Geography: Discipline, Profession and Subject since 1870

The GeoJournal Library

Volume 62

The titles published in this series are listed at the end of this volume.

Geography: Discipline, Profession and Subject since 1870

An International Survey

edited by

GARY S. DUNBAR

Professor Emeritus of Geography,
University of California,
Los Angeles, U.S.A.

A C.I.P. Catalogue record for this book is available from the Library of Congress

ISBN 1-4020-0019-7

Published by Kluwer Academic Publishers,
P.O. Box 17, 3300 AA Dordrecht, The Netherlands.

Sold and distributed in North, Central and South America
by Kluwer Academic Publishers,
101 Philip Drive, Norwell, MA 02061, U.S.A.

In all other countries, sold and distributed
by Kluwer Academic Publishers,
P.O. Box 322, 3300 AH Dordrecht, The Netherlands.

Printed on acid-free paper

Printed in the Netherlands.

CONTENTS

CONTRIBUTORS

Vincent R. H. Berdoulay, Professor of Geography, University of Pau, France

Anne Buttimer, Professor of Geography, University College Dublin, Ireland; President of the International Geographical Union, 2000-2004

Ilaria Luzzana Caraci, Professor of Geography and Pro-Rector, Third University of Rome, Italy

Gary S. Dunbar, Professor emeritus of Geography, University of California, Los Angeles, USA

Ingrid Hönsch, Chief (emerita) of the Central Geographical Library and the Archives of Geography, Institute for Regional Geography, Leipzig, Germany

David Hooson, Professor emeritus of Geography, University of California, Berkeley, USA

William A. Koelsch, Professor emeritus of History and Geography, Clark University, USA

Ben de Pater, Senior Lecturer in Human Geography, Utrecht University, The Netherlands

Bruno Schelhaas, Scientific Collaborator, Institute for Regional Geography, Leipzig, Germany

Paul Simpson-Housley, Professor of Geography, York University, Canada

John Warkentin, Professor emeritus of Geography, York University, Canada

Charles W. J. Withers, Professor of Historical Geography, University of Edinburgh, Scotland

1 Introduction

Gary S. Dunbar

"Pro-fesh-yon-al-i-zay-shun . . . pro-fesh-yon-al-i-ZAY-shun! . . . I do not think I like that word . . . I am quite certain I do not like that word!," said A. F. Martin of Oxford University in 1980 when I mentioned to him that I was making a study of the professionalization of geography in Europe and North America from about 1870 to 1930. I should not have been surprised by his reply, because Mr. Martin, a highly regarded lecturer, tutor, and university man, was in some ways a typical example of his generation of British geographers, in that he did not have an advanced degree and his publication record was meager. Except for a single remarkable essay (Martin 1951), he was not well known in the field of geography, at least not outside the United Kingdom. Was he, then, a *professional* geographer? Yes, indeed, because he was a full-time scholar in a distinguished university, even though he lacked the sort of credentials that have been demanded of subsequent generations of scholars, at Oxford and elsewhere.

I certainly do not mean to denigrate Martin or his ilk; I am simply using this anecdote to launch a book on the emergence of modern geography. I thought long and hard about Martin's remarks and then abandoned my project altogether in 1985 when I met Professor Horacio Capel of the University of Barcelona, who gave me a rather blunt assessment of the likelihood of success for a solo enterprise of this nature. But, then, in 1998, after I had been retired for a decade, I succumbed to the flattering proposal of Petra van Steenbergen of Kluwer Academic Publishers that I might write a book on the history of geography. It occurred to me that I might be able to salvage some of the material that I had gathered in the 1970s and 1980s by *editing* a book on the history of academic geography in Europe and North America, with the chapters written by true experts from the various countries. The terminal date that I had chosen for my original project—1930—had a certain logic for a book by a single author, because it would have enabled me to avoid the complexities of the recent past, but in 1998 I thought that a more important book

G.A. Dunbar (ed.), Geography: Discipline, Profession and Subject since 1870, 1–7.

would be produced by several authors, who could bring the story down to the present and perhaps even take a peek into the near future. In any event, Kluwer gave me the green light, and I proceeded to line up authors, mostly friends or friends of friends, to write the chapters on the emergence of modern geography in their homeland—or, in two cases, Sweden and Russia, countries on which the authors had vast knowledge, even though they were not natives or long-term residents. I assigned the purely arbitrary date of 15 January 2000 as the deadline for submitting chapters to me in either British or American English. Two chapters actually came in before that date, and I was pleasantly surprised that the last paper to arrive was less than seven months late—quite remarkable for an edited work of this sort. I had heard horror stories about other editorial projects that suffered long delays or even failed utterly because of a single weak link.

There are several ways in which a book of this sort can be arranged. A chronological treatment with, say, ten chapters covering European and American geography for each decade of the 20th century might have a certain logic, but it would probably be the least desirable strategy. A thematic arrangement would be better, but the country-by-country treatment of the present book is perhaps the most practical solution; certainly it has the advantage of attracting possible authors and maybe a larger readership as well. Mark Bassin has recently noted that "with remarkably few exceptions, historians of geography tend to have little serious interest in anything outside their own national traditions" (Bassin 2000, 476), but we hope that this collection, despite its structure, might be counted as one of those exceptions. The authors herein seem to be unusually careful to point out the faults, as well as the virtues, of the various national traditions. We are trying to approach David Livingstone's concept of *situated* knowledges—situated historically as well as geographically. As Livingstone has reminded us, "Geography has meant different things to different people in different places and thus the 'nature' of geography is always negotiated. The task of geography's historians, at least in part, is thus to ascertain how and why particular practices and procedures come to be accounted geographically legitimate and hence normative at different moments in time and in different spatial settings" (Livingstone 1992, 28-29).

I had hoped that the chapters would be of approximately equal length and would not have to be cut or expanded greatly. In fact, one of the chapters had to be reduced by more than 50% to bring it more or less in line with the others, but I implored the author to save all the original material and add to it in order to produce a book on the subject at some future time. Indeed, I hope that all the authors will expand their essays to book length. It is usually easier to expand chapters than to reduce them, abstracts may be even harder to write than a page of text, and titles can sometimes create the greatest anguish of all for an author. No title can perfectly convey the full character of any work, but it is hoped that the present title, suggested by William Koelsch, will win the approval of all but the most contrary readers.

Originally, I suggested that the authors should try to emphasize ideas and institutions over facts and individuals and to go light on bibliography and long quotations. If the chapters seem still to be a little too heavily factual for a reader's taste, it is only what one could reasonably expect from dedicated authors whose previous works have been models of historical scholarship. We cannot eliminate all "dead white men" from a book such as this in order to avoid the charge of political incorrectness. The lack of illustrations is due partly to my inability to handle such materials on my computer and to my desire to keep the price of the book as low as possible but mostly to my belief that they simply are not appropriate in a book of this nature. It would be easy to include photographs of well-known geographers (again, mostly "dead white men") or of buildings such as Oxford's School of Geography or Paris' Institut de géographie, but that would run counter to the intention to emphasize themes or concepts over individuals or monuments.

The order of the chapters might appear to be somewhat arbitrary. A strictly alphabetical order would put Canada first and the United States last, but I thought that it would be better to put the two North American nations together, following the seven European countries. Since Germany had the first modern research-oriented universities, it would be proper to place that nation first, but the subsequent order of European countries might be a matter of debate. Readers will all have different opinions on this matter, based on their own ethnic or national backgrounds, but they will agree that the modernization of geography, like most of the traditional academic disciplines, began in the last third of the 19th century in all the countries under discussion and then spread to the rest of the world. The development can be seen most readily in the appearance of university institutes or departments and in the establishment of professional journals. It is certainly no coincidence that these phenomena are linked to the urbanization and industrialization of Europe and North America.

After about 1870 a profound change came over universities in the western world. Enrollments expanded, students were allowed greater choice in selecting courses, libraries and laboratories grew, and the teaching staffs saw great increases in both numbers and quality. Professors were expected to have post-graduate degrees themselves and then to train younger scholars. The academic disciplines were professionalized and institutionalized so rapidly that universities of the World War I era bore little resemblance to their condition only fifty years earlier.

What is a "profession," and what is Geography's claim to that title? "The world's oldest profession" was established so early in pre-classical times that standards of entrance, certification, and practice were minimal or non-existent, but subsequent aspirants to the title of "professional" have had to endure a long apprenticeship followed by certification. From the Middle Ages onward, until the last quarter of the 19th century, the professions were limited to the priesthood (ministry), law, university teaching, and perhaps medicine, but the urban-industrial age brought a great clamor for recognition from all manner of skilled workmen who saw that professional status usually conveyed greater prestige, if not wealth. There was even a period of deprofessionalization in the second quarter of the 19th century when, for various reasons, professional standards, never very high, were relaxed considerably. In the 20th century so many groups have sought professional status that one American sociologist posed the query, "The professionalization of everyone?," and then answered his own question by denying entry to most of the occupations that "engage in heroic struggles for professional identification" (Wilensky 1964, cf. Goode 1969).

Geography is undeniably a profession, or a profession within a profession, in that geographers are university-trained, usually to the level of the doctorate, and then pursue a lifelong career, usually in university teaching. Before the almost universal modernization of universities in Europe and North America in the 19th century, beginning first in Germany with the founding of the University of Berlin in 1810 but not appearing elsewhere much before 1870, professors and students were few in number and only minimal standards were observed. Attendance was casual, and professors did not always think of their job as a career or lifelong pursuit. University libraries and laboratories were meagerly equipped, and there was little incentive to engage in research; indeed, most of the research work was produced outside the academy. In the late 19th and early 20th centuries university departments or institutes were small—so small, in fact, that professors had to spread themselves thinly over the broad spectrum of geographical themes. Doctoral theses often tended to be regional geographies covering both the physical and the human geography of a fairly large area, in order to demonstrate the author's ability to teach a wide variety of courses. As departments grew, each member of the teaching staff was still called upon to cultivate two or more regional and topical specialties. Students who went on to specialize in human geography were expected to take courses in physical geography, often for no better reason than that "it was good for their soul."

By the late 1950s and early 1960s, several profound changes occurred in the traditional university curriculum. The old rationale for a generalist education in the post-graduate training of geographers no longer held up. It became unreasonable to expect a doctoral candidate to hold forth learnedly on topics as far apart as plate tectonics, vegetation regions, and demographic patterns. (I remember with horror my own Ph.D. oral examination in 1955 when the obligatory geologist on the committee asked me about the mineralogical composition of black beach sand and a visiting instructor showed slides of tropical plants and expected me to identify them.)

In the 1960s there was concern, at least in the United States, that Regional Science might make inroads on the subject matter of Geography or even replace it altogether, but that proved to be an evanescent fear. Then, in the 1990s there was some apprehension about the possibility that Geographic Information Systems (GIS) might break away from Geography or perform some sort of matricidal act. At this moment, surveyors and engineers are threatening Geography's position as the main supplier of GIS training and accreditation. The rise of GIS has, paradoxically, been accompanied by decline in the use of maps in geographical research and publication. Ron Martin recently spoke of "the demise of the map" and called for the restoration of "the map as a central component of our theoretical and empirical work on space and place" (Martin 2000, 4-5).

Field studies in Geography declined in the latter part of the 20th century, just as in Geology and Biology. Indeed, the latter disciplines have undergone rapid change, with old-style Geology being phased out in some universities in favor of Geophysics, and the biological sciences have shifted in focus and scale, away from the old field-based studies in Botany and Zoology to an emphasis on Biochemistry and Microbiology. In the Social Sciences, fieldwork gave way after 1950 to the indoor processing of numerical data using the improved technology (computers) then coming into use. One hopes that there will always be a place for field studies in Geography, although geographers will need to make a more profound assessment of what it is that they are trying to accomplish through fieldwork (Driver 2000). Driving through the countryside at 100 kilometers an hour does *not* constitute fieldwork!

Other recent changes illustrate the vitality of Geography. Cultural geography, a German-American subject that found its greatest representation in the so-called "Berkeley School," has been given a very different orientation ("the cultural turn") by recent British scholars. In the last twenty years, feminist concerns have had a great effect on the practice of human geography generally, and not simply in discussion of topics of direct concern to women. The topics of colonialism and imperialism, which have been themes of perennial concern to geographers, have increased in importance in the recent past (Bassin 2000, 477-478). Even the study of the History of Geography has changed—and for the better—chiefly through the greater concern for establishing context and for the greater emphasis on archival research. All these new-old concerns are vital to the re-energizing of the field of Geography.

One of the problems of writing histories of Geography has always been that, with few exceptions, geographers have usually focussed narrowly on their own discipline without being sufficently mindful of the wider historical context. It would be salutary for them to consider the histories of other disciplines, particularly the "nearest neighbors" in the earth and social sciences. To take just two recent historical essays, one on Anthropology and the other on Sociology, we can see how exposure to such works could definitely broaden the perspective of geographers. George Stocking's "Delimiting Anthropology" is concerned with Anthropology's boundaries and the creation of "adjectival anthropologies" (sub-disciplines or specialties) in the Anglo-American world since the formation of the American Anthropological Association in 1904, but many of its themes have a more universal significance. He begins with the erstwhile geographer Franz Boas and gives a passing nod to Auguste Comte. Since the Association of American Geographers was also founded in 1904, Anthropology and Geography have had many similar experiences (Stocking 1995). Immanuel Wallerstein's "The Heritage of Sociology" could also be read with great profit by geographers. To Wallerstein, "Sociology as a discipline was an invention of the late 19th century, alongside the other disciplines we place under the covering label of the social sciences," but after 1970 they "have ceased to be disciplines, because they no longer represent obviously different fields of study with different methods and therefore with firm, distinctive boundaries." The names and organizational structure of these disciplines still remain, however, in the form of professional organizations, journals, university departments, and degrees, so that Wallerstein is chiefly concerned with "sociology as a culture, that is, as a community of scholars who share certain premises" (Wallerstein 1999, 2-3). These are but two of the recent works that could aid geographers in finding their place in the world of scholarship. If the present volume contributes to that understanding, we feel that our labors will not have been in vain.

References

Bassin, M. (2000) Studying Ourselves: History and Philosophy of Geography, *Progress in Human Geography* 24, 475-487.

Driver, F. (2000) Editorial: Field-work in Geography, *Transactions of the Institute of British Geographers* 25, 267-268.

Goode, W. J. (1969) The Theoretical Limits of Professionalization, Chapter 6 in A. Etzioni (ed.), *The Semi-Professions and Their Organization*, The Free Press, New York, 266-313.

Livingstone, D. N. (1992) *The Geographical Tradition: Episodes in the History of a Contested Enterprise*, Blackwell Publishers, Oxford.

Martin, A. F. (1951) The Necessity for Determinism: A Metaphysical Problem Confronting Geographers, *Transactions and Papers* (Institute of British Geographers)17, 1-12.

Martin, R. (2000) Editorial: In Memory of Maps, *Transactions of the Institute of British Geographers* 25, 3-5.

Stocking, G. W. Jr. (1995) Delimiting Anthropology: Historical Reflections on the Boundaries of a Boundless Discipline, *Social Research* 62, 933-966.

Wallerstein, I. (1999) The Heritage of Sociology, The Promise of Social Science, *Current Sociology* 47, 1-37, 39-41.

Wilensky, H. L. (1964) The Professionalization of Everyone?, *American Journal of Sociology* 70, 137-158.

Gary S. Dunbar
13 Church Street
Cooperstown, New York 13326
USA

2 History of German Geography: Worldwide Reputation and Strategies of Nationalisation and Institutionalisation

Bruno Schelhaas and Ingrid Hönsch

Introduction

The prestige of German geography up to the beginning of the 20th century and the high international position of German sciences have often been the focus of historical research. The rise of Prussia, the problematical nation-building, and the local diversity in the various regions of Germany have produced a very complex but exciting subject. The English-speaking outsiders' views of German geography, perceived by experts who visited Germany in the first half of the 20th century, explain the situation very well. As Thomas Elkins has pointed out, two anglophone geographers were of special importance: Richard Hartshorne, who published his influential *The Nature of Geography* (1939) after spending some time in Austria and Germany, and Robert E. Dickinson (Elkins 1990, 21-23). Another expert was Samuel Van Valkenburg, who had studied under Albrecht Penck in 1915-1916 at the University of Berlin. His paper on "The German School of Geography" (Van Valkenburg 1951) is still—a half-century later—a very readable historical review of 20th-century German geography.

Our present attempt to reconstruct the history of German Geography is the product of insiders' views, based on nearly twenty years of work on the historiography of the field. The main emphasis will lie on the various stages of institutionalisation, professionalisation, and nationalisation of the subject.

After the deaths of the two protagonists Alexander von Humboldt and Carl Ritter in 1859, which mark the end of the classical period, German geographers tried with varying degrees of success to professionalise their subject. The leading position of the geographical societies, above all the Berlin Society that had been founded in 1828, was increasingly eroded by the proliferation of university chairs and the rise of academic geographers. The stage of academicisation finally began in 1871, the year of German nation-building (Reichsgründung), and was finished before World War I.

G.A. Dunbar (ed.), *Geography: Discipline, Profession and Subject since 1870*, 9–44.

Ferdinand von Richthofen was the leading figure in German geography in the last quarter of the 19th century. His successor in the Leipzig chair was Friedrich Ratzel, whose concepts of human and political geography gave him a worldwide reputation. A number of international scholars (e.g., Ellen Churchill Semple) filled the lecture halls to listen to Ratzel's new geography. The German model of scientific geography was born, influenced by strong environmentalistic and geological elements but becoming increasingly accepted as an independent subject in its own right.

The nationalisation of German geography was, compared with that of Russia, France and Great Britain, dominated by contradictory developments. The political catastrophes that occurred in the 19th and 20th centuries, up to the end of the Cold War in 1989, were mirrored in the history of German geography. Often the link between geography and politics became dangerous, especially after World War I when a platform was given to revanchist, nationalist and, later, Nazi discourses. German geopolitics, military geography and applied geography since the 1920s illustrate that dark chapter of the history of geography. Germany's leading international position was already lost. Some prestige was restored in the post-World-War-II era with the international reception of Walter Christaller's Central Place theory and the recognition of such cosmopolitan German geographers as Carl Troll and Walter Manshard. After 1960 West German geographers made greater connections with the anglophone world, whereas East German geography looked eastward to the Soviet Union for direction. Some German geographical traditions have endured; a strict science system with sub-disciplines, regional geography (Länderkunde), applied geography and regional planning have all retained considerable importance. The post-war history of the two German geographical traditions has shown breaks and continuities in that project called German geography, creating a new and complex field of interest for any reconstruction of the history of science.

At the beginning of the institutionalisation of German geography the history of geographical thought was a well accepted subject. The classical works of the protagonists Humboldt, Ritter, Johann Gottfried Lüdde, Johann Eduard Wappäus and Johann August Zeune and the more specialised Oscar Peschel and Sophus Ruge illustrate that historical awareness in the early period. Historical research was done in every decade, but with different intentions and different methods. To a great extent historical reconstruction up to the present day has been connected with the writing of obituaries and Festschriften or pieces written on the occasion of jubilees or national and international meetings. Historiography has often been used to demonstrate the progress and national development of the subject. For a long time Hanno Beck was the only German professional who was devoted to fulltime research on the history of geography, although significant contributions to the subject were made by Ernst Plewe, Manfred Büttner, Walter Sperling, Josef Schmithüsen and Gerhard Engelmann.

Since 1980 a scientific turn has taken place. Earlier, the habilitation theses of Dietrich Bartels (1968) and Gerhard Hard (1970) had pointed the way, but the real breakthrough came in 1980 with the Ph.D. theses of Hans-Dietrich Schultz and Ulrich Eisel, which concentrated on the history of the changing paradigms and methodologies of German geography. The historic debates about the Nazi past (Historikerstreit) and the efforts within other science subjects have brought about a number of critical studies in the last twenty years (Sandner 1988; Wardenga 1995a).

The situation in GDR geography was different; the history of geographical thought was well supported, and numerous studies, mainly diploma and doctoral theses, were produced. The concentration on the history of German geography was to a high degree part of the creation of an independent German Marxist geography, illustrated by stressing the "positive" traditions (Humboldt, Ritter, Alfred Rühl and some others) and condemnation of geopolitics and regional geography (Länderkunde), which were classified as bourgeois, imperialist and partly pro-fascist.

In the united Germany specialists in the history of geographical thought have increased in number, and the subject has been a prominent topic in professional meetings. Its place in university curricula is still somewhat marginal, however, and there is no up-to-date German-language textbook covering the field.

Geographical Societies, from Bildungsbürgertum (educated bourgeoisie) to research organisations (Büdel 1982; Guntau 1986; Linke 1986; Schelhaas 1997)

The year 1828 marked an important step in the institutionalisation of German geography. The founding of the Berlin Geographical Society, seven years after its predecessor was established in Paris, resulted from the progressive mood of the Prussian capital towards the sciences and was connected with famous names such as Humboldt and Ritter. A framework for popular and exploration geography was born, a model of organisation with a worldwide reach.

Scientific societies had been founded in Europe beginning in the 17th century. Early examples of German associations are the Societas Ereunetica sive Zetetica in Rostock (1622) and the still-existing Leopoldina (The German Academy of Naturalists), constituted in 1652 at Schweinfurt and now situated in Halle.

In the burgeoning natural history societies (Naturforschende Gesellschaften) of the 18th century, geographical thought was mostly a side issue in the scientific debates. The sphere of activity was mainly local, and the academies often had a courtly or governmental character (e.g., Prussian Academy in Berlin, Saxonian Academy in Leipzig and Bavarian Academy in Munich). As with university geography some years later, the subject was mostly taught by non-geographers, most often by philologists, physicians or "universal scientists". At that time we can find no independent representation of geography in the modern sense. It was, rather, cartography and astronomy that formed organisations of their own. The Kosmographische Gesellschaft (Cosmographical Society), founded by Johann Michael Franz in Nuremberg (1746-1765), is the most famous forerunner of geographical societies in Germany. After the pre-classical period of German geography, which was concentrated in Nuremberg, Göttingen, Leipzig, Berlin and elsewhere, one can characterise the continuous attempts at a national concentration of geography as a long history of failed organisations (Linke 1986; Schelhaas 1997).

The Gesellschaft Deutscher Naturforscher und Ärzte (Association of German Naturalists and Physicians), founded in 1822 by Lorenz Oken in Leipzig, was an important science association in the early 19th century. In a special section, first called Geognosy (i.e., Geology) and Geography, later Geography and Hydrography, and since 1873 in an independent section of Geography, scientific geography had its first national arena, joined by some of the great names of classical German geography (e.g., Humboldt, Peschel, Karl Theodor Andree, Adolf Bastian, Georg Neumayer, August Petermann and Gerhard Rohlfs).

Before the rise of academic geography, the geographical societies were the most dominant power with regard to the publication and propagation of geography. In many German cities, even in those without any academic tradition, these societies developed into establishments for the middle classes (Bürgertum). As Ute Wardenga has stressed, the increasing German interest in Bildung (culture) in the 19th century was an important reason for the success of the societies. In addition, the local separatist movements in the various regions of Germany after the unification of the nation in 1871 and the resulting local and regional identities explain this institutional step (Wardenga 1999, 77ff.).

Berlin (1828), Frankfurt (1836) and Darmstadt (1845) established the oldest geographical societies, followed by Leipzig (1861), Dresden (1863), Munich (1869) and Bremen (1870). The lectures given at these societies were often events of great interest. The members were local businessmen, the higher civil servants and military officers. With the great number of European exploring expeditions in all parts of the world and their journalistic marketing in the form of journal articles and series, magazines, novels and popular lectures, the societies were very successful in their local areas.

The diverse nature of the German geographical societies reflected the political situation of the federalist and often splintered Germany. The non-existence of a nation-state created the special German situation of having no national geographical society but a network of local societies. The history of German geography is, compared with other disciplines or national traditions, a special case. German physicists founded their first national association in 1844, the geologists four years later and the chemists in 1867 (Guntau 1986). Again, German geography can be considered a belated development within the hierarchy of the sciences.

Here are some examples to show the different situation of geographical societies in the German states. The geographical society of Frankfurt, launched in 1836, was mainly a product of the commercial activities of the free imperial city (Freie Reichsstadt), a combination of bourgeois, commercial and scientific interests. Above all, two protagonists promoted that project: the publisher and cartographer August Ravenstein and the historian Georg Ludwig Kriegk (Fick 1986). In Leipzig, in those days a booming commercial centre, especially of the publishing industry, seventeen middle-class citizens founded in March 1861 the Leipziger Verein der Freunde der Erdkunde, including a Carl Ritter Foundation. One main reason for installing the society was to finance an expedition to find the missing explorer Eduard Vogel, who had wanted to supplement Heinrich Barth's African studies. Again, the social structure of the association was bourgeois, dominated by academics, civil servants, teachers, publishers, booksellers and bankers (Arnhold 1965; Gellert 1984; Grimm 1998). The Bremen society, founded in 1870 as the Verein für die Deutsche Nordpolarfahrt and reorganised six years later into a geographical society, was very much connected with the attempts to form a German Arctic expedition (Weißenborn 1938; Petri 1975).

Berlin is a special case (see Hellmann 1928; Quelle 1953; Lenz 1978; Nix/Wölke 1986). The University of Berlin, organised by Alexander von Humboldt's brother Wilhelm in 1810, the venerable Prussian Academy of Science and the popular and scientific fame of Alexander von Humboldt after his return to Berlin inspired institutionalisation in the Prussian capital. Already, two predecessors had shown the concentration process in early 19th-century Berlin: the Berliner Geographische Vereinigung, which was founded in 1809 under the leadership of Johann August Zeune but declined during the independence war; and Heinrich Berghaus' attempt at organising the Verein zur Beförderung der Erdkunde in 1822.

In 1828, to celebrate the 50th anniversary of Captain Reymann's greatest cartographic work, more than fifty famous scientists (including A. von Humboldt, Zeune, Leopold von Buch, H. Lichtenstein, A. von Chamisso and C. G. Ehrenberg) launched the Verein der Geographen Berlins, later renamed the Gesellschaft für Erdkunde zu Berlin. The first and long-time president was Carl Ritter. In the early stages of the society's existence the social aspect was dominant; the scientific work often consisted of merely reporting on French and British exploring expeditions. The belated colonial interest and modest research funding of the Prussian State gave the societies of Paris, London and St. Petersburg a decided advantage in the international competition of national geographical societies.

In 1849, the year of Heinrich Barth's first African expedition, a very successful stage of exploration and scientific work began. The Berlin society supported many expeditions (including those of Decken, Heuglin, Beurmann, Rohlfs, Nachtigal and Wissmann), at first mostly to Africa and then later into the polar regions. Research funding was guaranteed by the Carl Ritter Foundation (begun 1859) and successor foundations down to the present day. Reports were given at the Berlin society meetings and published since 1839 in the society's periodicals (Bader 1979). These journals were important instruments in the creation of an international geographical network through the exchange of publications, which, as a welcome side effect, filled the society's library. Along with the periodicals issued by publishing houses, such as Justus Perthes in Gotha, the Berlin Geographical Society's publications were important in the worldwide dissemination of geographical knowledge.

By the end of the 19th century, the Berlin Society's work became more and more scientific. As a result of the professionalisation of academic geography and specialization into geographical subdisciplines, e.g. geomorphology and economic geography, the meetings turned into scientific sessions. Ferdinand von Richthofen and Albrecht Penck were the most important representatives of this era. As the oldest, largest and most prestigious German society, the Berlin Geographical Society often tried to play a national role in the international scientific community, sometimes in conflict with the other societies (especially the big ones in Leipzig, Frankfurt, Hamburg and Munich). A highlight in the Berlin Society's history was the VIIth International Geographical Congress held in Berlin in 1899. Organised by the Berlin Geographical Society and under Ferdinand von Richthofen's chairmanship, this first German international meeting signifies the prominent position of German geography at that time.

During the period of growing colonial interest, African research aided the extension of the network among the geographical societies as well as to German geography in general, especially after the formation of the nation-state in 1871 (Bader 1978). Already, the African Association, founded in 1788, and its transformation into the Geographical Society of London in 1830, illustrates that close connection. Headed by the Berlin Geographical Society and initiated by Adolf Bastian, the societies of Berlin, Dresden, Frankfurt, Halle, Leipzig and Munich launched the Deutsche Gesellschaft zur Erforschungs Aequatorialafrikas in 1873. On an international level, the Association International Africain was founded on the occasion of the Brussels Congo Conference in 1876. The German explorers Lenz, Nachtigal, Richthofen, Rohlfs and Schweinfurth joined the new Association. In 1878 the Afrikanische Gesellschaft in Deutschland was formed through the collaboration of the geographical societies of Berlin, Dresden, Halle and Leipzig and several chambers of commerce and other economic interests. The life of this Society was very short. After supporting a number of African expeditions, beginning with Güßfeldt's Loango expediton in 1873, the African German Society broke up in 1899, mainly as a consequence of difficulties with the government and insufficiency of financial support.

Academicisation: geography chairs and colonial interest (Engelmann 1983; Gärtner 1992; Schulte-Althoff 1971)

The formation of a unified nation-state in 1871 had positive consequences for academic geography. The new centralised government had a great interest in specific knowledge about other countries and continents, mainly in order to use this information for the national political and economic aims. Germany wanted to demonstrate its world power, mainly as an answer to the political, military, economic and colonial developments in France and Britain. The belated German attempts to acquire colonies in Africa and other overseas territories assisted geography indirectly. The burgeoning German industries needed raw materials in order to expand and to remain competitive in world markets. The government supported expeditions to such territories, often in close connection with the geographical societies. The Landeskundliche Kommission zur Erforschung deutscher Schutzgebiete (Special Commission for Exploration in the German Protected Areas) was established to organise German administration in the colonies.

In spite of all this activity, there was a lack of specialists. There were no independent geography chairs and institutes before the 1870s. After Carl Ritter's death in 1859 the Berlin ("full") chair was vacant until 1874. During the 19th century geography was often a side issue in the curricula. In this situation several ausserordentliche (extraordinary) professors or private lecturers, often connected with other subjects like statistics or history, represented academic geography. For example, Georg Benjamin Mendelsohn in Bonn (from 1835) and Eduard Wappäus in Göttingen (from 1845) taught both geography and statistics (Brogiato 1998: 32ff; Engelmann 1983: 3-56).

In spite of the centralised government (Reichsregierung), education was organised on a federal basis. It was Leipzig University in Saxony that established a first full chair in geography in 1871, occupied by Oscar Peschel. Other universities followed. In Prussia new chairs for geography were created rapidly: Alfred Kirchhoff in Halle (1873); one year later Heinrich Kiepert in Berlin and Eduard Wappäus in Göttingen; Ferdinand von Richthofen in Bonn followed in 1875. Adalbert Falk, the Prussian Minister of Culture and Education, issued special instructions in 1875 to establish chairs in geography in all Prussian universities. Outside Prussia, only Munich, Dresden, Leipzig, and Strassburg had chairs in those days. The southern German universities founded geographical chairs after the turn of the century. But it was not easy to find genuine geographers to take these positions. It is not surprising that the new professors were often natural scientists, historians, journalists, philologists or even secondary school teachers. It was Ferdinand von Richthofen who succeeded in establishing a "school of geographers" (Richthofen-Schule). The young institutes had to organise the scientific staff, libraries and special collections of maps, rocks and minerals, and photographs. The Leipzig University Institute for Geography, for instance, was mainly influenced by Ratzel, who followed Oscar

Peschel, Otto Delitsch and Ferdinand von Richthofen in the chair. Ratzel's contributions to the Geographical Institute in its formative period were considerable. When he came to Leipzig in 1886, the Institute had only a few small rooms for its fifty students. In the period of Ratzel's directorship, the situation changed fundamentally. He succeeded in enlarging the Institute; in the 1890s more than 300 students, including some from abroad, attended his lectures each year. Ratzel founded an institute library, including a map collection and other teaching resources. He established a social bond between staff and students (including the young Alfred Hettner) by creating the Geographischer Abend, a forerunner of the Leipzig university geographical society. Apart from Ratzel's institutional innovations, his geographical concepts opened a new field of geography; he is often called the founder of anthropogeography. He succeeded in creating a deterministic blending of natural and cultural history. His ideas on the dependence of human life on natural conditions agreed with the colonial policy of the State. Besides Ratzel's institute there was another famous geographical institution in Leipzig: the Museum für vergleichende Länderkunde (Museum for Comparative Regional Geography) founded by the German geologist and vulcanologist Alfons Stübel in 1896. This specialised museum and the associated Archives for Explorers (now Archives for Geography) were unique in Germany.

Colonial geography was a new topic in those years. Hans Meyer, geographer and editor, founded a special chair for colonial geography at the University of Berlin out of his private assets. Fritz Jäger became the first holder of the chair in 1911.

Another Prussian ministry official, Friedrich Althoff, pressed for the academicisation of geography in Germany. In 1883 he called for greater representation of geography as an independent subject in the universities, including a more rigorous curriculum, cartography courses and satisfactory equipment of the geography departments and libraries (Engelmann 1983: 92ff).

By the beginning of World War I, 28 institutes had been established. This was an indication of the national reputation of geography. The full professors were very powerful. Their institutes developed into influential bodies with consequences for the propagation of geographical thought. The chairs at Berlin, Leipzig, Bonn, Munich and Heidelberg were the most respected ones (Penck, Ratzel, Philippson, Drygalski and Hettner, respectively). Speaking of this German phenomenon of a powerful Ordinarius (Full professor), Thomas Elkins has said: "Intellectually unchallenged within their own kingdoms, it is not surprising that some became dogmatic in the extreme" (Elkins 1990: 24).

Apart from the expansion of university teaching programs, large expeditions were organised. With financial help from the German government, Erich von Drygalski, who had been a student of Richthofen, organised the First German South Polar Expedition (1901-1903). He formed an international team of specialists in the field of polar research. Geographers such as Hans Meyer, Fritz Jäger and Eduard Pechuel-Lösche made important contributions to the geographical exploration and resource assessment in the new African colonies. Germany lost all its overseas territories after World War I, and geographical exploration was halted for several years.

The Deutscher Geographentag (annual national meeting of German geographers): nationalisation and demonstration of German geography (Brogiato 1995, 1998; Georgi 1968; Höhl 1981; Schelhaas 1997)

The lack of a national geographical society in Germany and the divergent developments in the German states led finally to the belated establishment of a national meeting of German geographers and national forms of publication. The unification of the German Reich in 1871 and the political and mental atmosphere during the Kaiserzeit (the German empire, 1871-1918) allowed for increased specialisation within the sciences.

In the spirit of the failed revolution of 1848, a first national meeting of geographers took place in Frankfurt in 1865 as a "meeting of masters and friends of geography" within the Freie Deutsche Hochstift (Georgi 1968). This foundation had been established in 1859 on the 100th anniversary of Friedrich Schiller's birth by Otto Volger as a liberal and bourgeois organisation of the educated classes and also as an expression of German historicism. The geographers' meeting, dominated by Volger and his consultant August Petermann, discussed programmatic topics concerning the future strategy of a professional German geography, including the realisation of an Arctic expedition (Deutsche Nordfahrt), establishment of a national geography meeting (expressed especially by Ferdinand von Hochstetter), initiation of a standardised meteorological logbook and promotion of a German oceanic research centre (Deutsche Seewarte, created in 1868 and existed from 1875 to 1945 in the form of a Reich institute). The expressed demands illustrate important steps on the way to a fully accepted and competitive science subject: nationalisation within the international competition of exploring the world, establishment of a national scientific forum, standardisation of empirical surveys and foundation of national research institutes. The ambitious project was doomed to fail. As a consequence of the German-Austrian war (1866) the free imperial city (Freie Reichstadt) of Frankfurt was annexed by Prussia. The liberal and prominent status was lost, and the geography meetings and the programmatic goals could not be realised.

It was August Petermann, the active geographer and cartographer from Gotha, who became the leading organiser of German geography in that era. He specified the Frankfurt plans for the Arctic expedition. In 1866 he wanted to launch a great German Geographical Society consisting of as many as 100,000 members in order to finance such expeditions. As the editor of *Mittheilungen aus Justus Perthes' Geographischer Anstalt*, which is still one of the world's oldest and most influential geographical periodicals (Bernhardt 1981; Köhler 1981), Petermann tried to combine his research reforms with expansion in the circulation of his publications. The project was thoroughly discussed by the experts but failed again in the turmoil of the Austro-Prussian War. Nevertheless, the first German Arctic expedition took place in 1868, led by Karl Koldewey, but the plans for the other German geographical projects remained dormant for a decade.

Again it was the Berlin Geographical Society that took the lead in the nationalisation of geography. Gustav Nachtigal now became the chief organiser. The Berlin Society planned a meeting in 1879 to celebrate the 100th anniversary of Carl Ritter's birth. Within the organising committee the old project of a national geographical society was revived. Precise suggestions by Nachtigal, Georg Neumayer, Friedrich Marthe, Ernst Behm and Karl Zöppritz were again refused, because the existing geographical societies, which were to be assimilated by the central society, fought for their independence and their own publications. Only one goal could be reached: the Deutscher Geographentag. During the Berlin anthropologists' meeting in August 1880, again under Nachtigal's direction, the delegates decided to arrange a general German geography meeting in Berlin in the next year. On 7-8 June 1881 more than 80 participants joined the first of what has now become 53 meetings. Only six lectures were given at that first meeting of the Geographentag; the selection of papers on a miscellany of topics illustrates the vague and rather marginal position of geography at that point in the 19th century. School geography was represented in a special section at the Berlin meeting, promoted by Alfred Kirchhoff and Hermann Wagner.

The annual meetings of the Deutscher Geographentag were very successful. In 1882, 434 participants joined the second congress in Halle, and the number of papers increased to 14. Frankfurt (1883) counted 504, Munich (1884) 345, Hamburg (1885) 633, Dresden (1886) 331 and Karlsruhe (1887) 402 participants. At the Munich meeting the delegates decided to install a Central Commission (Zentralausschuß des Deutschen Geographentages), the highest body of German geography, which arranged meetings in cooperation with the local committees down to 1934. The Deutscher Geographentag, with its commissions and sections, acted for a long time (and does still, to a certain extent) as the national representative of German geography, as a kind of umbrella association combining all geographical interest groups.

At the second meeting, in Halle (1882), another important and still functioning institution was born: the Zentralkommission für wissenschaftliche Landeskunde von Deutschland (Central Commission on the Regional Geography of Germany), initiated by Richard Lehmann and managed first by Friedrich Ratzel and later by Kirchhoff and Penck (Meynen 1948; Meynen/Richter 1982; Wardenga 1999). The first projects concentrated on a detailed bibliography of regional literature, and in 1885 publication of a bibliographical series was begun, Forschungen zur Deutschen Landes- und Volkskunde (Research Studies on the Regional Geography and Ethnology of Germany). In an article on the Commission, Ute Wardenga (1999, 77) identified three different periods of this body:

1. collaboration with amateur geographers organised in different associations (1882-1900)
2. more or less compulsory involvement of academic geographers (1900-1916)
3. declining adaptability to changing organisational structure (1916-1941)

Wardenga pointed out the difficult position of research on the homeland, the German landscape itself. Regional geography at that time mostly involved overseas research. Down to the 1920s, the Commission's work was dominated by the struggle between academic and amateur geographers.

In the beginning, the geographical societies, which were often the local organisers, dominated the meetings. Later, the assemblies came to be dominated by the university geographers—again an academisation of geography. Down to 1912, the year of the founding of the Verband Deutscher Schulgeographen (Association of German School Geographers), the Geographentag acted as a kind of official body also with regard to school geography, especially concerning discussions of curricula (Brogiato 1995, 1998).

Verband Deutscher Schulgeographen (Association of German School Geographers) (Heske 1988; Brogiato 1998)

The first national concentration in German geography was achieved by school geographers. Already within the early Deutsche Geographentage school geography had its own section, increasingly carried by the Ständige Kommission für den erdkundlichen Unterricht (Standing Commission on School Geography), initiated by Hermann Wagner and founded in 1901. Because of its ineffectiveness the Commission was dissolved during the 18th meeting of the Geographentag at Innsbruck in 1912. In its place, Hermann Haack and the Justus Perthes publishing house in Gotha developed a new vehicle for school geography. The *Geographischer Anzeiger*, edited by Haack and Heinrich Fischer since 1899, increasingly became the most influential periodical for school and, in part, academic geography (Brogiato 1998). After several appeals published in the magazine, Haack established the Association of German School Geographers in 1912. The new organization developed very well, counting 3,000 members by 1914.

The ensuing political disasters, beginning with World War I, did not cause any loss of prestige or influence. The Association and its organ, the *Geographischer Anzeiger*, joined in the patriotic and in part imperialist mood. Haack and his colleague Friedrich Knieriem prepared the Association's transition into the Nazi institutional structure. During two meetings at the Thuringian town of Eisenach, the assembled school geographers decided to continue the Association's work within the Nationalsozialistischer Lehrerbund (National Socialist Teachers' Association), a centralised and doctrinaire organisation comprising all school and, in part, academic affairs. The official execution of this momentous decision took place in 1934 at the first Geographentag after Hitler's seizure of power. The Sachgruppe Erdkunde (Geography Section), headed by Albrecht Burchard and, after his death (1939) by Knieriem, developed into the most powerful body in German geography in those years, organised in a very strict way (Führerprinzip) and divided into 33 regional sub-commissions, the so-called Gausachbearbeiter. Seven Reichsreferenten (special advisors) were appointed to coordinate scientific and technical matters. Of that number, three played a special role: Ludwig Mecking, who was responsible for scientific geography and also acted as the chairman of the Central Commission of the Deutscher Geographentag; Hans Mortensen, who was responsible for university geography and acted as the chairman of the Association of German Academic Geographers; and Karl Heck, the Association of German School Geographers' last chairman, who was responsible for curricular affairs. The position and influence of these Reichsreferenten in the prewar era is even now hard to determine. Between 1934 and 1942, four assemblies of this geographical section took place, dominated by Nazi geography debates, but activity ceased after that time.

After the war, it was Julius Wagner's role to initiate the institutional work of school geography. During the first postwar Geographentag, held in Munich in 1948, the delegates decided to reinstate the old Association of German School Geographers. On the occasion of the school geographers' meeting in the Hessian town of Jugenheim, the official reconstitution was accomplished in April 1949. The Association is today Germany's largest geographical organisation.

Ideologising and specialisation after World War I
(Fahlbusch/Rössler/Siegrist 1989; Heinrich 1990, 1991; Jüngst 1993; Mehmel 1995; Sandner 1994; Schultz 1977; Wardenga 1995b)

German geography became more and more political after the turn of the century. Geography developed into an applied subject as well as an academic and school discipline. After the filling of the chairs in geography at nearly all German universities and the establishment of geographical societies in several cities, the first institutional stage was completed. The following stage was dominated by theoretical, methodological and political discussions. With the growing nationalism in the period of industrial expansion from 1871 onward, the connection between geography and politics got very close. In general, the German-French War of 1870-1871, the new colonies, military geography, border studies and patriotic geography (vaterländische Geographie) were some of the main topics in German geography around the turn of the century (Sandner 1994). World War I, the consequences of the Versailles peace treaty, the reparations, the military restrictions, and the loss of the colonies and German territories were topics that almost all academic and school geographers refused to deal with.

In World War I geography very quickly became an important scientific as well as military subject (Mehmel 1995; Schultz 1977;

Wardenga 1995b). The improvement of topographical and geographical information for the military and the politicians, as well as for the masses, was the main reason for the new acceptance and support of geography. The discussions of war strategy were accompanied by a shift of interest from geomorphology to human geography, in the form of economic, political and military geography. Many unscientific and propagandistic studies dealing with the new territorial and political-geographical situation were undertaken, and articles on such topics filled the geographical periodicals.

Within the special commissions on regional geography (Landeskundliche Kommissionen) many geographers conducted research in the occupied areas. Wardenga (1999, 80) has pointed out the main goal of those commissions: "They were aimed at collecting abundant data concerning the geology, geomorphology, hydrology, pedology, and also the settlement structures, population and economy of the occupied countries in order to provide a basis for further decisions during peace negotiations". With governmental support, geographers discovered a new and worthwhile area of responsibility. Out of their empirical surveys came maps, handbooks, articles and Ph.D. theses, partly kept secret by the military. In 1921 the Deutscher Geographentag at Leipzig offered an opportunity to present some of the successful results of that research.

In the 1920s there were increasing numbers of studies in lebensraum research, Grenzraumforschung (border studies) and Volks- und Kulturbodenforschung (Ethnic German research). The Stiftung für deutsche Volks- und Kulturbodenforschung in Leipzig, founded in 1926 and encouraged by Albrecht Penck, developed into the most influential organisation of regional geography after World War I (Fahlbusch 1994). Already in former patriotic associations, such as the Verein für das Deutschtum im Ausland (Society of German Affairs in Foreign Countries), the Alldeutscher Verband, the Schutzbund für das Grenz- und Auslandsdeutschtum (Association of German Affairs in Border and Foreign Regions) and the Deutsche Akademie, the imperialist background had been prepared. With the new organisations, especially the Leipzig foundation, scientific proofs were advanced as a legitimation of the demand for world power. Managed by Wilhelm Volz and joined by a number of German geographers, the Leipzig foundation developed successfully, but it was disbanded in 1931. Conferences were organised, research studies—mainly in eastern Europe—were supported and publications were produced. A major publication was the *Handwörterbuch für das Grenz- und Auslandsdeutschtum*, an extensive dictionary of

German affairs, mainly produced by Emil Meynen and Friedrich Metz. After the mysterious liquidation of the foundation, institutionalisation of Volkstumsforschung (Ethnic German research) increased within the Volksdeutsche Forschungsgemeinschaften. Between 1930 and the end of World War II, ethnic German research was conducted in six regional sections (including the former colonies) and coordinated by the Berlin office under Meynen. After Hitler's seizure of power, these organisations became important arenas for Nazi ideology and settlement planning (Fahlbusch 1999).

The Deutscher Geographentag also became more propagandistic and revanchist instead of scientific and neutral. Not only the Versailles peace treaty, the loss of territory and the colonies and the hard retributions but also the fervent patriotism among German academics created a general impetus for geography. After Strassburg (1914) the next meeting of the Geographentag took place in Leipzig in 1921. The 20th meeting had been scheduled for 1916, but the war and the political disturbances thwarted these plans. Only a handful of German academic geographers showed up in Heidelberg in 1916. The Central Commission held a working session in Gotha in 1920 to prepare for the Leipzig assembly the next year as well as for German geography in general.

Under the topic "Geography and the World War" the Leipzig delegates passed resolutions whose intent was unmistakable. In his welcoming speech, the chairman of the Central Commission, the well known colonial geographer Hans Meyer, named the central themes of the congress:

1. What were the achievements of German geography during the war?
2. What kind of geographical research has been undertaken since the last meeting?
3. What are the consequences of the World War with regard to school geography?
4. What are the territorial consequences of the World War? (Meyer 1922, 17)

Another famous geographer, the Munich chair-holder Erich von Drygalski, filed an application for the unrestricted cartographic mapping of the lost territories, including the colonies, as a natural part of Germany. There were two reports on the regional geography of the war zone, and Norbert Krebs gave the keynote address on the "Territorial changes of the World War" (Krebs 1922).

Verband Deutscher Hochschullehrer der Geographie (Association of German Academic Geographers)

Since the acceptance of geography as a university subject and the establishment of chairs of geography, the academic geographers increasingly formed a social group of their own, often in conflict with school and amateur geographers. After World War I the Geographentag developed in an increasingly unsatisfactory direction and was dominated mostly by school geographers. It was Alfred Philippson who managed in 1925, on the occasion of the 21st meeting of the Geographentag, held in the Silesian city of Breslau, to establish the Association of German Academic Geographers. In the beginning, the Association was dominated by school and curricular affairs. The demand for a better representation of geography at all levels of the schools and universities was expressed by means of several resolutions addressed to the education authorities in charge.

In the 1930s Nazi ideology became more and more dominant and governed the Association's work. Above all, Albrecht Burchard, Ludwig Mecking, Hans Mortensen, Siegfried Passarge and Wolfgang Panzer represent this chapter of German academic geography. As we have seen in the history of the school geographers' association, the academic geographers also were unable to keep Nazi ideology out of their Association. The work of the Association of German Academic Geographers down to the end of World War II seems to be very convoluted. An official liquidation of the Association never took place. Immediately after the end of the war, some influential members tried to continue the institutional work. Above all, Carl Troll, Wilhelm Credner and Emil Meynen managed the reinstallation in those years. In 1947 the first postwar meeting of academic geographers was held in Bonn under Troll's chairmanship.

Research funding and geography

Before the 1920s research funding was guaranteed by several institutions, above all by the academies of science and the scientific associations. As previously noted, the geographical societies, in particular the Berlin society, supported research, expeditions and publications since the middle of the 19th century. The most influential research network was the Kaiser-Wilhelm-Gesellschaft, which provided support for the sciences, especially the booming field of physics. Centralised government funding was institutionalised by means of the Notgemeinschaft der Deutschen Wissenschaft, which was known as the Deutsche Forschungsgemeinschaft (German Research Council) after 1920 (Hammerstein 1999). This organisation, constituted two years after the end of World War I, was headed by the former Prussian Minister of Culture and Education Friedrich Schmidt-Ott and included the German academies of science, 35 universities, the technical universities, the Kaiser-Wilhelm-Gesellschaft and some other research institutions. The Council's goal was to guarantee scholarships, support scientific publications and libraries and, last but not least, fund specific research studies. Under the custody of the Ministry of the Interior and controlled by the Reichstag, the body was divided into special commissions for several science subjects, among them geography.

From 1921 to 1927 Alfred Philippson was the German Research Council's commissioner for geography. Erich von Drygalski assumed that position in 1927. A sub-commission on expeditions (Reisekommission), made up of Drygalski, Max Friedrichsen, Norbert Krebs, Albrecht Penck and Philippson himself, was installed for the Geographentag meetings in Leipzig in 1921. After 1925 the official duties were assigned to a new commission: Philippson (chairman), Krebs (regional geography) and Leo Waibel (human geography). Within German Research Council's special commissions, endowed with extra research funding, large research projects were carried out. Among the geographical projects were the following:

1. Atlas of German folklore
2. Pre- and early history of eastern Germany
3. Research in the Saar region
4. German historical geography

The Research Council was dominated by the Nazis after 1933. In 1934 Schmidt-Ott was urged to step down, and he was replaced by the physicist Johannes Stark. Several institutions provided research funding in Nazi Germany. These institutions had changing and confusing responsibilities and were often in competition with each other (Rössler 1990, 77). After the outbreak of World War II another organisation became influential, the Reichsforschungsrat, a council that was established in 1937 to help fulfil the Vierjahresplan (the Nazi four-year plan to expand the German economy in preparation for the war effort).

Eckart Ehlers has described the post-World-War-II funding of geography in West Germany (Ehlers 1992, 19-25). In East Germany geographical research was partly connected with political and economic projects, often in order to fulfill the Socialist Party's resolutions and economic plans. Apart from research conducted by the East German university institutes and planning authorities (especially the Forschungsleitstelle für Territorialplanung), the Academy of Science with its Geography Department had considerable strength and backing. Geographical research in East Germany was somewhat restricted, however. It was competitive in the international scientific community only in some fields—e.g., geo-ecology, cartography and research on urban agglomerations and settlement systems.

German geography and National Socialism (1933-1945)

(Black/Smith 1946; Heinrich 1990a, 1991; Heske 1986, 1988; Rössler 1987, 1989, 1990; Schelhaas 1997; Troll 1947, 1949; Wissmann 1949)

The year 1934 marked the establishment of national-socialist geography in Germany. Hitler's seizure of power in 1933 did not mean a drastic change in the contents of the sciences but, rather, in the institutions. There was an expansion of these institutions, but the new regime brought forced conformity, censorship and a strict ideological science policy. Several opportunistic geographers saw that they could establish a successful career by adapting to the new order.

During the 25th Deutscher Geographentag at the small spa town of Bad Nauheim in 1934, an agenda was set for a national-socialist geography that fulfilled the NSDAP party doctrine. Vienna was originally designated to organise the meeting but had to cancel because of political disturbances in Austria. The delegates passed some landmark resolutions in Bad Nauheim. Accommodation to the Nazi science policy became an act of affirmative acceptance for the majority of German geographers. The papers read at the meetings reflected the political doctrine—e.g., Hans Schrepfer's lecture on "Raum, Rasse und Volk" ("Space, Race and Nation," 1935) and Ludwig Mecking's keynote speech on the future of national-socialist academic geography (1935).

The Jena meeting in 1936 was the last official meeting until after World War II. Nazi ideology was again well represented. The main topics were spatial research (Raumforschung) as a new and applied science subject, which profited from the Nazi policy, and also school geography and its new challenges. Two local meetings took place before the outbreak of war in 1938—the Bremen meeting for the northern German geographers (Niederdeutscher Geographentag) and the Reichenau meeting for the southern geographers (Oberdeutscher Geographentag). The larger meeting that was planned for the Sudeten spa town of Teplitz-Schönau was cancelled.

The war brought the military conscription of many geographers but also the expansion of several planning institutions. Geographical activities did not rest during the war. Four meetings of German geographers were arranged in 1940 and 1941 (Kiel, Leipzig, Vienna and Prague). Institutional alignment with the Nazi strategy, essential for the war effort, was forced onto these assemblies. Oskar Schmieder and Paul Ritterbusch became the dominant organisers in those years. Geography was integrated into the large-scale project Einsatz der Geisteswissenschaften im Krieg (Input of the humanities in the war effort), a well supported working group headed by Ritterbusch that was to fit the humanities (including geography) into the war strategy, as was happening with the natural sciences. Oskar Schmieder, a colleague of Ritterbusch at the University of Kiel, took advantage of this special situation. He managed to gain a very powerful position in German geography between 1939 and 1945. During the Kiel meeting Schmieder took over the chairmanship of the Central Commission of the Deutscher Geographentag, and Heinrich Schmitthenner became his deputy.

The old plan of installing one umbrella organisation over German geography was achieved for the first time in 1941, mainly under Schmieder's direction (Schelhaas 1997). Within the Deutsche Geographische Gesellschaft (German Geographical Society) all geographical societies of the Reich, the German Institute for Regional Geography at Leipzig and the Abteilung für Landskunde (Meynen's Department of Regional Geography) were collected under a strict organisational body. The German Geographical Society must be regarded as an ambivalent product of the concentrating and controlling of sciences in Nazi Germany. This first "successful" umbrella organisation was concerned with war-related research, and in its short life only a few activities were organised. Lebensraumfragen europäischer Völker (European lebensraum research) was the most important project of those years, and it attracted many geographers. Three published volumes, with some strict Nazi chapters, illustrate the tenor of German geography at that time. The editors, Karl Heinz Dietzel, Schmitthenner and Schmieder, set the tone in the preface of the first volume: ". . . the root of all the other problems that mankind is suffering from ultimately stems from a geographical matter, the glaring imbalance between peoples and their lebensraum . . . That is why German geographers have decided to take a stand on the most important current issues" (Dietzel/Schmitthenner/Schmieder 1941).

The second highlight in the German Geographical Society's work was the European geographical meeting at Würzburg in March 1942. After the expulsion of Germany and Austria from the International Geographical Union in 1925, German-speaking geographers were disadvantaged in international scientific competition. Attempts at reconciliation from 1934 onward were not very successful. The German participation in the International Geographical Congress at Warsaw (1934) and Amsterdam (1938) was strictly regulated and dominated by nationalist geography (Robic/Briend/Rössler 1996). The Würzburg meeting was to be the beginning of a new international geographical platform, dominated by the fascist states, as a kind of counterpart of the IGU. The assembled Italian, Spanish, Finnish, Bulgarian and German geographers decided to meet on a regular basis, but further meetings were prevented by the events of World War II. The ambivalent and mysterious life of the German Geographical Society is difficult to comprehend. Its formal liquidation took place in the 1950s.

A less academic but still important field of German geography in those years was the booming field of regional planning. The institutionalisation of the relevant geographical organisations and their integration into Nazi and war policy has been described by Mechtild Rössler (1987, 1989, 1990). Lebensraum research, often in the occupied territories in the East (Ostforschung), was carried out in several governmental and university departments—e.g., the Reichsstelle für Raumordnung (Agency for Regional Policy, 1935) and the Reichsarbeitsgemeinschaften für Raumordnung (Working groups on spatial research, 1935) (Vennhoff 2000). German geographers were involved in the heartless work of resettlement planning, the degrading expulsions and the support of genocide. Christaller's Central Place theory must be reexamined in light of its political and military ramifications (Rössler 1989).

In the final stages of World War II the connections between geography and war strategy grew even greater. Beside the activities within the Reichsforschungsrat and the Wehrmacht, many university geographers were engaged in military research, which was above all concerned with the occupied eastern territories. In 1943 Hermann Göring appointed a special commissioner for geographical research (Sonderbeauftrager für erdkundliche Forschung), headed by the young explorer and military man Otto Schulz-Kampfhenkel. Many German geographers were involved with military geography, often at the front lines, and were also concerned with new research instruments such as air photo interpretation and spatial analysis.

Continuity after 1945 (Black/Smith 1946; Böhm 1991; Sandner 1995)

Immediately after the end of the war some geographers were able to continue their work and to find new fields to explore, often in cooperation with the Allies. Between March and June 1945 the American geographers Thomas Smith and Lloyd Black visited numerous German geographical institutions, such as university departments, civil and war agencies and private institutions. They interviewed a number of geographers and assessed the status of past and present German geography. Black and Smith ended their report with the sober statement: "The German geographers are confronting so many difficulties that their return to professional usefulness will be slow. Many of these [difficulties] are physical. . . . The prime problems, however, are still deeper. If the geographers are to succeed, they must eliminate from their ranks all those who supported the Nazi regime or who were passively sympathetic to it; they must reorient their research toward the internal problems of Germany; and they must revise the entire Nazi philosophy of geographical education to eliminate, especially, ideas of nationalistic expansion" (Black/Smith 1946: 408). Samuel Van Valkenburg came to a similar conclusion but was more pessimistic. He saw a special trait in the German geographers: "their desire was to clear themselves from the blame of Nazi contacts, or if they had been members of the Party to bring out the fact that the membership was made under pressure" (Van Valkenburg 1951, 114).

In addition, the FIAT-Review of German Sciences with its geographical section compiled by Hermann von Wissmann (1949) and Carl Troll's review of German geography from 1933 to 1945 (1947, 1949) are important published sources on the immediate postwar period.

Subsequently, three geographers and their networks became very important: Carl Troll, Wilhelm Credner and Emil Meynen. The little town of Scheinfeld in Franconia, the home of Troll's mother, developed into the most influential centre of German geography for a short time. The Geographical Institute of Bonn University was evacuated to Scheinfeld under fortunate circumstances in the winter of 1944, managed by Troll and Wolfgang Hartke (Böhm 1991, 303ff). Meynen's Abteilung für Landeskunde (Department of Regional Geography) made an odyssey at the end of the war and during the first postwar days from Berlin to the Thuringian town of Worbis and then, after the Soviet occupation of Thuringia, to Scheinfeld in June 1945 (and later to Landshut, Remagen and finally to Bonn). The

United States Army was interested in having the Department, with its extensive collections of maps, aerial photographs and statistics, in its own military territory (Böhm 1991, 309).

Another organisation, the Reichsarbeitsgemeinschaft für Raumforschung, was able to resume its work a few days after the capitulation under the old chairman, Kurt Brüning, without grave problems. The name of the body was changed a year later to Akademie für Raumforschung und Landesplanung (Academy for Spatial Research and State Planning).

In the Black/Smith and Van Valkenburg reports one major problem of German society in the recovery years was brought out: the different strategies of denazification. As we have seen, the majority of geographers were involved with Nazi policy; opposition to the cruel system was rather minimal. The year 1945 did not bring a sharp personnel break in the western part of Germany. Biographies of many of the geographers show them involved in planning for the reconstruction of the destroyed towns in that period. Meynen's department was a sort of reservoir of geographical talent in those chaotic times.

After the Potsdam resolutions (i.e., resolutions of the Allied leaders at the Potsdam Conference in 1945) and the stabilisation of the occupied zones, academic geographers tried to organise a basic network, supported by representatives from the zones (Behrmann for the Soviet zone, Troll for the British and Credner for the American and French). Böhm has published some interesting documents regarding the wartime and postwar situation in the University of Bonn and Carl Troll's influential activities (Böhm 1991, 301ff). The assembly of the academic geographers in Bonn in 1947 was their first organisational success.

In a recent review, Gerhard Sandner has described the immediate postwar situation. He distinguished between two institutional steps: first, consolidation (1945-1948), which ended with the Geographentag at Munich in 1948; and second, stabilisation of geographical and institutional frameworks (1949-1951), based on the first periodicals, the foundation of the Zentralverband der Deutschen Geographen and the resumption of research funding by the German Research Council (Sandner 1995: 144). The development within the Soviet zone and GDR was very different.

West German postwar geography (Ehlers 1992; Manshard 1976)

In the late 1940s the geographical institutions were mostly fit for work. Geographical periodicals resumed publication in 1947. Again Carl Troll was a driving force. His new journal *Erdkunde* developed into an internationally successful periodical (Böhm/Ehlers 1996). An internal network was constituted by the *Rundbrief* (since 1948) and the *Geographisches Taschenbuch* (since 1949), both edited by Emil Meynen. The *Rundbrief* carried internal news about research projects and academic affairs, including a calendar of events, as a kind of official newsletter of the Zentralverband down to 1991. The *Taschenbuch* contained lists of names and addresses of geographers and their institutions and in the first volumes included statistics, regional geographic essays, and review articles with a didactic purpose. The traditional periodical of the Berlin Geographical Society, which was declared the official organ of the German Geographical Society during the last years of the Nazi period, was re-installed in 1949 under the new name *Die Erde*. In the same year, the school geographers established the *Geographische Rundschau*, managed by Julius Wagner, and it is today the periodical with the highest circulation. The *Geographische Zeitschrift* resumed publication in 1963.

At the Geographentag meeting in Frankfurt in 1951, the Zentralverband der Deutschen Geographen, the umbrella organisation of all German geographical associations down to 1995, was launched. The academic geographers again played the dominant rôle. Only university professors were allowed to gain membership in the Association of German Academic Geographers. After the 1970s the organisational structure of the Zentralverband became quite convoluted, as Ehlers has pointed out: "Traditions, the perseverance of hierarchical thinking and order and certainly also the special German fondness for joining organizations have contributed to an organizational structure in German geography that is probably unique in its complexity and sometimes also ineffectiveness"(Ehlers 1992, 16).

Beside the traditional Association of German School Geographers (re-organised in 1949) and the Association of German Academic Geographers (re-organised in 1948), a third organisation was established: the Verband Deutscher Berufsgeographen (Association of German Professional Geographers), founded in 1950 and mainly supported by Walter Christaller, Paul Gauss and Emil Meynen. The aim of the professional association was to create a platform for applied geography and regional planning. In 1980 the association changed its name to Deutscher Verband für Angewandte Geographie (German Association of Applied Geography).

The Geographentag at Kiel in 1969 provided a lasting break, a real revolution within West German geography. The younger geographers, especially the students, proclaimed the establishment of a new and scientific subject in the schools and universities to replace the unsatisfactory older variety. The old Länderkunde (regional geography) was rejected. A general reform of university education and institutional structures was demanded.

Since the 1960s emphasis shifted to social geography and landscape ecology, but the traditional institutional powers could still defend their status. The old Association of German Academic Geographers prevented a more democratic and pluralistic structure from becoming dominant. The consequence was the "complexity and ineffectiveness" that Ehlers spoke of.

In 1970 some academic geographers who could not become members of the Association of German Academic Geographers because they had not gone through the habilitation process launched a new organisation, the Verband Deutscher Hochschulgeographen (Association of German University Geographers), mainly in order to "strive for better cooperation between all university geographers" (Ehlers 1992, 18). A year later, a fifth association was born: the Hochschulverband für Geographie und ihre Didaktik (University Association for Geography and Its Teaching Methods). All discussions on cooperation or merger of the various groups were bound to fail. Down to 1990 and, to some extent, even to the present day, it has been the traditional geographers with their Association and their powerful university chairs who have mainly determined science policy.

University geography has expanded rapidly since the end of the 1960s. The departments were enlarged to accommodate the increasing numbers of students. Geography, like other disciplines, developed into an overcrowded subject. Some new universities were founded, often with a reforming educational impetus—e.g., the university at Bochum, with its well known geography department headed by Peter Schöller. Almost all German institutions of higher learning, including the technical universities and teachers colleges, now have geography sections. In the 1980s applied geography became more and more relevant. Professional geographers could now graduate with a diploma degree.

East German Geography (F. Hönsch 1978; Schmidt/Richter 1995; Schelhaas 2000; Sperling 1977)

The segregation of the two German geographical developments finally began in the 1950s. The first years after the war were marked by destroyed cities and universities and the total loss of ideology. Nevertheless, there was a strong will to overcome these difficulties. After a brief presence of the American Army in Thuringia and parts of Saxony, the Soviet Army began to organise civilian life in the summer of 1945. In the first period (parallel to the Western development) it was necessary to reconstruct the university buildings, to organise the return of evacuated documents, instruments and books and—most important—to denazify the teaching staff. Johannes Gellert, Rudolf Käubler, Ernst Neef and Joachim Schultze were suspended from their duties at the universities. Some leading geographers, such as Heinrich Schmitthenner and Karl Heinz Dietzel of Leipzig University, were evacuated by the U. S. Army to a camp near the Hessian town of Weilburg in order to protect them from the Soviet Army and to use their special knowledge. Schmitthenner got a chair in geography at the University of Marburg. The "brain drain" of German academics, instigated by both the Western Allies and the Soviets, is a fact of German history that has not always been acknowledged.

The Soviet military administration demanded the reopening of the schools and universities as early as possible. The old traditional institutes of geography resumed their work in the autumn of 1945, in spite of a great lack of qualified teaching staffs. Retired professors, such as Hassert, Schlüter and Krebs, returned to their old chairs. Jena was the first university in the Soviet sector to be reopened, followed by Berlin, Halle, Leipzig, Greifswald and Rostock.

The first university reform in 1946—the so-called "Antifascistic-Democratic" reform—in East Germany was directed at the abolition of the higher education privilege (Brechung des Bildungsprivilegs), to the training of new teachers (Neulehrer) who were not burdened with Nazi ideology, and to the establishment of new forms of education and teaching at the universities. To prepare people who did not have the proper educational background, Arbeiter-und-Bauern-Fakultäten (Faculties for the working class) were founded at all universities. The proletarian pre-studies faculties were to train the students in Marxist philosophy and also educate the urgently needed teachers and cadres.

The famous publishing house of Justus Perthes at Gotha continued its work without any break. The Soviet military administration brought the aged Hermann Haack out of retirement to organise the publishing of maps, serials and teaching materials. Haack, who had been heavily involved in Nazi geography, was a corresponding member of the Russian Geographical Society. His helpful contacts with the Soviets, both military men and geographers, allowed the continuation of the former Nazi-Musterbetrieb (Nazi model plant). In 1948 the famous periodical *Petermanns Geographische Mitteilungen* was revived. The periodical then put additional emphasis on physical geography, cartography and Soviet geography, but the quality remained on a high international level. The nationalised publishing house, renamed VEB Hermann Haack, developed into the largest institution in the GDR for the publication of geographical and cartographic materials.

Another important periodical was begun in 1949, the *Zeitschrift für den Erdkundeunterricht* (School Geography Magazine). This journal developed into a highly influential institution for the propagation of Marxist geography in the schools but also to some extent in the universities. After the foundation of the Geographical Society in the GDR, the journal *Geographische Berichte* (Geographical Reports) was made the official organ of GDR geography in 1956.

The Second University Reform in the GDR (1951/1952) covered more ideological goals and demanded the introduction of Marxist ideas in all sciences. For geography it meant the division of the field into physical and economic geography. The latter (Politische und Ökonomische Geographie) needed new specialists. A small group of young geographers, trained especially by the economist Heinz Sanke, a protagonist of communist ideology, gained positions of leadership in all geographical institutes and governmental authorities (e.g., Horst Kohl, Helmut Riegraf and Alfred Zimm).

The university institutes developed in a stable manner. Beside the old ones in Berlin (Humboldt-University), Halle, Greifswald, Leipzig, Jena, Rostock and Dresden (Technical University), new chairs were founded at the Pädagogische Hochschule Dresden (teacher training college), the Verkehrshochschule Dresden (Traffic and Transportation University), the Brandenburgische Landeshochschule in Potsdam and the Hochschule für Ökonomie in Berlin.

Aside from university geography, a number of geographical working groups (Arbeitskreise für Erdkunde), which became part of the socialist mass organization Kulturbund in 1949, tried to carry on the old ideas of the traditional geographical societies. Their lectures and field trips were given not only by East German geographers but also by West German and (infrequently) foreign geographers. After the foundation of the GDR in 1949, science policy and national education became increasingly centralised and strictly organised.

A team of active geographers launched the centralised Geographical Society in (later "of") the GDR in 1953, led by the first president, Ernst Neef. The Society organised cooperation between East German and foreign geographers. The national association was divided into specialised sections for physical and economic geography, cartography, school geography and applied geography. The local sections (often the former local working groups) had more popular functions, above all in lectures and in-service training of teachers. In the first years annual meetings were organised (Jahreshauptversammlung), later general scientific meetings (Wissenschaftliche Hauptversammlung) at intervals of two years, and from 1975 to 1989 Geographical Congresses (Geographenkongress). Contacts with geographers in West Germany were alternated with contacts with colleagues in socialist countries. Both organisation and curricula were reformed after the Soviet model.

On the occasion of the 19th International Geographical Congress, held in Stockholm in 1960, GDR geographers were accepted by the International Geographical Union as equal partners within the international organisation. This decision was supported by the delegates from socialist countries and also by the IGU president Dudley Stamp, who had visited the GDR in 1959. From 1959 to 1990 two National Committees for Geography existed in Germany. After the construction of the Berlin Wall in 1961, East German geographers were increasingly separated from developments in West Germany, and finally there was a total breaking off of all relations, with a few exceptions in the form of private contacts and regular meetings between Western and Eastern teachers before 1965.

In the 1960s conformity between the sciences and party policy became very close. The main topics of research were the problems of economy and social structure on the one hand and of landscape ecology on the other. Within the State Planning Commission (Staatliche Plankommission), an Economic Research Institute was founded, later called the Forschungsleitstelle für Territorialplanung (Research headquarters for territorial planning), which had close contacts with geographers. All scientific activities were planned and ordered by governmental and party authorities. The Ministry of Education and the Ministry of Academic Affairs played an important rôle in formulating curricula and research plans. The party sub-commissions had geographical sections, which were primarily concerned with the propagation of Marxist geography. In spite of everything, the importance of geography was perhaps greater than in West Germany. The high degree of organisation and the cadre system brought a great sense of certainty, but sciences in the GDR were part of the socialist and totalitarian social system, which brought injustice and partial isolation within the international scientific community.

In the 1960s a division of research took place. The Greifswald Institute, for instance, was specialised in coastal research and recreation geography; Leipzig and Dresden became centers for geoecology, above all driven by Ernst Neef and his students; and Berlin developed a nucleus of economic and political (Marxist) geography under Heinz Sanke's leadership. The 3rd Academic and University Reform in 1968-1969 was directed to the incorporation of the sciences, including geography, into the social system of socialism. Both teaching and research were to promote the socialist economy and policy of the GDR. The institutional consequences were very great:

1. Geographical teaching and research were to be concentrated in a few large institutes. The traditional institutes in Leipzig, Jena, Dresden and Rostock were closed, and their staff members were transferred to the larger institutes in Halle, Berlin and Greifswald.
2. Traditional university faculties and institutes were transformed into so-called sections.
3. Within the Academy of Sciences in Leipzig there was established a special research institute for geography, the Institute for Geography (and later) Geoecology, in place of the old Deutsches Institut für Länderkunde (German Institute for Regional Geography).
4. The Geographical Society of the GDR became affiliated with the Academy of Sciences.

This restructuring enabled the Sozialistische Einheitspartei Deutschlands (Socialist Unity Party of Germany) to gain greater control of the management of the institutions. The concentration represented a loss, but also an opportunity, for geography. Special working groups for cooperation in the fields of environmental problems, landscape ecology and settlement structure of the GDR were established, organised by the State Planning Commission and the Ministry of Science and Technology.

A main topic of the cooperation between geographers and cartographers in the 1970s was the creation of a national atlas (*Atlas DDR*) under the leadership of Edgar Lehmann and Heinz Sanke. The Institute for Geography and Geoecology at Leipzig successfully worked in the areas of environmental evaluation, spatial organisation of economy and settlement structure of the GDR, as well as in the methodology of remote sensing and monitoring of the earth. All research projects operated at a high international level (I. Hönsch 1995, 1996).

Some trends after the reunification of Germany

The political changes in Eastern Europe since 1989 have had consequences for the institutional structure of German geography. The attempts to reconcile GDR and FRG traditions caused intensive discussions among the geographers. The Geographentag at Basel (1991) was a platform for these discussions. The two Associations of Academic Geographers were combined to form the Verband der Geographen an Deutschen Hochschulen (Association of Geographers at German Universities); the habilitation was no longer a prerequisite for membership. On the occasion of the first Geographentag held in the former GDR (Potsdam 1995), the delegates launched a new organisation, the Deutsche Gesellschaft für Geographie (German Society of Geography), which replaced the old Zentralverband.

After the reunification of Germany, geography in the Eastern part went through a new phase. At first there was a debate about the correction of the 3rd University Reform. As a result, the university institutes in Leipzig and Jena were reopened. After a successful evaluation by the Scientific Council of the FRG, the former Institute for Geography and Geoecology in Leipzig was transformed into a smaller Institut für Länderkunde (Institute for Regional Geography). New geography chairs were created in Erfurt, Frankfurt/Oder and Chemnitz. Another product of the reunified German geography is the present project of the German National Atlas, combining geography, cartography and other sciences—once again a belated development in the process of nationalisation and institutionalisation within German geography.

References:

Arnhold, Helmut (1965) Die Pflege der Geographie in Leipzig 1860-1870, *Leipziger Geographische Beiträge*, 7-16.

Bader, Frido J. W. (1978) Die Gesellschaft für Erdkunde zu Berlin und die koloniale Erschließung Afrikas in der zweiten Hälfte des 19. Jahrhunderts bis zur Gründung der ersten deutschen Kolonien, *Die Erde* 109, 36-48.

Bader, Frido J. W. (1979) Rückblick nach 100 Jahren der Zeitschrift der Gesellschaft für Erdkunde zu Berlin, *Die Erde* 110, 93-117.
Bartels, Dietrich (1968) Zur wissenschaftstheoretischen Grundlegung einer Geographie des Menschen, *Geographische Zeitschrift, Beihefte, Erdkundliches Wissen* 19, Wiesbaden.
Bernhardt, Peter (1981) "Petermanns Geographische Mitteilungen" und die deutschsprachigen geographischen Zeitschriften bis zum Ende des 19. Jahrhunderts, *Petermanns Geographische Mitteilungen* 125, 167-182.
Black, Lloyd D. (1947) Further notes on German geography, *Geographical Review* 37, 147-148.
Black, Lloyd D. and Thomas R. Smith (1946) German geography, *Geographical Review* 36, 398-408.
Böhm, Hans, ed. (1991) Beiträge zur Geschichte der Geographie an der Universität Bonn, *Colloquium Geographicum* 21, Bonn.
Böhm, Hans and Eckart Ehlers (1996) Erdkunde - 50 Jahrgänge "Archiv für wissenschaftliche Geographie". Ein Rückblick und ein Ausblick, *Erdkunde* 50, 360-379.
Brogiato, Heinz Peter (1995) Die Schulgeographie im Spiegel der Deutschen Geographentage, *Geographische Rundschau* 47, 484-490.
Brogiato, Heinz Peter (1998) "Wissen ist Macht - Geographisches Wissen ist Weltmacht". Die schulgeographischen Zeitschriften im deutschsprachigen Raum (1880-1945) unter besonderer Berücksichtigung des Geographischen Anzeigers, *Materialien zur Didaktik der Geographie* 18, Trier.
Büdel, Julius (1982) Die Deutschen Geographischen Gesellschaften als Bewahrer der Welt- und Landeskunde in den Stufen der großen Bürgerzeit 1828-1982, *Mitteilungen der Geographischen Gesellschaft Lübeck* 55, 7-22.
Dietzel, Karl-Heinz, Oskar Schmieder, and Heinrich Schmitthenner, eds. (1941) *Lebensraumfragen Europäischer Völker*, Vol. 1, *Europa*, Leipzig.
Ehlers, Eckart (1992) German geography 1945-1992: Organizational and institutional aspects, in Eckart Ehlers (ed.), *40 Years After: German Geography. Developments, Trends and Prospects 1952-1992*, Bonn.
Eisel, Ulrich (1980) Die Entwicklung der Anthropogeographie von einer 'Raumwissenschaft' zur Gesellschaftswissenschaft, *Urbs et Regio* 17, Kassel.
Elkins, Thomas H. (1990) Human and regional geography in the German-speaking lands in the first forty years of the twentieth century: an outsider's view, in Eckart Ehlers (ed.), Philippson-Gedächtniskolloquium, *Colloquium Geographicum* 20, Bonn, 21-33.
Engelmann, Gerhard (1983) Die Hochschulgeographie in Preußen 1810-1914, *Erdkundliches Wissen* 64, Wiesbaden.
Fahlbusch, Michael (1994) "Wo der deutsche . . . ist, ist Deutschland!" Die Stiftung für Volks- und Kulturbodenforschung in Leipzig 1920-1933, *Abhandlungen zur Geschichte der Geowissenschaften und Religion/Umwelt-Forschung*, Supplement no. 6, Bochum.
Fahlbusch, Michael (1999) *Wissenschaft im Dienste der nationalsozialistischen Politik? Die "Volksdeutschen Forschungsgemeinschaften" von 1931-1945*, Baden-Baden.
Fahlbusch, Michael, Mechtild Rössler, and Dominik Siegrist (1989) Conservatism, ideology and geography in Germany 1920-1950, *Political Geography Quarterly* 8, 353-367.
Fick, Karl E. (1986) 150 Jahre Geographische Gesellschaft zu Frankfurt am Main 1836-1986, *Frankfurter Geographische Hefte* 55, 3-106.

Gärtner, Rainer W. (1992) Die Entwicklung der wissenschaftlichen Geographie in Sachsen unter dem Einfluß von Oscar Peschel, Otto Delitsch und Ferdinand von Richthofen, Unpublished dissertation, University of Halle-Wittenberg, Halle.

Gellert, Johannes F. (1984) Die Gesellschaft für Erdkunde zu Leipzig 1861-1945. Historische Analyse und Erinnerung, *Arbeitsblätter für Wissenschaftsgeschichte der Martin-Luther-Universität Halle-Wittenberg* 14, 42-57.

Georgi, Johannes (1968) "Die Allgemeine Deutsche Versammlung von Freunden der Erdkunde" in Frankfurt a.M. 1865 und ihre Bedeutung für die Geographie, *Petermanns Geographische Mitteilungen* 112, 104-111.

Grimm, Frank-Dieter (1998) Zur Geschichte und Gegenwart der Geographischen Gesellschaft zu Leipzig, in Alois Mayr and Helga Schmidt (eds.), *Geographie in Leipzig. Tradition und Neubeginn*, Leipzig, 94-100.

Guntau, Martin (1986) Wissenschaftliche Gesellschaften und disziplinäre Erkenntnis in der Geschichte der Wissenschaft, *Rostocker Wissenschaftshistorische Manuskripte* 14, 7-23.

Hammerstein, Notker (1999) *Die Deutsche Forschungsgemeinschaft in der Weimarer Republik und im Dritten Reich. Wissenschaftspolitik in Republik und Diktatur*, München.

Hard, Gerhard (1970) Die 'Landschaft' der Sprache und die 'Landschaft' der Geographen, *Colloquium Geographicum* 11, Bonn.

Hartshorne, Richard (1939) The nature of geography: a critical survey of current thought in the light of the past, *Annals of the Association of American Geographers* 29, 171-658.

Heinrich, Horst-Alfred (1990a) Der politische Gehalt des fachlichen Diskurses in der Geographie Deutschlands zwischen 1920 und 1945 und dessen Affinität zum Faschismus, *Geographische Zeitschrift* 78, 209-226.

Heinrich, Horst-Alfred (1990b) Der propagandistische "Mißbrauch" geographischer Fachzeitschriften bei der Annexion Österreichs 1938, *Petermanns Geographische Mitteilungen* 134, 193-200.

Heinrich, Horst-Alfred (1991) Politische Affinität zwischen geographischer Forschung und dem Faschismus im Spiegel der Fachzeitschriften. Ein Beitrag zur Geschichte der Geographie in Deutschland 1920-1945, *Gießener Geographische Schriften* 70, Gießen.

Hellmann, Gustav (1928) Aus der Geschichte der Gesellschaft für Erdkunde zu Berlin im zweiten halben Jahrhundert ihres Bestehens (1879-1928), *Zeitschrift der Gesellschaft für Erdkunde zu Berlin, Sonderband zur 100-Jahr-Feier*, 1-14.

Heske, Henning (1986) German geographical research in the Nazi period: a content analysis of the major geography journals, 1925-1945, *Political Geography Quarterly* 5, 267-281.

Heske, Henning (1988) "*... und morgen die ganze Welt...*" *Erdkundeunterricht in Nationalsozialismus*, Gießen.

Höhl, Gudrun (1981) 100 Jahre Deutscher Geographentag 1881-1981, in Ingrid Dörrer (ed.), Mannheim und der Rhein-Neckar-Raum. Festschrift zum 43. Deutschen Geographentag in Mannheim, *Mannheimer Geographische Arbeiten* 10, 13-28.

Hönsch, Fritz (1978) 25 Jahre Geographische Gesellschaft der DDR, *Geographische Berichte* 23, 87-98.

Hönsch, Ingrid (1995) Kontinuität und Diskontinuität geographischer Forschungs- und Sammeltätigkeit am Länderkunde-Institut Leipzig, in Ingrid Hönsch and Ute Wardenga (eds.), Kontinuität und Diskontinuität der deutschen Geographie in Umbruchphasen. Studien zur Geschichte der Geographie, *Münstersche Geographische Arbeiten* 39, 9-18, Paderborn.

Hönsch, Ingrid (1996) Die Entwicklung des Instituts 1896-1989, in Alois Mayr, Frank-Dieter Grimm, and Sabine Tzschaschel (eds.), 100 Jahre Institut für Länderkunde 1896-1996. Entwicklung und Perspektiven, *Beiträge zur Regionalen Geographie* 40, 11-33, Leipzig.

Jüngst, Peter (1993) Deutsche Geographie und Nationalismus - ein historisches Thema?, *Geographische Zeitschrift* 81, 69-81.

Köhler, Franz (1981) 125 Jahrgänge von Petermanns Geographischen Mitteilungen. Wandlungen im Profil der Zeitschrift, *Petermanns Geographische Mitteilungen* 125, 1-10, 109-115.

Kost, Klaus (1988) Einflüsse der Geopolitik auf Forschung und Theorie der Politischen Geographie von ihren Anfängen bis 1945. Ein Beitrag zur Wissenschaftsgeschichte der Politischen Geographie und ihrer Terminologie unter besonderer Berücksichtigung von Militär- und Kolonialgeographie, *Bonner Geographische Abhandlungen* 76, Bonn.

Kost, Klaus (1988) The conception of politics in political geography and geopolitics in Germany until 1945, *Political Geography Quarterly* 8, 369-385.

Krebs, Norbert (1922) Die territorialen Veränderungen des Weltkrieges, in Walter Behrmann (ed.), *Verhandlungen des 20. Deutschen Geographentages zu Leipzig*, Berlin, 196-217.

Lenz, Karl (1978) 150 Jahre Gesellschaft für Erdkunde zu Berlin, *Die Erde* 109, 67-86.

Linke, Max (1986) Deutsche Geographische Gesellschaften von den Anfängen bis in das erste Jahrzehnt der Einrichtung geographischer Lehrstühle an deutschen Universitäten, *Petermanns Geographische Mitteilungen* 130, 247-255.

Manshard, Walter (1976) Recent trends in German geography, *Geoforum* 7, 383-385.

Mecking, Ludwig (1935) Die Geographie an den Hochschulen, in Albrecht Haushofer (ed.), *Verhandlungen und wissenschaftliche Abhandlungen des 25. Deutschen Geographentages zu Bad Nauheim 1934*, Breslau, 161-170.

Mehmel, Astrid (1995) Deutsche Revisionspolitik in der Geographie nach dem Ersten Weltkrieg, *Geographische Rundschau* 47, 498-505.

Meyer, Hans (1922) Ansprache des Vorsitzenden des Geographentages, in Walter Behrmann (ed.), *Verhandlungen des 20. Deutschen Geographentages zu Leipzig*, Berlin, 14-18.

Meynen, Emil (1948) Die Tradition des Amtes für Landeskunde. Ein Rückblick auf die Zentralkommission für wissenschaftliche Landeskunde von Deutschland 1882-1941, *Berichte zur deutschen Landeskunde* 5, 2-33.

Meynen, Emil (1966) Institut für Landeskunde. Das erste Vierteljahrhundert seiner Tätigkeit 1941-1966, *Berichte zur deutschen Landeskunde* 36, 145-206.

Meynen, Emil and Gerold Richter (1982) Der Zentralausschuß für deutsche Landeskunde. Ein Bericht über die Jahre 1941-1982, *Berichte zur deutschen Landeskunde* 56, 63-87.

Nix, Bernd and Tobias Wölke (1986) Die Gesellschaft für Erdkunde zu Berlin, *Berichte der Humboldt-Universität Berlin* 6, no. 4 (*Berliner Geographische Arbeiten Sonderheft* 3), 56-72.

Petri, Hans-Hermann (1975) Die Geographische Gesellschaft in Bremen und die Völkerkunde, *Jahrbuch der Wittheit zu Bremen* 19, 121-141.

Quelle, Otto (1953) *125 Jahre Gesellschaft für Erdkunde zu Berlin*, Berlin.
Richter, Hans (1998) Umfeld, Umbrüche und wissenschaftliches Profil des Geographischen Instituts der Universität Leipzig von 1871 bis 1996, in Alois Mayr and Helga Schmidt (eds.), *Geographie in Leipzig. Tradition und Neubeginn*, Leipzig, 18-49.
Robic, Marie-Claire, Anne-Marie Briend and Mechtild Rössler, eds. (1996) *Géographes face au monde. L'Union Géographique Internationale et les Congrès Internationaux de Géographie*, L'Harmattan, Paris and Montréal.
Rössler, Mechtild (1987) Die Institutionalisierung einer neuen Wissenschaft im Nationalsozialismus. Raumforschung und Raumordnung 1935-1945, *Geographische Zeitschrift* 75, 177-194.
Rössler, Mechtild (1989) Applied geography and area research in Nazi society: central place theory and planning, 1933 to 1945, *Environment and Planning D* 7, 419-431.
Rössler, Mechtild (1990) "Wissenschaft" und "Lebensraum". Geographische Ostforschung im Nationalsozialismus, *Hamburger Beiträge zur Wissenschaftsgeschichte* 8, Berlin and Hamburg.
Sandner, Gerhard (1983) Die "Geographische Zeitschrift" 1933-1944. Eine Dokumentation über Zensur, Selbstzensur und Anpassungsdruck bei wissenschaftlichen Zeitschriften im Dritten Reich, *Geographische Zeitschrift* 71, 65-87, 127-149.
Sandner, Gerhard (1988) Recent advances in the history of German geography 1918-1945. A progress report for the Federal Republic of Germany, *Geographische Zeitschrift* 76, 120-133.
Sandner, Gerhard (1994) In search of identity: German nationalism and geography, 1871-1910, in David Hooson (ed.), *Geography and National Identity*, Blackwell, Oxford (England) and Cambridge (Massachusetts), 71-91.
Sandner, Gerhard (1995) Die unmittelbare Nachkriegszeit: personelle, institutionelle und fachlich-inhaltliche Aspekte 1945-1950, in Ute Wardenga and Ingrid Hönsch (eds.), Kontinuität und Diskontinuität der deutschen Geographie in Umbruchphasen. Studien zur Geschichte der Geographie, *Münstersche Geographische Arbeiten* 39, 151-160, Paderborn.
Schelhaas, Bruno (1997) Die Deutsche Geographische Gesellschaft. Ein Beitrag zur Verbandsgeschichte der deutschen Geographie, Unpublished Diplomarbeit (Master's thesis), University of Münster.
Schelhaas, Bruno (2000) DDR-Geographiegeschichte - Problemlagen und Zugriffe einer wissenschaftshistorischen Rekonstruktion. Einige Aspekte aus der Frühphase der DDR-Hochschulgeographie, *Hochschule Ost* 9, 161-178.
Schmidt, Helga and Hans Richter (1995) Entwicklung der Politischen und Ökonomischen Geographie in der DDR, *Geographische Rundschau* 47, 506-512.
Schrepfer, Hans (1935) Raum, Rasse und Volk, in Albrecht Haushofer (ed.), *Verhandlungen und wissenschaftliche Abhandlungen des 25. Deutschen Geographentages zu Bad Nauheim 1934*, Breslau, 65-84.
Schulte-Althoff, Franz (1971) Studien zur politischen Wissenschaftsgeschichte der deutschen Geographie im Zeitalter des Imperialismus, *Bochumer Geographische Arbeiten* 9, Paderborn.
Schultz, Hans-Dietrich (1977) Die Situation der Geographie nach dem Ersten Weltkrieg. Eine unbekannte Umfrage aus dem Jahre 1919, historisch kommentiert, *Die Erde* 108, 75-102.

Schultz, Hans-Dietrich (1980) Die deutschsprachige Geographie von 1800 bis 1970. Ein Beitrag zur Geschichte ihrer Methodologie, *Abhandlungen des Geographischen Instituts, Anthropogeographie, Freie Universität Berlin* 29, Berlin.

Schultz, Hans-Dietrich (1989) Die Geographie als Bildungsfach im Kaiserreich. Zugleich ein Beitrag zu ihrem Kampf um die preußische höhere Schule von 1870-1914 nebst dessen Vorgeschichte und teilweiser Berücksichtigung anderer deutscher Staaten, *Osnabrücker Studien zur Geographie* 10, Osnabrück.

Schultz, Hans-Dietrich (1995) "Was ist des Deutschen Vaterland?" - Geographie und Nationalstaat vor dem ersten Weltkrieg, *Geographische Rundschau* 47, 492-497.

Sperling, Walter (1977) *Geographie und Geographieunterricht in der DDR*, München.

Troll, Carl (1947) Die geographische Wissenschaft in Deutschland in den Jahren 1933 bis 1945. Eine Kritik und eine Rechtfertigung, *Erdkunde* 1, 3-48. English version (1949) Geographic science in Germany during the period 1933-1945. A critique and a justification, *Annals of the Association of American Geographers* 29, 99-137.

Van Valkenburg, Samuel (1951), The German school of geography, in Griffith Taylor (ed.), *Geography in the twentieth century: a study of growth, fields, techniques, aims and trends*, Philosophical Library, New York; Methuen, London, 91-115.

Vennhoff, Michael (2000), Die Reichsarbeitsgemeinschaft für Raumforschung (RAG) und die reichsdeutsche Raumplanung seit ihrer Entstehung bis zum Ende des Zweiten Weltkrieges 1945, *ARL-Arbeitsmaterial* 258, Hannover.

Wardenga, Ute (1995a) Geschichtsschreibung in der Geographie, *Geographische Rundschau* 47, 523-525.

Wardenga, Ute (1995b) "Nun ist alles, alles anders!" Erster Weltkrieg und Hochschulgeographie, in Ingrid Hönsch and Ute Wardenga (eds.), Kontinuität und Diskontinuität der deutschen Geographie in Umbruchphasen. Studie zur Geschichte der Geographie, *Münstersche Geographische Arbeiten* 39, Paderborn, 83-97.

Wardenga, Ute (1999) Constructing regional knowledge in German geography: the Central Commission on the Regional Geography of Germany 1882-1941, in Anne Buttimer, Stanley Brunn, and Ute Wardenga (eds.), Text and image. Social construction of regional knowledges, *Beiträge zur Regionalen Geographie* 49, Leipzig, 77-84.

Weißenborn, Johannes (1938) 50 Jahre Geographische Gesellschaft in Bremen 1877-1927, *Deutsche Geographische Blätter* 40, 249-253.

Wissmann, Hermann von, ed. (1949) Geographie, *Naturforschung und Medizin in Deutschland 1939-1946. Für Deutschland bestimmte Ausgabe der FIAT Review of German Science* 44-47, Wiesbaden.

Bruno Schelhaas and Ingrid Hönsch
Institut für Länderkunde
Schongauerstrasse 9
04329 Leipzig
Deutschland (Germany)

3 Geography in France: Context, Practice, and Text

Vincent Berdoulay

From eulogy in the history of geographical thought to partial oblivion in subdisciplinary assessments, a wide range of opinion has characterized international scholarly opinion on French geography. Obviously, contrasting views have also existed, and still exist, within France, including those held by geographers themselves. Laudatory, critical, or biased opinions, as well as scientific debates, have to be understood in relation to the necessity for this geography to get recognition in the national and international market of ideas. For achieving this status, it depends upon institutional support, which has largely been dependent on national context. It should be noticed, however, that this context is not something given that more or less determines the evolution of geographical thought. Of course, the context has a conditioning rôle, but it also results itself from the initiatives taken by geographers. In addition, detailed historical research easily demonstrates that geographers usually do not form a community—i. e., a unified group that is able to act collectively. The dominant situation is a competitive one, each group taking inspiration and support from non-geographers. Thus, the circulation and elaboration of ideas is best understood by focusing on "circles of affinity" that largely cut across disciplinary boundaries (Berdoulay 1995). They are part of the process of institutionalization, as they help to clarify the differential success achieved by various ways of geographical thought and practice. Most of all, they are the major mediations between text and context. Because of their relative stability, academic institutions, as well as teaching, research, and fieldwork practices, also play a significant mediating rôle. Less noticed, but equally important, discursive genres and rhetorical practices can bring an even longer stability in the mediation between text and context (Berdoulay 1988).

A particular attention paid to major societal issues helps to disclose the various ways in which geographers have interacted with context. The success of French geography is closely linked to the remarkable conjunction that it achieved with cultural, political,

G.A. Dunbar (ed.), Geography: Discipline, Profession and Subject since 1870, 45–78.

institutional and scientific initiatives that played a major rôle in the history of France. Particularly striking at the time of Vidal de la Blache, this phenomenon has repeatedly occurred over the last two centuries. There was something like a favorable ecology for the development of geographical thought in this country, with opportunities that geographers could tap and, in turn, produce. Many publications in several languages have appeared on the subject of French geography. Recently, Paul Claval has contributed a very comprehensive history of French geography from 1870 to the present, with numerous contextual elements and bibliographical references (Claval 1998). Thus, the focus here will be on the selection of a few internationally important contributions by French geographers, with brief reference to their favorable contextual ecology. This concise overview of the history of French geographical thought will be particularly attentive to cultural aspects. Since it is authored by a French geographer, this selective presentation does not eschew subjectivity. But it is informed by a distanced view, on two counts, both because of a long student and professional experience in North American geography and a parallel long involvement in research on international and French geographical thought.

Nineteenth-century France as a seedbed of academic geography

The pioneering contributions of the German geographers Carl Ritter and Alexander von Humboldt in the first half of the 19th century, as well as the upsurge of geography in French universities after 1870, have overshadowed the rich, but diverse and at times unrelated, geographical activities that took place in France before that year. They provided a background knowledge to geographical issues that gave depth to the ideas developed later on. They accustomed people to geographical issues and, whether or not they were developed in opposition to previous ones, they gained legitimacy from the existence of this intellectual field of curiosity.

Actual geographical activities continued in part what the innovative construction of the French state and society had already started to induce. At the end of the 18th century, the French Revolution showed how far this construction had progressed, as the idea of Nation came to the foreground of political conceptions and practices. Given the difficulties of exporting these views far from their place of institutionalization, the universalistic ideas of nation rapidly took a particular cultural and territorial dimension. But the universalistic thrust within the approach to state building, to which 18th-century science contributed, clearly continued in geographic activities.

The same universalistic thrust is felt in geographical exploration, as well as in the growing importance of rivalries among nation-states (Broc 1981, Puyo 1999, Godlewska 1999). More important is the number of private or semi-private initatives for exploring places abroad. The establishment of the Société de Géographie de Paris in 1821 highlights the beginning of this movement. The co-presence in its activities of scientists, travellers, and other laymen provided the ground for a continued geographical inquiry on the diversity of the world's regions. Most of all, through its *Bulletin*, it created the first full-fledged geographical periodical, where reports from the wide world were made available after some form of debate among the members of the Société. In spite of the poorly peer-controlled information that was published, the impact of this pre-institutionalization of geography should not be minimized, as it bred many significant debates. Many of the early contributions to geography by Élisée Reclus were developed within the Société (Dunbar 1978), and that was also the case with many other authors. Most notably, Jules Verne was an active member. Although his *Géographie de la France* was written in the standard descriptive form of the day, the active rôle played by Jules Verne in the Societé de Géographie is illustrative of the breadth of interests of many of its members. The narrowly conceived scientific conceptions that presided over the university institutionalization of geography unfortunately restricted these views, and geography was quite effectively separated from literature for a long time.

Other significant initiatives were taken by various members of the Société. Among them, Jules Duval (1813-1870), who was also close to the Le Play circles, argued for developing the complementarities between geography and political economy. He particularly stressed the benefits that the colonization of previously uncultivated lands would bring to humanity. Indigenous populations were expected to share this additional wealth, evolve, and then gain political emancipation. His personal involvement in colonization heralded the importance that this issue took shortly aferwards in the development and institutionalization of geography. Duval was the editor of an economic periodical and a contributor to a few others at the same time that he was the active vice-chairman of the Société de Géographie de Paris. Actually, with Prosper de Chasseloup-Laubat (1802-1873) as its president, the Société prepared the ground for future colonial propaganda by promoting exploration and by insisting on the civilizing mission of France abroad.

These colonization initiatives are to be included within the growing concern of the French élites for the relative decline of their national strength—i. e., a loss of power as measured in demographic and industrial productivity terms in relation to Prussia, England, and the United States (Claval and Nardy 1968). The "nationalization" of political and social scientific thought, which was also occurring in other European countries, nurtured many debates in France, where the universalistic bias was strong, and this provided the impetus for geography-related attention to place, territory and resources. The social and political unrest, illustrated by the revolutions of 1830 and 1848 or by the coup of Louis Napoléon Bonaparte in 1851, highlighted the need for a better understanding of its underpinnings and for adequate theory to cope with it. The dire living conditions of the industrial proletariat laid the ground for accepting the pressing need for social and political reforms.

Opposed by those who resisted change or by those who became advocates of complete change through revolution, French reformers had a difficult path to trace. Often with a strong Catholic or Protestant background, the reformers incrementally succeeded in introducing improvements by legal means and through the creation of associations. Among them, Frédéric Le Play (1806-1882) played a remarkable rôle, bridging a strong concern for both the advancement of positive scientific knowledge and the making of policy. Through his own publications and through numerous associations that he helped to found in order to promote his views, he managed to influence public and private policy in the realms of rural and urban affairs. This key figure of 19th-century reformist thought is now receiving the attention that he and his disciples deserve (Kalaora and Savoye 1989, Berdoulay and Puyo 1997). They were also among those reformers who actively tried to solve the many problems raised by the dramatic events that struck French society in 1870-71. These events acted as the catalyst of the new geography that was incubating in the preceding years.

Geographical concerns and national reconstruction

After a period of relative peace and internationalism in Western Europe, the year 1870 brought traumatic experiences for French society, as well as the opportunity to fundamentally restructure its institutions: the military defeat by Prussia, the collapse of the

Imperial regime, the uprising and bloody repression of the Paris Commune, and the aborted monarchist restoration. The progressive establishment of the Third Republic had to come to terms with the consequences of these events in a new international context marked by the rise of nationalism. As there were many paths to national reconstruction, the rôle devoted to geography was to be varied. And so were the circles of affinity that suppported these geographical concerns. It is well known that the school led by Paul Vidal de la Blache came to monopolize the geography chairs and positions in the university system: by World War I they were all held by students of Vidal. This school was, however, not the only one to try to promote its views. Other circles of affinity produced strategies of institutionalization and (often high-quality) scientific works. Among them, one can distinguish historical geographers (who focused on past political and administrative areas), the makers of the inventory of the earth (linked to the geographical societies), the circle of Ludovic Drapeyron (editor of *La Revue de Géographie*), the circle of Emile Levasseur (close to the liberal Le Playist periodical *La Réforme sociale*), the circle of *La Science sociale* (around Edmond Demolins and Henri de Tourville, who advocated a stricter scientific obedience to Le Play), the Durkheimian social morphologists (who contested the field of human geography with the Vidalians), and individual geographers who worked in peripheral institutions (especially the cousins Franz Schrader and Élisée Reclus). Although all succeeded in securing strategic and social or institutional positions, the Vidalians emerged as the ones who were most in harmony with the ruling political groups and who managed to gain the key places in the educational institutions (which were almost all under the centralized control of the State). A brief summary of a few major social-contextual issues is enough to show how the relative positioning of the circles of affinity in relation to those issues bred a differentiated approach to geography (Berdoulay 1995).

Essentially the creation of Prussian initiatives, the new German state posed a challenge to the French that extended far beyond power politics. The military defeat and the ensuing annexation of Alsace-Lorraine by Germany induced a feeling of humiliation, which triggered varied reactions. The challenge thus extended from politics to psychology, but also to most aspects of the cultural, social and institutional organization of France. The fact that Germany had become a leader in many economic, industrial and scientific fields had to be addressed. The response ran from a conservative turn among the monarchists and religious-clerical groups to the deliberate policy of selective imitation of German approaches to science and institutions. The latter attitude was taken by the supporters of the new republican regime, often with quite insistent rhetoric. Most scientists knew German, made pilgrimages to scientific centers in Germany, and worked at importing German ideas and educational institutions. The French university was largely reorganized after the German model, with required attendance of students and the establishment of seminars and laboratories. In this context, it turned out that the Vidalians were those who demonstrated greatest admiration for German geography. The Durkheimians came second but focused more on the works of German sociologists and psychologists rather than geographers. Among the least interested in German ideas were the most ideologically conservative; as they did not secure significant institutionalization, they became relative outcasts of the established intellectual scene.

Thus, German science set the standards of scholarship, measured by the German yardstick. The formation, institutionalization and professionalization of Vidalian geography followed the German model. The Vidalians, who kept abreast of most German geographical advances, were accorded recognition from the international geographical community, which was then dominated by the Germans. They gained prestige both nationally and internationally. Thus, it would be erroneous to consider their approach to geography as a nationalistic project, since its international relevance was at its core. The German rôle in legitimizing an internationally recognized academic discipline also set the bias in favor of a cultural-historical definition of geography. This may explain the relative separation that occurred between academic geography and previous French research where the concern for history and the humanities was not so central. Nevertheless, it should be pointed out that the German standards were accepted selectively. References to Ritter, Humboldt, Oscar Peschel and Friedrich Ratzel largely dominated in French works (Mercier 1995). This shows that the German influences were not accepted blindly and completely; rather, there existed a demand for particular ideas, which were drawn upon more than others.

It is in relation to this issue of national reconstruction that the importance of the colonial movement is to be appraised, and not in relation to the growth of capitalistic imperialism as is so often heard. French capitalist ventures abroad went mostly to non-French-controlled countries, such as Russia, and private business investments in the colonies never became important. Furthermore, the French colonial empire was then conquered anew before protectionist tariffs were instituted. In view of the French indifference, if not frank hostility, toward colonial expansion, pressure groups worked at modifying governmental and public attitudes. The turning point of 1871 helped them in that the acquisition of colonies came up as one of the solutions to strengthen the country by compensating for its territorial losses. But, as the need for the ownership of new resources was not backed by the business and industrial world, colonial expansion was largely promoted with arguments involving a mixture of nationalist pride (to be restored), humanitarianism and economic utopias. Many geographical societies were thus founded after the 1870-71 humiliation, and they served as pressure groups. Again, geography became central in this societal issue. Through geographical publications, lectures and exploration subsidies, these societies tried to propagandize the colonial idea. Here again, not all geographical circles of affinity were associated with this ideology. The Vidalians were among the strongest supporters. The Durkheimians, with their socialist leanings, rarely supported colonial expansion; however, even those who were the most reluctant did not put forth a coherent anti-colonialist therory, leaving the colonial field little tilled by their own approach. Colonial geography became a new subject of study, for which geographical societies and the government created several lectureships. If the colonial movement legitimized the institutionalization and professionalization of geography, its actual impact on the theoretical content of the discipline remains a difficult question (which will be dealt with below).

This reinforced what was observed in other countries, that is, the perception by the public and the government of the usefulness of geography for the education of citizens in a modern country. As the schooling of all children became compulsory and free at the beginning of the Third Republic, the secularization of education brought additional demands on geographers concerning the content of curricula and textbooks. In a sense, the political and social demand for geographical education literally burst out in the 1870s and '80s, and its institutionalization almost preceded what the (often self-improvised) geographers could offer. The competition for these institutional opportunities became quite strong. Many geographers, such as Levasseur, Drapeyron, Schrader and the Vidalians, worked in close association with rival publishing houses and competed for the production and use of textbooks, atlases and wall maps. The competition extended to the training of teachers: in this respect, the Vidalians succeeded best, because they managed to control the École normale supérieure (ENS), which was the seedbed for the educational élite. Vidal was in charge. Furthermore, when Vidal succeeded

Louis-August Himly at the Sorbonne in 1898, he directed all the doctoral dissertations, which were to open the doors of all university positions.

The progressive preeminence of the Vidalians was also felt in other societal issues. These include all academic and educational institutional reforms, the regional reorganization of France, and the establishment of the foundations of a lay morality that could be taught in schools. The involvement of geographers in these issues is mentioned below.

Vidal and his circle of affinity

Often referred to as the "founding father" of modern French geography, Paul Vidal de la Blache (1845-1918) was its major figure, the leader whose thought has served as an inescapable reference point for most of the 20th century. Like most founders of new academic fields, he advocated ideas whose main characteristic is their progressive elaboration, even though his followers tended to take them as "ready made". Actually, Vidal kept modifying his ideas and approach throughout his life. There obviously was continuity and centrality in its core problematics of geography, which has made its success long-lasting, but it is also true that his approach remained evolving and quite open. This is why his thought can profitably be revisited in light of more recent issues. A confusing image in the understanding of his thought is produced by the fact that, long after his death, many French geographers have felt the need to position themselves vis-à-vis the founding father, even though they had little knowledge of his work. They actually did so in order to counter the positions of others. The history of the interpretations of Vidalian thought is thus full of amazing stereotypes! They extend from a purely regional or descriptive geography to a nationalistic ideology.

This stereotyping is also the case with the interpretations of Vidal's character. He can be cast as the pure product of the rising class of provincial origin through the service to the state (his father reached the level of provincial inspector of education). Finishing the preparation of his doctorate in history at the time of the Paris Commune, which he completely ignored, Vidal demonstrated a remarkable dedication to hard work, making of himself the best image of what was to be called "republican élitism"; the new regime was supposed to give recognition to hard work and education rather than to social or economic heritage. His very sober lifestyle, quite in line with the growing and demanding ethic of the bourgeoisie, added to his coldness in social, professional and family relations (Sanguin 1993) He lived for a long time in the buildings of the ENS, before his late marriage with another representative of the new republican state élite, a daughter of the historian Alfred Rambaud. But this apparent self-righteousness has to be understood in the context of an authentic, sincere and full commitment to the improvement of current social, economic and intellectual life. This is why one cannot attach the label of conservatism to him. Quite the contrary, his whole professional life has an intrinsic innovative and progressive thrust. His evident love for his homeland went hand in hand with a respectful approach to other cultures, a global view and a forcefully international (especially Germanic) appraisal of scientific advancement. Clearly, this made him a strong supporter of the republican reformers after 1870, as well as an admirer of German achievements who was eager to carry French efforts beyond them.

Through his teachings and actions, Vidal knew how to inspire a school of thought. He loosely structured it, however, giving it much less social binding and fewer theoretical norms than did other "founding fathers," such as Émile Durkheim for his own group. This openness, in his thought and in his organizing of his school, probably extended the significance of his initiatives, in his own time as well as later. At the same time, his institutional strategic options have proved remarkably clever. His group benefited from the positions he held in the ENS and the Sorbonne, the two key institutions in the republican reform of the educational system. The *Annales de Géographie*, founded in 1891 to function as a research organization for Vidal's group and a medium to diffuse its doctrine, was made possible by the interest and help of the publisher Armand Colin, himself in close association with the republican reformists. At the same time contributions by foreign authors and extensive reviews of literature published in other languages fully inscribed the Vidalian approach in international scientific research.

It is noteworthy that Himly, Levasseur, Reclus and members of *La Science Sociale* never published in the *Annales*. Even if, from the beginning, this journal seemed more open to specialists of other disciplines than geography, they were carefully selected—e. g., Charles Flahault and Gaston Bonnier in botany, and Alfred Angot in climatology. The efforts of the Vidalians to get closer to the physical and biological sciences is reflected by the fact that, until 1905, there were twice as many articles on physical as on human geography. The latter, especially through regional studies, took on importance largely at the expense of the old historical geography, which, little represented in the first issues, almost disappeared later on.

Nearly all contributors to the *Annales* were professors. But, while keeping their academic image (they very rarely published accounts of their travels), the Vidalians tried to maintain good public relations, keeping in contact with geographical societies. The most important link was Emmanuel de Margerie, a co-director of the *Annales* and one of the animators of the *Bulletin* of the Paris Geographical Society. The Vidalians also maintained a presence in the more traditional, government-backed Commission on Historical and Descriptive Geography. Their strong involvement in the colonial and regionalist movements and in the World War I peace negotiations gave them a firm position in contemporary societal issues, and their penetration of the educational system at all levels (including the training of élites at the ENS and the Ecole Libre des Sciences Politiques) provided them with a stable grounding in French high culture. With the election of Vidal to the Académie des Sciences Morales et Politiques in 1906, his concept of geography was recognized in one of the highest honorific institutions.

Most of the legitimacy of the group was established by Vidal himself. His circle included influential historians, such as Ernest Lavisse, who reigned over the Sorbonne for a long time, quite in agreement with the reformers of education and of Republican policies. Vidal made sure to establish his competence in the established field of historical geography (Vidal 1880). He kept distant but good relations with a few other circles; for instance, he gave a lecture at the Le Playist Société d'Économie Sociale and two others at the École des Hautes Études Sociales. Most noteworthy, he belonged to the Union pour la Vérité, founded in the early 1900s, an association whose aim was to debate live issues from a viewpoint leaving partisanship aside. There he could meet and discuss with prominent figures such as Émile Durkheim, Lucien Lévy-Bruhl, Félix Pécaut, Ferdinand Buisson, and Léon Brunschwig and demonstrate the value of his geographical approach.

Although this school of thought was unquestionably led by Vidal, there were differences in themes and approaches. One can distinguish two "generations" of Vidalians, the age distinction coinciding with a difference in scholarly interests. Among the first generation—that is, Vidal's earliest students—Marcel Dubois (1895), Lucien Gallois (1908) and Pierre Camena d'Almeida (1893) are best known. Gallois, upon whom Vidal relied from the beginning, was his master's successor at the ENS and as the head of the Vidalian school of geography. The common intellectual bond among these first disciples was that they witnessed the evolution of geography from the old-style historical geography toward physical, human and regional studies. They had to do their theses, however, under Himly in historical geography or the history of geography, and they seem, like Vidal, to have kept an interest in these fields for the rest of their lives. They can thus be distinguished from the second generation of students who did their theses later under Vidal. The latter usually demonstrated their qualifications by writing regional monographs. These works, and others, made them famous, in France and abroad: Jean Brunhes (1902, 1912), Camille Vallaux (1925), Albert Demangeon (1905), Emmanuel de Martonne (1902, 1909), Raoul Blanchard (1906, 1911), Jules Sion (1909) and Maximilien Sorre (1943-52). It is this generation that came to occupy most positions in the universities, either to newly created chairs or as replacements for retiring members of the historical geography circle of affinity. The first generation tended to have philosophical views in line with the dominant neo-Kantianism that was established early by the republican reformers. The other generation, born around and after 1870, also built on this philosophical background but were probably more sensitive to, or ready to integrate, the new philosophical developments, such as those of Henri Bergson.

The apparent homogeneity of Vidal's group should not be accepted uncritically. Actually, two trends did not harmonize well with the major Vidalian stream. The first was almost a sub-group around Dubois. Although he had taken only one course from Vidal (in 1877, when Vidal had just been appointed at the ENS), Dubois kept in close contact with him and contributed to the founding of the *Annales de Géographie* as its co-director. He taught very successful seminar-like classes at the Sorbonne, before being named to the chair of colonial geography as soon as it was created in 1893 (when Himly was the only other professor of geography). This made Dubois a leader in the training of geographers who would not, or could not, integrate with the ENS, and there was some rivalry between the two groups. For reasons of ill health, Dubois and his main disciple, Henri Schirmer, progressively withdrew from the academic scene before World War I. The tension between Dubois and the main Vidalian stream has strong significance for understanding controversies about the approach to planning (a theme examined below). The second trend to depart from the major Vidalian current was originated by Jean Brunhes, who, while keeping close relations with Vidal and other members of the group, had only formal relations with others, such as Demangeon, who dominated human geography after the War. Strongly interested in ethnography, quite involved in social Catholicism, and partially atracted by non-neo-Kantian philosophers such as Bergson and Maurice Blondel, Brunhes was impressive by the breadth of his knowledge and concerns. Having stayed for sixteen years at the University of Fribourg in Switzerland until he obtained the newly created chair of human geography at the prestigious Collège de France in 1912, he did not benefit from an institutional setting fostering the training of disciples. He did, however, inspire several scholars (notably Pierre Deffontaines) and provided geography with a different image among the public.

Thus, the Vidalians shared enough common ground to impress others by the coherence of their geography, at the same time as it had enough potential diversity for securing its future. However, a closer look at two main areas of Vidal's contribution will demonstrate that the challenge of keeping alive the whole scope of his legacy has not been met.

A possibilist theory grounded in conventionalism

When historian Lucien Febvre, who wanted to show his appreciation of Vidal's view of the relationships between nature and society, referred to it as possibilism, he probably did not anticipate the success of the term (Febvre 1922). This is because environmental determinism has so often biased geography that its practitioners welcomed such a congenial motto. The majority of geographers, however, have tended to view possibilism as the antithesis of any kind of environmental determinism and, sometimes, of any kind of determinism at all. Nothing is more remote from Vidalian possibilism. According to Vidal, geography is a science that involves some form of scientific determinism, and it includes the study of environmental influences among its tasks. This points to the interrelations of two issues for clarifying what possibilism was about: Vidal's scientific epistemology, and the theoretical scope attributed to possibilism (Vidal 1902, 1903a, 1911, 1913, 1922).

Actually, when Vidal elaborated his views on geography, French society was highly concerned with the foundations of the moral principles that should be favored in the establishment of the Republic. Most of all, the issue had a very practical and urgent aspect: what foundations of morality should be taught in the new republican, secularized schools whose student body was expected to be somewhat heterogeneous? Although treated in often polemical terms, this involved deep philosophical issues. The established Catholic Church had earlier ceased its outright support of the French conservative forces, and Republican politics became associated not only with the secularization of public institutions but often with anticlericalism. Concerning education, what foundations of morality could be taught, once religious and aristocratic references had been chased out of the schools? Two solutions were offered. One was a turn toward Kant's moral philosophy, which did not rest on religious foundations, but instead on rationalism, individualism, responsibility and cooperation. This call for Kant's philosophy was particularly convenient since it did not contradict the traditional values taught for centuries by Christianity. The early Republicans, who formed the mainstream of Republican politics, favored this idealist moral philosophy, which widely diffused in the population through public schools and colleges. As the tutor of these institutions, the university was equally penetrated by this philosophy. The second solution was of a more positivist nature. Sociologists, and especially the Durkheimians, wanted to establish the foundations of morality in the scientific study of society. Given a particular society, the best moral rules can be identified and then established as the norm. This solution remained confined to the most radical Republicans and did not sweep the board as neo-Kantianism did.

Quite naturally, this neo-Kantian orientation extended to other aspects of philosophy, especially with respect to science and knowledge, which were referred to under the label of conventionalism. Most philosophers who supported this line of thought also belonged to the mainstream of Republican ideology. They distanced themselves from the positivist criteria of science (about laws, facts, theories, proofs, etc.) by stressing the cognitive dimension of scientific activity. Challenging the idea that theories directly result from facts, they insisted that the mind imposes its categories on nature, that is, an interpretation of the perceived reality. Thus, laws and theories are conventions; they are neither right nor false. Rather, they function as categories or frameworks, which allow for the registering and classification of empirical information, which, in turn, may modify them. Through this interactionism and constructivism, French conventionalist philosophers such as Henri Poincaré (1902, 1905) or Pierre Duhem (1906) inspired epistemological attitudes among those who shared their circles of affinity, and their critical position vis-à-vis positivism was internationally diffused and was later picked up by others in the English-speaking world (e. g., the Americans W. V. Quine and Thomas Kuhn).

Surely, the neo-Kantian conventionalist epistemology of Vidal and his first disciples is manifest. It accounts for the now obscure polemics between the Vidalians and the Durkheimians (or social morphologists). As the latter had more positivist leanings, their respective concepts of laws, theories and scientific facts were quite different, and this hindered communication between the two groups (Berdoulay 1978). In their writings, the early Vidalians demonstrated that they were fully aware of the relativity of their explanations, preferring to talk of "notions", "theories" or "principles", rather than "concepts" or "laws"(e. g., Vidal 1896, 1911b). In fact, it is in this conventionalism that the epistemological foundations of Vidalian possiblism should be understood. It provided some form of epistemological operationality to the broader neo-Kantian philosophy in its view of the human relation to nature. Clearly, the ecological viewpoint of Ratzel provided the empirical dimension to fieldwork, but the epistemological and theoretical framework was derived from neo-Kantian moral philosophy and conventionalism.

As a free moral agent, the individual is viewed as realizing his "intentions" in relation to his environment. In other words, values come from the individual himself, as an ontologically free agent, or, better, as a free user of the earth. In this respect, no determinist position can be taken: values cannot stem from the contemplation or imitation of the order of nature. In their writings, especially in the textbooks that they authored, the Vidalians stressed human creativity as well as the virtues of the prolonged contact with, and transformation of, nature. Hard work, initiative, association and social solidarity were extolled, as was the case in mainstream Republican ideology. The possibilist approach to the human relationship to the earth then takes on a deeper significance than a mere reaction to environmental determinism. It implies a whole philosophy of these relationships. Actually, it comes out as the broad theoretical framework for studying them.

The approach is focused on the study of mediations—that is, on what emerges from the human interaction with nature, regions, pays, genres de vie, but also field morphology or patterns, landscapes, etc. (Vidal 1903b). This interactionist and constructivist approach was characteristically included in the neo-Kantian conventionalist view of epistemology: the focused-upon mediations were discussed in terms of "notions", themselves referring to resulting forms of interactions among logically independent causal series. The Vidalian approach insisted on the contingent emergence of phenomena and on their explanation through the identification of series of cause-and-effect relations, rather than through the identificaton of a universal law. If human freedom was constrained by nature, and correlatively by the stabilizing "force of habit" once solutions had emerged, it nevertheless could strive for taking initiatives. In this respect, the Vidalian approach took inspiration from neo-Lamarckism, which corresponded in France to a strong antimechanistic interpretation of evolution.

Interest in naturalistic theories of evolution had gained momentum after the middle of the 19th century in France, and the polemics surrounding Darwin's views had much to do with it. Among them, the revival of Lamarck's contribution played a great rôle, as it came also from abroad, from Darwin's own work and the equally known contributions from Ernst Haeckel. August Weismann's criticisms (in the 1880s) of the inheritance of acquired traits did not temper neo-Lamarckism for several decades. This was partly because the criticisms did not hit at the core of what neo-Lamarckism was about for many of its supporters. What was important for them was the search for a non-mechanistic interpretation of evolution, whereby the evolutionary process resulted from the interaction of an active environment and an equally active organism. Adaptation was to be studied as a reciprocal process. Neo-Lamarckism had the merit of looking critically at the idea of natural selection (where the individual organism is passive) and it drew attention to the initiating rôle of movement and form in the course of evolution. These views extended from naturalistic science to social theory, as they were congruent with the above-mentioned neo-Kantian conceptions of the human relationships to nature (Berdoulay and Soubeyran 1991). Effort and will, and therefore finality, could bring into existence new patterns in the human relationship to the environment. Physical geography, "geography of life", and human geography were thus intimately associated.

However, the problem of instrumentalizing the possibilist theoretical framework remained a difficult task. The neo-Kantian epistemology helped in this respect, as it allowed for a nuanced utilization of, and discourse on, empirical observations. But it stopped short of providing the appropriate methodological tools; the language of science was still embedded in positivistic analytical methodologies. Thus possibilism suffered most from this lack of an adapted scientific language. If natural language and its associated rhetorical devices (Berdoulay 1982) helped for setting the problem, they provided no methodological help for gaining advantage from this theoretical breakthrough. Certainly, this is one of the reasons for the later decline in interest in the possibilist stance.

Milieu, landscape, and ways of seeing

French geography displayed great sensitivity to the actual societal demand for a new geographical understanding of the world. Geographers took up the challenge and reinforced their concern for the modes of representation of their results and observations. One of the most challenging aspects was to induce and convey a new way of seeing, i. e., a new, more holistic way of looking at the environment and representing it. Landscape became a focus of attention, as it is the most visible part of what results from the complex human interactions with the earth. But the very complexity that produces landscapes is best referred to as *milieu*. As it includes both objective and subjective aspects as well as holistic concerns, the challenge has encouraged varied approaches (Buttimer 1971). The Vidalians were among those who tackled the issue, but other geographers contributed interesting attempts in this respect, such as Franz Schrader, who tried to combine his skills in cartography, topography and artistic representation.

Generally speaking, French geography echoed the 19th-century primacy devoted to the eye in all spheres of knowledge. This attitude, which French geographers favorably noticed in Humboldt's work, associated art, science and literature as valuable means for understanding the world. Although declining through time, the weight of the Romantic, antimechanistic and holistic view of nature never quite disappeared. In fact, this view was just a particular manifestation of a more enduring philosophical concern (Da Costa Gomes 1996, 2000). Organicism then represented a concern in early French geography. It should not, however, be taken as an analogical model; rather, it is best characterized as a metaphor—that is, as a rhetorical device to induce a new way of seeing (Berdoulay 1982). The use of the organic metaphor had the advantage of breaking away from sectorial approaches in favor of stressing the dependency between each part of a geographical complex and the whole; it stimulated both synthetic and analytical research on emergent mediations such as those already mentioned (regions and genres de vie). However, the tendency of some authors to slide from metaphor (with a cognitive aim) to analogy (an analytical and explanatory model), and the correlative overlooking of dysfunctional phenomena and the introduction of teleological reasoning, explain why the younger Vidalians (as well as many other social scientists) showed a definite reluctance to use organicist rhetoric. This did not prevent them from building on what the Vidalian-backed organic metaphor had helped them to envision; and terms such as "interconnexion" and, later, "complexes" and "systems", substituted for it.

The concern for finding other ways of seeing also manifested itself in the receptivity to the neo-Kantian and neo-Lamarckian views of the world. Their acceptance of the inner finality that may be at work within living wholes, and the related antimechanistic attitude, laid the ground for a definitely humanistic approach. This is also why there was a strong interest in Henri Bergson's philosophy. In spite of his criticisms of both Darwinism and Lamarckism for their monist approach, Bergson can be noted for his appreciation of the neo-Lamarckian stress on effort and will as forces in evoluton (Bergson 1907). He distinguished matter from a vital or life force (élan vital) that provided species with their own powers of invention to overcome the obstacles they might face. His view of evolution was not linear, but, rather, it developed in divergent directions, each species being collectively affected. It is because life "invents" that species appear, survive, and evolve. Thus the theme of creativity is emphasized, especially that of the human being, whose evolution is viewed as the highest realization of life force, and who is able to perceive the latter through intuition. As a participant in the life force, one gains one's freedom through action, through creating when dealing with the natural conditions in which the life force operates. There is no place for teleological explanation, because the process of evolution is regarded as creative and spontaneous. In addition, this approach stressed constant change and readjustment in nature and the primacy of life over matter. Although Vidal's writings echoed some of these themes in a rather neo-Lamarckian manner, Brunhes' reflected the early, direct influence of Bergsonian ideas, while Jean Gottmann may be the latest to refer to them explicitly (Brunhes 1912, Gottmann 1947). Bergson's view of a creative evolution responsible for a state of perpetual change, adjustment, and redefinition is similar to Brunhes' "principle of activity". A key notion of his human geography, the "psychological factor", was given a very Bergsonian definition, since it "might be called, according to an expression dear to Henri Bergson, 'the direction of the attention'" (Brunhes 1912, 731). Consequently, creative evolution and, especially, human initiative could not be accounted for by simple mechanistic science; it reinforced the epistemological tendencies that have been mentioned above, especially as regards contingency and the emergence of new wholes. For instance, the notion of genre de vie is an attempt to integrate man's creative evolution into a scientific framework. Lucien Febvre understood just this when he stated that human creativity could be given a place in science in the same way that others had given a place in biology to "spontaneity", the "faculty to create something new", that is, to what "Bergson named . . . the impulse (élan) and creative power of life" (Febvre 1922, 445).

Through this kind of concern, the Vidalian idea of milieu could not be a mere juxtaposition of sectorial analyses. The way of seeing that it conveyed could not be entirely satisfied through the classical canons of actual science. This explains the repeated temptations for other cognitive tools or languages. Literature and painting came up as the most solicited, in order to compensate for the actual failures of science to approach the very complexity—both objective and subjective—of milieu (Berdoulay 2000). As well illustrated by their contemporary Marcel Proust (e. g., Proust 1918), the early French geographers could seek to compensate for the shortcomings of positivistic science by turning toward the discursive resources displayed in literary expression as well as in innovative paintings. Most noteworthy, the apparent lengthy descriptions should in fact be considered as key elements of an overall narrative framework and world view. This point should not be overlooked: narrativity can be placed at the core of the Vidalians' epistemology, because it accounts for their view of the relationships among time, space and action, and because it enabled them to draw upon natural language resources. Their interest in painting did not go as far as with others, especially with Franz Schrader. This cousin of Élisée Reclus was a non-academic self-made geographer, who worked with the publisher Hachette but also had numerous associative activities related to geography, cartography, art and mountain hiking (Berdoulay and Saule-Sorbé 1998, 1999). He very well illustrated his contemporaries' wish to see the world differently—that is, to explore it from new points of view. He was particularly interested in taking advantage of the mobility of the eye looking at the landscape. Through technical devices of his own making, such as the orograph, he tried to instrumentalize new perspectives, especially in mountainous terrain. Camille Vallaux' "tours d'horizon" are reminiscent of this concern (Vallaux 1925). Furthermore, Schrader paralleled most developments in landscape painting of his time. When photography became fully competitive for the accuracy of the representation of morphology, Schrader turned away from the topographical use of painting in favor of more impressionistic research. In the manner of Claude Monet, he tried to grasp the "enveloppe" that one may feel within a particular milieu. In so doing, he clearly wanted to complement the actual failure of scientific language to come to terms with the full significance of milieu.

The mobility that was encouraged in the geographical ways of seeing echoed a concern for the benefits of the geographer's distancing himself from others and from his own discourse. This relationship to alterity (or otherness), once transposed into particular textual strategies, can express the diversity of the experience of place. This is what makes the enduring quality of Vidal's texts, especially in his *Tableau de la géographie de la France* (1903), where the distance between the author and his object of study is not fixed. His capacity to multiply the points of view and to adopt a mobile way of seeing introduces an openness to alterity, as he sees both with and like the eyes of the Other. While this plurality of the ways of seeing gives relativity to the geographer's eye, it also does not question its discursive, scientific identity. Through the textual opening to otherness, Vidal's geographical discourse demonstrated the possibility of offering complex representations of reality (Laplace-Treyture 1998).

All this shows that the new vistas opened by the early French geographers did not call for a separation among the various modes of knowledge, be they scientific, narrative, literary or artistic, even though scientific understanding defined their overall purpose. Landscape and milieu were two key referents of this attempt to understand geographical complexity. If these concerns, choices and attitudes provided long-lasting inspiration, however, they were hard to hold together, as French geographers kept plowing more and more diverse fields.

Dispersed legacies and the diversity of commitments

For about a half century after the death of Vidal in 1918, French geography went almost full cycle, from the international assertion of a self confident school in the 1920s and '30s to a growing isolation in the '50s and a final internal crisis in the late '60s (before the recent reconstruction and renaissance). All through this cycle, there was an apparent paradox: the same unitary set of references to Vidal's thought were proclaimed, whereas a profusion of research directions gave evidence to the contrary. In fact, while French geography indeed witnessed a multiplication of new themes of research, there was no theoretical renewal. Furthermore, inherited theoretical foundations were reduced to a minimum, because of a quasi-complete retreat from epistemological issues and debates. Dogmatism substituted for epistemological and theoretical research. Thus, the so-called Vidalian references offered a very truncated image of what the master had envisioned. They also took away from his contribution the constant inquietude that made him modify his own recommendations. In other words, the French school extended a few of Vidal's contributions to new fields of study—e. g., rural, industrial, transport, urban and tropical geography and various aspects of physical geography. They are too numerous to be accounted for here, especially as they have been examined in various reviews (L'Information Géographique 1957, *La pensée géographique* . . . 1972, Meynier 1969, Claval 1998, Claval and Sanguin 1996). Suffice it here to put in perspective a few revealing characteristics of this history.

The formal legacy was held by students of Vidal himself. Lucien Gallois, who succeeded him at the Sorbonne, continued the project launched by Vidal of a new, but collective, *Géographie universelle* (1927-1948). Orthodoxy was also guaranteed by Emmanuel de Martonne (Vidal's son-in-law) in regional and physical geography, Albert Demangeon in human geography, and Raoul Blanchard in regional and physical geography. Jean Brunhes, from his peripheral but prestigious position at the Collège de France, illustrated human geography to a wider public until his death in 1930. Under their auspices, a wide array of topics were approached from their simplified, but effective approach: city evolution, political boundaries, ethnic impact on landscapes, housing forms and functions, natural spatial ecology, landscape morphology, regional contexts in Latin America, Canada and Africa, etc. (Augustin and Berdoulay 1997, Claval 1998, 177-213). The use of topographic and geological maps and the practice of fieldwork was thought sufficient to keep alive the concern for seeing the landscape from various points of view (Nozawa 1996).

But, beyond this façade, a few tendencies that were perceptible earlier gained a strong momentum. The marginalization of research trends that differed from orthodoxy now affected members of the school itself, in that their ideas were not incorporated into mainstream geography—e. g., Maximilien Sorre's work on ecological, social and medical geography. Strikingly, the intrusion of geomorphology within the scope of physical geography kept growing to a point where it almost substituted for the whole field with respect to research. And possibilism was progressively reduced to a simple ideological stance against environmental determinism—i. e., to a position completely forgetful of its underlying theoretical and philosophical foundations. All these observations that evidence an important retreat from epistemological and theoretical issues should be related to two processes. One is the classic formation of a "normal", routinized science, where paradigmatic security provides for the full extension of powerful analytical devices. The other is more contextual, as the process was not limited to geographical thought. In this respect, it should be mentioned that the innovative epistemological research that existed in France before World War I did not extend beyond this period. Most of the great conventionalists were dead by then. This epistemological vacuum was progressively occupied by authors who, for the most part, were active scientists or those who were directly inspired by their practice. A "soft" positivism gained ground, which legitimized a radically empirical attitude based on defiance of theory. Furthermore, the concern for a "universitarian" style made most of these geographers shy away from explorations at the interface with other modes of knowledge, including fine arts and literature. Finally, an institutional conservative bias was introduced through recruiting practices. The "agrégation", a national, highly competitive examination created for the hiring of secondary-level teachers, became a "normal" step in the curriculum of the aspiring university professor. As the jury was (and still is!) renewed largely on the basis of cooptation by its own members, this examination progressively became a powerful means of enforcing strict orthodoxy. It remained until recently the major institutional device for a chronic epistemological conservatism (Claval 1998). It imprinted on the students' minds encyclopedic, conservative and empiricist standards for what geography was supposed to be.

The Second World War considerably modified the conditions and the issues that geographers had to work with. This context, characterized by doubt and exploration, opened up and provided opportunities for the emergence of a host of new themes. Most of the geographers who took up the challenge had not directly received the teachings of Vidal de la Blache. Consequently, their contribution was even more remote from his views, even though they kept invoking his authority. By so doing, they accentuated the impoverishment of his legacy, which became completely stereotyped around a persistent empirical attitude and a shallow possibilism. However, their exploration of new themes made geographers concerned with most of the issues of spatial modernization and global distributions of wealth, poverty and resources, and the image they gave of geography was that of a discipline open to current issues but dependent on theoretical explanations provided by others (Claval 1998, 241-294). This is when the former holistic foundations were reduced to a definition of geography as a "science of observation" and/or a "science of synthesis".

Beyond the increase in diversity of the treated themes, and beyond the continued empiricism, two contextual changes (among others) should be noted, because they had a long-lasting influence on geographical research. They both relate to the privileged interest given to permanent or stable structures, in spite of an apparent concern for change. They indeed correspond to the way structuralism was then understood. The first one concerns the weight of Marxism in postwar French society. The prestige of the victorious Soviet Union and the place taken in the French Resistance by the Communist Party contributed to a widespread, albeit often edulcorated, Marxist view of society. The academic milieu was particularly receptive, and the new generations of geographers increasingly became sympathetic to Marxism, whether or not they registered as members of the Communist Party. A few leading geographers, such as Michel Rochefort, Jean Tricart, Jean Dresch and Pierre George, were known for these ideological leanings. One of the most striking aspects of Marxist ideology, however, was its relatively limited impact on geographical thought itself. To be sure, it impacted greatly on academic life in terms of careers and related professional strategies, but the conceptual core of the discipline did not incorporate the major foundations of Marxism. This should be related to the empirical, atheoretical conservatism already mentioned. What Marxism most contributed to alter was subject matter and categories of observation, such as social classes, capital appropriation, urban domination of the countryside, and unequal development. Pierre George was instrumental in renewing geographical themes of research. He encouraged the development of social, demographic, industrial and economic geography, together with the study of Eastern Europe and underdevelopment. Furthermore, he actively participated in the research and discussions that took place on the modernization of France. In all these respects, his importance to French geography is considerable (Claval 1998, 268-284). And so was Marxism. But the structuralism that it induced dissolved into the more classic search in the French school for permanences, which were preferably to be found in physical geography and traditional or routinized behavior. The Marxist impulse stopped short of theoretical and epistemological reformulations of geography, including a prospective and planning thrust, as will be discussed below. Rather, most of its present conceptual legacy is a critical attitude toward existing power relationships (Lacoste 1976), as well as an impassioned categorization of geography as a social, rather than natural science, thus substituting one type of determinism for another.

The geographical structuralism that reduced Marxism to a classic approach also contributed to the biased relationship of geographers with historians. Because of the privileged association of the two disciplines in primary and secondary education, most students of geography studied history, and vice versa. The founder of the "Annales School" of historical thought, Lucien Febvre, insisted on his debt to Vidalian geographers, notably to Vidal himself, whom he had known while a student. Febvre's book *La terre et l'évolution humaine* (1922) praised geography against Durkheimian sociology in order to justify his own views on a new approach for history. By praising certain aspects of geography, however, this successful book locked the discipline within specific boundaries—that is, the identification of genres de vie in well defined regional contexts—leaving to history the concern for social and economic change as well as the analysis of mentalités (cultural representations). This "vampire's kiss" was repeated after World War II, when Fernand Braudel assumed the leadership of the Annales School. Actually, he also took a strong leadership over most human sciences until the 1970s, thanks to the initial help of the Ford Foundation and the creation of the Maison des Sciences de l'Homme in Paris (Coutau-Bégarie 1983). Braudel kept praising Vidal and the first Vidalians, clearly implying that contemporary geographers were not dealing with the right topics and that they deserved the inferior ranking they had in academia (which he controlled to a great exent). For Braudel and leading historians, geography should deal with the more stable structures in society—i. e., those structures that had emerged from ancient, long-lasting human relationships with the natural environment. The influence of the Annales School was far from what foreign geographers could recently get from reading of its accomplishments! Again, geography was confined (this time by others) to the identification of stable structures inscribed in the land.

Thus, consciously or indirectly driven away from some of its most promising developments, French geography had to assume a relative decline of audience. This occurred nationally, but also internationally because the epistemological and theoretical conservatism of post-World-War-II French geographers made them lose the habit of discussing foreign contributions. By the late 1960s, French geography entered two decades of major conceptual, theoretical, epistemological and methodological turmoil. The classic schemes collapsed in favor of a plurality of approaches. The logic of debates has necessarily changed, as the numbers of geographers involved have completely changed in scale. Whereas academic geographers numbered 40 in 1920, there were 71 in 1955, 550 in

1971, and about 1000 in 1995 (Claval and Sanguin 1996, 8). Paradoxically, with the help of the agrégation, the conservative approach still counts many practitioners. Most notewothy, the quantitative revolution, as such, did not much change the conservative views of geography. What was taken from the possibility of handling a great quantity of data was often more a reinforcement of the traditional empiricist approach than a turn toward theorizing or modelbuilding. Nevertheless, thanks to its plurality, French geography has now renewed international contacts, and it fully participates in international thematic and theoretical challenges (Claval 1998). Among its various contributions, two can be emphasized here, as they illustrate older, innovative concerns that were poorly institutionalized but are now in the process of gaining a renewed interest. First, we shall consider the increasing concern with planning and applied geography, and then we shall turn to the growing emphasis on the social science and humanistic aspects of geography.

From colonization to planning

One of the distinctive features of the institutionalization of French geography has been the repeated concern of its practitioners for the planning relevance of their work. Surely, this was not the case with all individuals, many of whom gave the impression of retreating into the university ivory tower, but planning created an overall significance for geographical thought. This has to be related to the notable sensitivity of French geography to its societal context. Live issues inspired geographers, and, in turn, they tried to contribute to their solutions, at least by providing elements for understanding. Paradoxically, in spite of much research that cut across the boundary between the educational and applied aspects of geography, the resulting impression is that of a series of only occasional events. The two World Wars, but also live issues such as regionalization and decentralization, gave impulse to these boundary crossings (Robic 1996). Geographers seemed satisfied with demonstrating the broad social relevance of their science, with occasional implications for live issues or emergencies, while remaining dedicated to a non-applied academic discourse. This was more than a matter of style; there definitely was an epistemological "original sin", if not a "hidden murder", in their history that prevented their methodology from freely crossing the boundaries of academic standards toward planning. Colonial geography was the scene of the drama.

Colonial expansion had provided geographers with new territories in which theories and concepts could be tested, not only through observation but also through experimentation. Colonies to be developed became potential incubators of modernity, because unequal power relationships allowed the colonizers to test new approaches of social and spatial organization. To a great extent, French colonial ventures in lands such as Morocco, Tunisia or Indochina gave theoretical and practical foundations to the main approaches of urban and regional planning that were later developed in metropolitan France (Rabinow 1989, Marié 1989). The goal of understanding and developing colonial lands could be but a stimulus to a geographical approach that would be geared to planning ("l'aménagement"). Marcel Dubois unquestionably came out as the intellectual leader of those who took up the challenge.

Dubois clearly set "colonial geography" against a "geography of colonies". While the former elaborates methods of its own in order to develop a country along rational lines, the latter simply consists of the local use of the traditional geographical approach (Dubois 1894). In other words, colonial geography was planning-oriented. A possibilist framework made Dubois argue against the exploitation of conquered people, in favor of social and economic development through the transformation of the resource base that will eventually lead to political emancipation from the mother country (Dubois 1895). By favoring voluntary transformations, however, Dubois questioned the actual conception of the discipline. He forcefully required a conception of geography as an "open" or "applied" science. In doing so, he entered into confict with other French geographers, most notably Lucien Gallois, who advocated the formulation of a well-knit and easily reproducible paradigm for geography, which he saw in the regional monograph. On the contrary, Dubois criticized the static and bounded regional geography of those colleagues, and he favored general geography as best able to take into account the more fundamental globalization of contemporary economic relations. For various reasons, Dubois ended up losing this "battle", as he withdrew from the co-direction of the *Annales de géographie* (Soubeyran 1989, 1997). And, by World War I, as mentioned above, his sub-group among the Vidalians faded away. Thus, under the leadership of Gallois, the above-mentioned reduction of the Vidalian approach imposed a regional method at the expense of a more planning-oriented epistemology. Mainstream geography had no logical, inner relevance to planning. The latter was possible, but only as an aside, as the punctual "application" of an otherwise non-planning-oriented science. And, for part of the world, colonial geography was easily transformed into "tropical geography" (e. g., Gourou 1971).

The dilemma was recurrent all through the institutionalization of French geography. It explains its relative weakness in non-educational professionalization. The case of urban planning is evident. Referred to in early 20th-century France as "urbanisme", urban planning was defined as "the science of cities". In other words, fundamental and applied knowledge would not be distinguished. Its proponents worked in the footsteps of, and in association with, Le Playist circles of affinity. Its Musée social was the nursery of most urban planners. Some of them demonstrated more affinity with the Vidalians, who were also at the forefront of the institutionalization of urban planning at the end of World War I (Robic 1996). By the late 1920s, however, in spite of the use of geographical input, urban planners had structured their professionalization by way of the exclusion of geographers, to the benefit of historians, engineers and architects. Clearly, in this process, the relative epistemological and methodological closure of geography did not help its case.

Similar shortcomings limited the long-term impact of Michel Phlipponneau's crusade in favor of "applied geography". The content and the success of his 1960 manifesto demonstrated the interest that French geographers had for the modernization of their country, as well as their already significant professional implication in planning (Phlipponneau 1960). It also demonstrated how the dominant empirical (atheoretical) attitude prevented geography from providing general principles for giving foundations to the proposed solutions, rather than case-by-case particular treatments, at a time when competition from other disciplines was strong. The critical reaction to applied geography that Pierre George and his group published in 1964 fed a lasting debate (George et al. 1964). Their Marxist-inspired criticisms concerned the use that could be made of the proposed solutions, and they preached a critical attitude toward those who tried to transform the life conditions of disadvantaged social groups. But it is significant that these geographers did not criticize (or did not perceive) the lack of theoretical content of the geography that Phlipponneau wanted to apply.

The issue changed with the rise of theoretical concerns in the 1970s and later. Roger Brunet was the prominent leader of the movement. His founding of the periodical *L'Espace géographique* in 1971 and his close ties with governmental planning agencies (e. g., DATAR [Délégation à l'aménagement du territoire et à l'action régionale]) made him highly instrumental for giving visibility to the rôle of geography and geographers in planning at all territorial scales. Also influential in secondary-level educational reform, he significantly modified curricula, giving his geographical approach a widespread popularity. He introduced the term "choremes" for elementary spatial structures and produced a "table" of their diverse combinations (Brunet 1980). They serve for giving some analytical power to the graphic models that he proposed for representing geographical organization. Their simplification of otherwise complex phenomena has been criticized, but they possess a powerful communication efficiency in planning and pedagogical matters. But, more fundamentally, what is behind this approach is the same structuralist search for relatively stable patterns. Old biases are still alive. Whatever the future of this chorematic approach, it is but one sign of the extremely active and diverse involvement of French geographers in planning issues. The present societal demand for regional planning is strong, because of the French tradition of state involvement in public affairs and because of the contemporary readjustments that Europeanization and globalizaton are imposing on territorial policies.

In this respect, environmental issues have increased the concern for finding new ways for planning. If they have to integrate new types and scales of data, they also have to cope with the increased demand for local and participatory democracy (Berdoulay and Soubeyran 1993, 2000). French geographers have managed to become heavily involved in the training of professional planners. However, as this is usually achieved outside university geography departments, the recurrent epistemological issue of the relationship between the fundamental and applied aspects of the discipline is still alive.

Humanistic significance

Another lasting contribution of French geography has been its humanistic significance (Buttimer 1971). At a time of positivistic retreat, environmental crisis, renewed territorial identifications, and other destabilizations of beliefs and values, there is a strong societal demand for a better order of the spatial inscription of human life. Because of its definitely modernist foundations in political and cultural styles, French society has been relatively little affected by postmodernism, especially as far as gender and ethnicity are concerned. However, other issues, and especially various territorial identifications, have sprung up again. Public policies are incorporating part of them. For instance, since 1992, planning laws have allowed the possibility of creating various "pays" (thus recognized by the State), as far as its major socio-economic and public agents find themselves in agreement. As recent literature on the topic was lacking, one witnessed a rediscovery of many Vidalian writings. This renewed professional relevance of Vidalian thought met up with the more enduring academic interest in Vidal's humanistic approach, which encompassed human creativity and a sense of place and otherness. His books on France, Eastern Europe and human geography have all been reprinted, some of them several times in the recent past, for the benefit of a wide public. The same phenomenon is affecting some of Élisée Reclus' publications. It is not so much the radical or revolutionary fiber that is solicited from his work; rather, it is the humanistic flavor that comes out of the reading of his geographical descriptions. In fact, however confusingly, what may be attracting the readers is Reclus' deep sense of the potential that each individal possesses. His faith in the individual ability to generate a better world paralleled his tentative, albeit little explicated, formulation of an anarchist geography, resting on the individual as a creative and moral agent.

Present-day epistemological and theoretical debates remarkably echo these types of concern. The retreat from structuralism, where the individual conceptually had no initiative, has now swept across most social sciences. For some, the retreat is partial. For Pierre Bourdieu, for instance, his insistence on habitus leaves the individual strongly constrained in the logic of his decisions. In spite of Bourdieu's disdain for geography, his thought has encouraged a few Marxist-inspired geographers to espouse a more balanced view of socio-economic determinants. Other sociologists, such as Alain Touraine and François Dubet, have worked at better anchoring their approach of individual action in the autonomy of the Self. In philosophy, the last "return to Kant", with the contributions of Alain Renaut and Luc Ferry, goes in the same direction, that of a human subject who is acting as the author of his own life.

French geographers have gone along similar pathways. The major turning-point—if there ever was any—is less what many claim it to be, i. e., a shift from a so-called naturalistic view of geography to a social-scientific one (e. g., Lévy 1999), than a grounding of geography in a fully experiential perspective (Claval 1998). In recent times, the first well known attempt was the notion of lived space (espace vécu) that Armand Frémont illustrated in various publications (Frémont 1976). Interestingly, it was his own intellectual background and context that yielded this notion, and not a specific borrowing from foreign philosophical or scientific research. Existentialism had indeed penetrated many ways of thought in postwar French intellectual life. With the 1968 upsurge of positivistic and Marxist approaches, the apparently more objective notion of "territory" attracted a lot of attention. The critical interest in power relations was made easier with this notion. The reading of authors such as Michel Foucault encouraged the focus on territory. But at the same time, and increasingly, it referred to the subjective ties that an individual or a group may weave with the environment. Representation, rather than perception, became another notion that addressed the same theme. The analysis of spatial ideologies, with their strategic aspect, has helped geography to get closer to human initiative and creativity (Berdoulay 1985). All these attempts, however, have stopped short of fully encompassing the experiential viewpoint of an active subject within his or her environment.

At the same time, the study of the environment has benefited from a renewed interest in the landscape. Conceptualized through the ideas of "geosystems", "interactive space", "information", or "visibility", the landscape idea yielded a lot of research where physical and human aspects are usually combined (Bertrand 1968, Berdoulay and Phipps 1985, Rougerie and Béroutchachvili 1991). This wllingness to integrate social-scientific data with naturalistic ones is another enduring legacy of early French geography. However, the rôle of the subject is still to be clarified in this type of approach, and this is where the notion of milieu has been called forth again. In a very Heideggerian way, Augustin Berque is among the very few geographers who have addressed the issue, through the elaboration of a whole new conceptual framework, which revolves around the notions of milieu and "mediance" (Berque 1990, 1996). This approach nevertheless remains little concerned with the active part of the subject in relation to milieu.

This is where the older notion of place, once instrumentalized through the attention devoted to narrative, insists on the persistence of the ties that contemporary agents may keep with their milieu. Actually, place and subject cannot be viewed separately; they feed each other. The subject constructs place through the narratives that give meaning to his or her relationship to surrounding people and objects. Culture and materiality tend to stabilize these constructions (Berdoulay and Entrikin 1998). What is relevant for the study of modern life is that place does not refer a priori to a particular scale or to specific boundaries. Furthermore, by insisting on the active, auto-constructive implication of the subject, the notion of place is useful to approach the issue of citizenship and democracy (Berdoulay

1997). Whatever the future of these research vistas on place and active citizenship, they demonstrate the concern of French geographers for an experiential approach to contemporary cultural and political issues. The increased number of publications on cultural and political geography bear witness to this flourishing trend (Claval 1995, Lévy 1991, Lussault 1993).

Institutionalization in perspective

When compared with that of other countries, the institutionalization of geography in France occurred rather early (second only to Germany), and it has kept its momentum in the educational system down to the present day. Beyond the apparent unity of a well known school of thought, debates and conflicts have nurtured diversity. It would thus be erroneous to give an account of its history by molding it in a succession of paradigmatic changes. There is no more sense in dividing this history according to a Vidalian period followed by a post-Vidalian period, than in opposing a naturalistic geography to one that would be a social science, or than in espousing the succession of paradigms of Anglo-American presentations of geography.

Another distinctive trait of French geographers was their involvement in social issues. Although a modest discipline on the Parisian intellectual scene, geography has played a significant rôle in shaping the evolution of French society. Geographers have greatly participated in providing the image that this society holds of itself. The societal inscription of French geography is certainly one of its greatest achievements.

Nowadays, if professionalization does not call for much worry at the secondary and university levels, it is still not quite secure—although significant—in planning training and practice. The institutional aspects of this issue will have to be met by the clarification of its epistemological biases.

Through its stronger ties with francophone countries and geographers all around the world, and through its renewed relationships with other linguistic traditions, including in the context of the International Geographical Union (where French is co-official with English), French geography has managed to benefit from an institutional setting that should allow the furthering of creative contributions.

References:

Augustin, J.-P. and V. Berdoulay, eds. (1997) *Modernité et tradition au Canada: le regard des géographes français jusqu'aux années 1960*, L'Harmattan, Paris and Montréal.
Berdoulay, V. (1978) The Vidal-Durkheim Debate, in D. Ley and M. Samuels (eds.), *Humanistic Geography: Prospects and Problems*, Maaroufa Press, Chicago, 77-90.
Berdoulay, V. (1982) La métaphore organiciste. Contribution à l'étude du langage des géographes, *Annales de Géographie* 91, 573-586.
Berdoulay, V. (1985) Les idéologies comme phénomènes géographiques, *Cahiers de Géographie du Québec* 29, 205-216.
Berdoulay, V. (1988) *Des mots et des lieux. La dynamique du discours géographique*, Editions du CNRS, Paris.
Berdoulay, V. (1995) *La formation de l'école française de géographie (1870-1914)*, Editions du CTHS, Paris (1st ed. 1981).
Berdoulay, V. (1997) Le lieu et l'espace public, *Cahiers de Géographie du Québec* 41, 301-309.
Berdoulay, V. (2000) Le milieu entre description et récit. De quelques difficultés d'une approche de la complexité, in V. Berdoulay and O. Soubeyran (eds.), *Milieu, colonisation et développement durable*, L'Harmattan, Paris and Montréal, 25-37.
Berdoulay, V. and J. N. Entrikin (1998) Lieu et sujet. Perspectives théoriques, *L'Espace géographique* 27, 75-89.
Berdoulay, V. and M. Phipps, eds. (1985) *Paysage et système. De l'organisation écologique à l'organisation visuelle*, Editions de l'Université d'Ottawa, Ottawa.
Berdoulay, V. and J.-Y. Puyo (1997) La pensée géographique de Le Play, à propos *Des Forêts*, *Les Etudes sociales* 126, 19-36.
Berdoulay, V. and H. Saule-Sorbé (1998) La mobilité du regard et son instrumentalisation. Franz Schrader à la croisée de l'art et de la science, *Finisterra* 65, 39-50.
Berdoulay, V. and H. Saule-Sorbé (1999) Franz Schrader face à Gavarnie, ou le géographe peintre de paysage, *Mappemonde* 3, 33-37.
Berdoulay, V. and O. Soubeyran (1991) Lamarck, Darwin et Vidal: aux sources naturalistes de la géographie humaine, *Annales de Géographie* 100, 617-634.
Berdoulay, V. and O. Soubeyran (1993) Retour de la prospective et crise de la modernité, *Espaces et Sociétés* 74/75, 3-10.
Berdoulay, V. and O. Soubeyran, eds. (2000) *Milieu, colonisation et développement durable*, L'Harmattan, Paris and Montréal.
Bergson, H. (1907) *L'évolution créatrice*, Alcan, Paris.
Berque, A. (1990) *Médiance, de milieux en paysages*, GIP RECLUS, Montpellier.
Berque, A. (1996) *Etre humain sur la terre*, Gallimard, Paris.
Bertrand, G. (1968) Paysage et géographie physique globale. Esquisse méthodologique, *Revue géographique des Pyrénées et du Sud-Ouest* 39, 249-272.
Blanchard, R. (1906) *La Flandre. Etude géographique de la plaine flamande en France, Belgique et Hollande*, Danel, Lille.
Blanchard, R. (1911) *Grenoble. Etude de géographie urbaine*, A. Colin, Paris.
Broc, N. (1981) Les grandes missions scientifiques françaises au XIXe siècle (Morée, Algérie, Mexique) et leurs travaux géographiques, *Revue d'Histoire des Sciences* 3, 319-358.

Brunet, R. (1980) La composition des modèles en analyse spatiale, *L'Espace géographique* 9, 253-265.

Brunhes, J. (1902) *L'irrigation. Ses conditions géographiques, ses modes et son organisation dans la Péninsule ibérique et dans l'Afrique du Nord*, C. Naud, Paris.

Brunhes, J. (1912) *La géographie humaine. Essai de classification positive, principes et exemples*, Alcan, Paris (1st ed. 1910).

Buttimer, A. (1971) *Society and Milieu in the French Geographic Tradition*, Rand McNally, Chicago.

Camena d'Almeida, P. (1893) *Les Pyrénées, développement de la connaissance géographique de la chaîne*, A. Colin, Paris.

Claval, P. (1995) *La géographie culturelle*, Nathan, Paris.

Claval, P. (1998) *Histoire de la géographie française de 1870 à nos jours*, Nathan, Paris.

Claval, P. and J.-P. Nardy (1968) *Pour le cinquantenaire de la mort de Paul Vidal de la Blache*, Les Belles Lettres, Paris.

Claval, P. and A.-L. Sanguin, eds. (1996), *La géographie française à l'époque classique (1918-1968)*, L'Harmattan, Paris and Montréal.

Coutau-Bégarie, H. (1983) *Le phénomène 'Nouvelle Histoire'. Stratégie et idéologie des nouveaux historiens*, Economica, Paris.

Da Costa Gomes, P. C. (1996) *Geografia e modernidade*, Bertrand Brasil, Rio de Janeiro.

Da Costa Gomes, P.C. (2000) Milieu et métaphysique: une interprétation de la pensée vidalienne, in V. Berdoulay and P. Soubeyran (eds.) *Milieu, colonisation et développement durable*, L'Harmattan, Paris and Montréal, 55-72.

Demangeon, A. (1905) *La Plaine picarde: Picardie, Artois, Cambrésis, Beauvaisis. Etude de géographie sur les plaines de craie du nord de la France*, A. Colin, Paris.

Dubois, M. (1894) Leçon d'ouverture du cours de géographie coloniale, *Annales de géographie* 4, 121-137.

Dubois, M. (1895) *Systèmes coloniaux et peuples colonisateurs, dogmes et faits*, Masson & Plon, Paris.

Dunbar, G. (1978) *Elisée Reclus, Historian of Nature*, Archon Books, Hamden, Connecticut.

Duhem, P. (1906) *La théorie physique, son objet et sa structure*, Chevalier et Rivière, Paris.

Febvre, L. (1922) *La terre et l'évolution humaine*, La Renaissance du Livre, Paris.

Frémont, A. (1976) *La région, espace vécu*, P.U.F., Paris.

Gallois, L. (1908) *Régions naturelles et noms de pays. Etude sur la région parisienne*, A. Colin, Paris.

George, P., R. Guglielmo, B. Kayser, and Y. Lacoste (1964) *La géographie active*, P.U.F., Paris.

Godlewska, A. (1999) *Geography Unbound. French Geographic Science from Cassini to Humboldt*, University of Chicago Press, Chicago.

Gottmann, J. (1947) De la méthode d'analyse en géographie humaine, *Annales de Géographie* 56, 1-12.

Gourou, P. (1971) *Leçons de géographie tropicale*, Mouton, Paris and The Hague.

L'Information Géographique (1957) *La géographie française au milieu du XXe siècle*, Baillière, Paris.

Kalaora, B. and A. Savoye (1989) *Les inventeurs oubliés. Le Play et ses continuateurs aux origines des sciences sociales*, Champ Vallon, Seyssel.

Lacoste, Y. (1976) *La géographie, ça sert, d'abord, à faire la guerre*, Maspero, Paris.
Laplace-Treyture, D. (1998) Ecriture savante et relation au voyage, *Finisterra* 65, 75-82.
Lévy, J., ed. (1991) *Géographies du politique*, Presses de la Fondation nationale des Sciences politiques, Paris.
Lévy, J. (1999) *Le tournant géographique. Penser l'espace pour lire le monde*, Belin, Paris.
Lussault, M. (1993) *Tours: Images de la ville et politique urbaine*, Université François-Rabelais/Maison des Sciences de la Ville, Tours.
Marié, M. (1989) *Les terres et les mots*, Méridiens Klincksieck, Paris.
Martonne, E. de (1902) *La Valachie. Essai de monographie géographique*, A. Colin, Paris.
Martonne, E. de (1909) *Traité de géographie physique*, A. Colin, Paris.
Mercier, G. (1995) La région et l'état selon Friedrich Ratzel et Paul Vidal de la Blache, *Annales de géographie* 104, 211-235.
Meynier, A. (1969) *Histoire de la pensée géographique en France (1872-1969)*, P.U.F., Paris.
Nozawa, H. (1996) L'école vidalienne et l'excursion géographique. Une note préliminaire, in H. Nozawa (ed.), *Social Theory and Geographical Thought*, Kyushu University, Fukuoka, Japan.
La pensée géographique française contemporaine. Mélanges offerts à André Meynier (1972), Presses Universitaires de Bretagne, Saint-Brieuc.
Phlipponneau, M. (1960) *Géographie et action. Introduction à la géographie appliquée*, A. Colin, Paris.
Poincaré, H. (1902) *La science et l'hypothèse*, Flammarion, Paris.
Poincaré, H. (1905) *La valeur de la science*, Flammarion, Paris.
Proust, M. (1918) *A l'ombre des jeunes filles en fleurs (A la recherche du temps perdu)*, N.R.F., Paris.
Puyo, J.-Y. (1999) La science forestière vue par les géographes français, ou la confrontation de deux sciences 'diagonales' (1870-1914), *Annales de Géographie* 108, 615-634.
Rabinow, P. (1989) *French Modern*, M.I.T. Press, Cambridge, Massachusetts.
Robic, M.-C. (1996) Les géographes français et l'organisation de l'espace (1914-1950), in V. Berdoulay and J. A. van Ginkel (eds.), *Geography and professional practice*, Nederlandse Geografische Studies, Utrecht.
Rougerie, G. and N. Beroutchachvili (1991) *Géosystèmes et paysages. Bilan et méthodes*, A. Colin, Paris.
Sanguin, A.-L. (1993) *Vidal de la Blache, 1845-1918. Un génie de la géographie*, Belin, Paris.
Sion, J. (1909) *Les paysans de la Normandie orientale*, A. Colin, Paris.
Sorre, M. (1943-1952) *Les fondements de la géographie humaine* (4 vols.), A. Colin, Paris.
Soubeyran, O. (1989) La géographie coloniale, un élément structurant de l'école française de géographie, in M. Bruneau and D. Dory (eds.) *Les enjeux de la tropicalité*, Masson, Paris, 82-90.
Soubeyran, O. (1997) *Imaginaire, science et discipline*, L'Harmattan, Paris.
Vallaux, C. (1925) *Les sciences géographiques*, Alcan, Paris.
Vidal de la Blache, P. (1880) *Marco Polo, son temps et ses voyages*, Hachette, Paris.
Vidal de la Blache, P. (1896) Le principe de la géographie générale, *Annales de Géographie* 5, 129-142.
Vidal de la Blache, P. (1902) Les conditions géographiques des faits sociaux, *Annales de Géographie* 11, 13-23.

Vidal de la Blache, P. (1903a) La géographie humaine, ses rapports avec la géographie de la vie, *Revue de Synthèse historique* 7, 219-240.
Vidal de la Blache, P. (1903b) *Tableau de la géographie de la France*, Hachette, Paris.
Vidal de la Blache, P. (1911) Les genres de vie dans la géographie humaine, *Annales de Géographie* 20, 193-212, 289-304.
Vidal de la Blache, P. (1913) Des caractères distinctifs de la géographie humaine, *Annales de Géographie* 22, 289-299.
Vidal de la Blache, P. (1922) *Principes de géographie humaine*, A. Colin, Paris.

Vincent R.H. Berdoulay
Laboratoire SET-CNRS (UMR 5603)
Domaine universitaire
64000 Pau
France

4 A Partial Biography: The Formalization and Institutionalization of Geography in Britain since 1887

Charles W. J. Withers

> We have been told by Mr Chamberlain to think imperially; but in order to do this we must first of all learn to think geographically (*The Times*, 19 November 1904)

Introduction: writing geography's history

In the academic year 1999-2000, over 21,000 students in 109 universities and other institutions of higher education in Britain were registered for geography or geography-based honours degrees. At A-level (the principal advanced school qualification in England and Wales), geography is the fifth-most popular of all subjects. Geography is a required part of the national curriculum at GCSE-level (the mid-range school qualification in England and Wales). In 1996, nearly 1,050 professional geographers in all departments and schools of geography in 68 institutions of higher education submitted their research work to peer review as part of the 1996 Research Assessment Exercise, itself overseen by the four national funding councils responsible for managing the monies allocated to higher education by the government of the United Kingdom.

A hundred years ago, such popularity, numerical strength and institutional diversity in British geography would have been unimaginable. Until 1887 (in Oxford) and 1888 (in Cambridge), there was not, anywhere in the United Kingdom, a formally recognised department of geography with staff paid exclusively to teach that subject to students examined in it. The first Chair of Geography in a British university was not established until 1903 (in University College, London). Honours classes leading to formal qualifications in geography only began in 1917 (in Liverpool).

G.A. Dunbar (ed.), Geography: Discipline, Profession and Subject since 1870, 79–119.

To be sure, there were academics who taught geography in British universities well before the later nineteenth century. There was even a Chair of Geography of sorts in Captain Alexander Maconochie's inactive incumbency at University College, London, between 1833 and 1836. There were, of course, geographical societies, notably the Royal Geographical Society begun in 1830 (hereafter the RGS), the Royal Scottish Geographical Society (RSGS) founded in Edinburgh in 1884 (with its 'Royal' epithet from July 1887 and branches in Aberdeen, Glasgow and Dundee), as well as geographical societies in Manchester, Liverpool, Newcastle, Hull and Southampton by the later 1800s. Such popular and civic interest in the subject as a matter of thinking imperially was closely connected with the academic establishment of the subject. Geography's emergence as a subject of higher education also reflected contemporary concerns about the woeful state of school education in geography. Yet the fact remains: geography's present central place on the map of knowledge in Britain is the result of a history, which, in the general sense of formal institutionalized activity, is little more than a century old.

Recognizing the recency of geography's modern history is not to be presented, however, with a clear agenda as to how to recover and to write that history. It is even less clear how we should explain and understand it. There are no guidelines for undertaking disciplinary biography. Attention to seemingly straightforward questions demands that we address not just a range of work about the discipline itself and about its practitioners but also that we note changing fashions of writing about the history of geography and make mention, too, of the scholarship that has considered geography in its wider social, political and intellectual context.

Consider, for example, simple *when* questions: when was geography first taught in British universities? When was the first Honours class? When was the first Ph.D. in geography awarded? There are many other such questions. One answer to the first query, in the sense of geography understood as a formal discipline with students gaining recognised qualifications in it, has already been given, for Oxford and for Cambridge at least. But as Lesley Cormack has shown, geography in a variety of forms was taught, widely-read and understood as a basis to courtly culture and to political activity in Cambridge, Oxford and Gresham College, London, in the late sixteenth century (Cormack 1991, 1997; see also Baker 1963, 119-129). In Edinburgh, following the foundation in 1583-1584 of 'The Toun's Colledge' (the University of Edinburgh), the first Principal,

Robert Rollock, was required to teach chorography, essentially regional geography.

By the later eighteenth and early nineteenth centuries, geography was taught in each of Scotland's universities in association, variously, with mathematics, natural history, geometry, astronomy and geology (Withers 1996, 2000). The teaching staff were no less professional in their approach and in the seriousness with which they and their students took geography than their dedicated geographical counterparts at the end of the nineteenth century. Explaining the differences to such 'when' questions demands knowing the different purposes to which geography was put, how the subject has been differently understood over time and who, exactly, has been the subject's audience in given academic contexts.

Consider, too, *what* questions. With what other subjects was geography associated in its several 'originating' contexts? What relation did the geography taught and practised by professionals in academic institutions have with the popular interest in that subject shown by civic bodies? What connections are there between the geography taught in schools and university geography? Robert Mayhew, for example, has considered geography's place in eighteenth-century England. As was the case in Scotland then, geography was taught in schools in relation to mathematics and to civil history and was a matter of considerable interest in the public sphere where, amongst dozens of private teachers of the subject, some university professors taught geography in order to supplement their university income (Mayhew 1998a, 1998b; Withers 1999). In these terms, the foundation of formal departments ought not to be seen as recent, but rather, perhaps, as late events in an altogether more complex history of geography, one that places the subject's modern academic origins in much wider social and geographical contexts since geography has always been undertaken away from the sites of academe in public halls, in 'the field' and in private homes,

And *how* should we write our history of geography? As David Livingstone has shown, 'telling geography's story' has been undertaken in several ways: as 'textbook chronicles' presented chronologically; as Whiggish and normative internalist accounts designed to support particular views as to what geography is (and always has been); or as presentist arguments designed to see the past only in terms of the present and never in its own terms as a matter of difference, contradiction—'situated messiness' as Livingstone puts it—and debate (Livingstone 1992, 28). One important tendency within recent writing on the history of geography has been a move away from studying the history of *geography* as a discipline and towards the study of the histories of *geographical knowledge*, understood as that set of discursive practices and different social meanings that at any time were in use in making sense of the world geographically (on fashions in writing the history of geography, see Driver 1992; Driver and Rose 1992; Driver et al. 1995; Livingstone 1992, 1-31; Stoddart 1986).

Recognising these issues of content, context and historiography, this essay does not pretend to completeness. The principal institutions considered are of higher education. The essay does not document in detail geography's quantitative history (numbers of students taking geography, or staff teaching it, for example: useful outlines in these terms are offered by Johnston and Brack 1983; Stoddart 1967, 1981a, 1983). No attempt has been made either to date the first establishment of geography in the curriculum of all 109 higher education institutions currently teaching geography in Britain or to document the research profiles of those 68 departments registering their research quality in the subject in 1996. Developments within the discipline in Britain—so far as they should be separated from developments elsewhere—are not chronologically itemised. The essay is written by someone with a professional engagement with historical geography and currently more interested in recovering the longer-run place of geographical knowledge in academic and popular spaces than in establishing an unproblematic and progressive genealogy for the discipline's recent history.

The essay is, then, both partial and personal. What follows is, in part, a summary review of known work. In part also, the essay makes reference to unpublished manuscript evidence, notably to personal correspondence, to illustrate something of the personalities, places and politics behind geography's professionalization. The essay as a whole is much shorter than is merited by the subject matter available. References have been kept to an effective minimum. My focus is on Britain, on the United Kingdom, and where I separately discuss England or Scotland or Wales, I mean to do so: Ireland is not considered (Alexander and Gillmor 1989; Fahy 1981).

My starting point, as a working beginning, is the formalization and institutionalization of geography in Oxford and in Cambridge in the late nineteenth century. Three topics form the focus to this partial biography: the emergence of the discipline from the 1880s to the early 1930s; issues of institutional activity and disciplinary credibility from 1933 (the foundation of the Institute of British Geographers) to the present; and current debates about the 'quality' of British geography in teaching and in research.

Questions of disciplinary emergence, 1887-1933

The establishment of geography as a university subject in Britain is a story of overcoming others' disciplinary prejudices—notably, of geologists, who felt threatened by geography's engagement with the natural world and who could see no distinctive 'core' to the subject that was not already within their remit. It also meant overcoming doubting institutional authorities, unsure exactly what benefits the study of geography would bring, and, like geologists, how exactly it related to 'established' disciplines (like history). It is also a study of individual vision and institutionalized endeavour—from men like Halford Mackinder and John Scott Keltie and the RGS—and of energetic appeals to the utility of the 'new' geography as a science, a means to popular knowledge and to the education of the imperial citizen. Keltie's 1885 *Report* on the state of geographical education and the advantages to Britain of improving it and Mackinder's 1887 paper on the 'new' geography are both now recognised as defining moments in the emergence of geography as a modern discipline (Keltie 1886; Mackinder 1887; Stoddart 1986).

Outside Oxford and Cambridge, the first teaching position in England was in Manchester in 1892. In Scotland, Hugh Mill had lectured in physiography in Perth in 1887, at the invitation of Patrick Geddes, as part of the University Education movement. From 1888 to 1892, Mill lectured in physiography and in commercial geography at Heriot Watt College in Edinburgh. Most teaching posts in geography came in the early twentieth century: Aberystwyth (1906), Reading (1907), Edinburgh, Sheffield, and Liverpool (1908), Glasgow and Leeds (1909), Southampton (1913), the London School of Economics and Political Science (hereafter LSE) (1918), Aberdeen (1919). At St Andrews the first teaching post was established in 1935, yet the University had professors of law teaching the subject as part of mathematics classes in the early 1580s (Withers 1996).

Even by 1920, these were only six Chairs of Geography: in addition to University College, London (with Professor L. W. Lyde, in 1903), these were in Reading (1907), Oxford (1910), Aberystwyth and Liverpool (1917) and at Birkbeck College, London (1920). Honours classes leading to formal qualifications were, generally, even later: in the LSE and Aberystwyth in 1918, in Cambridge in 1919, Manchester in 1923 and Sheffield in 1924, and, in 1933, some thirty-four years after the foundation of the School of Geography there, in Oxford. Further growth, beyond Oxbridge, Scotland and the 'Redbrick' universities, occurred with the development of the technical colleges from the late 1940s and in the diversified curriculum of many teacher training colleges from the 1960s and of former polytechnics after 1992. As Stoddart has remarked, 'The recency of many of these developments . . . is salutary, and the protracted history of developments at most of these institutions gives some indication of the difficulties which the subject often faced' (Stoddart 1986, 46; see also Unwin 1992, 83-85).

Explaining the recency of geography's foundation as a recognised discipline has, for England and Wales anyway, much to do with the relative recency of the institutions themselves. Outside Cambridge and Oxford, only University College, London, was in existence at the foundation of the RGS in 1830. King's College, London, opened in 1831, Durham in 1832 and other civic universities were established in the later nineteenth century: Owens College, Manchester (1851), Southampton (1862), the College of Natural Science at Newcastle (1871), the Yorkshire College of Science at Leeds and Firth College in Sheffield (1874), the College of Science, Bristol (1876), Mason College, Birmingham (1880), and university colleges at Liverpool and Nottingham in 1881 and at Reading and Exeter in 1892 and 1895 respectively (Stoddart 1986, 46-58).

The argument of institutional youth and the prejudice of other, more strictly 'scientific' disciplines, does not directly work for Scotland. Evidence that geography was taught in the universities of Edinburgh, Glasgow and St Andrews in the 1580s and that it was widely used in both King's and Marischal Colleges in eighteenth-century Aberdeen means the formal departmental recognition of the subject in those places only from 1908 onwards is both recent *and* late and, further, without easy explanation. Answers are partly to be found in the suspicions of geologists and others—and, in that case, there are parallels with England even if leading geologists like Archibald Geikie were strong advocates of geography's professionalisation—and, partly, in the sense that, for some members of Scotland's professoriate, geography understood as the location of natural phenomena was anyway so central to their own work that its separate identification was unnecessary (Withers 1996, 2000).

These issues of institutional uncertainty, disciplinary 'turf' and discursive credibility may be illustrated by reference to the two 'founding' departments at Cambridge and Oxford and to the departments of geography in Scotland's universities. This is not to suppose, however, that geography's foundation in those places was the model for later experiences elsewhere.

1 Modern beginnings: the establishment of geography at Oxford, Cambridge and in Scotland's universities

There was a paid lecturer in geography, in Magdalen College, Oxford as early as 1541 (one Baldwin Norton). The origins of geography's modern disciplinary foundation at Oxford date from reorganisation of the University's Honours Schools in 1850, from the concerns of Thomas Butler, a fellow of Exeter College, who had lectured in physical and historical geography in Oxford in 1853, and from the promptings of the RGS in 1871 and 1874 (Baker 1963, 14-32, 119-129; Scargill 1976). A further RGS memorandum in 1879 considered geography's integrative rôle and the prospective duties of professorships in geography at both Oxford and Cambridge:

> It is through geography alone that the links can be seen that connect physical, historical, and political conditions; and it is thus that geography claims the position of a science distinct from the rest, and of singular practical importance . . . The duties of such a Professor . . . would be, first, to promote the study of scientific geography as defined above; and, secondly, to apply geographical knowledge in illustrating and completing such of the recognized University studies as require its aid (RGS Archives, Supplementary Papers I, 519).

Further RGS proposals in 1886 laid out the financial conditions by which posts would be supported. In January 1887, the University informed the RGS that it would fund a Reader in geography for five years at an annual salary of £300: this was soon reduced to £150 from the University, with the RGS paying the balance. H. J. Mackinder was appointed to the position in June 1887.

Mackinder's contribution to the nature of 'modern' geography is well known (Blouet 1987; Parker 1992). Probably less well known are the difficulties he faced in establishing his subject. For one thing, geography was not recognised as a subject in its own terms within Oxford: physical geography had been removed as a separate subject in the Natural Science School from 1872, and although lectures in geography were given by historians such as H. B. George and E. A. Freeman, they received little attention since the subject had almost no status in examinations. For another, Mackinder had no dedicated teaching equipment, no special room and little money for books. And it may well be that geography's development at Oxford was hindered by Mackinder's principalship of the university extension college at Reading in 1892 (which position he held alongside his Oxford post) where he was responsible for history and geography, and, later, by his lectureship in economic geography at LSE (at which institution he gave evening lectures from 1895).

The School of Geography at Oxford was founded in 1899. Mackinder's plan for such a school—modelled partly on European departments and partly on what he had observed at Harvard—was to site it in London. This was not possible and, indeed, rather more than the University, the RGS, through Sir Clements Markham, took the lead in establishing the School at Oxford. The School began its Diploma in Geography on 1 October 1899. *The Times*, on 29 March 1899, certainly saw the intended School as a welcome imperial venture: 'The serious mistakes made and the risks of war incurred by geographical ignorance have often been referred to; with the establishment of this school there will be no excuse for such ignorance among those who have the conduct of the Empire's affairs'. Other staff appointed were A. J. Herbertson, who had earlier taught at Heriot Watt College in Edinburgh and at Owens College in Manchester, H. N. Dickson and G. B. Grundy, both part-time, who lectured in physical geography and in ancient geography respectively.

Mackinder resigned from Oxford with effect from 1905. There can be no question concerning his zeal for geography. Yet doubts must remain over the degree of his commitment to Oxford. Mackinder had unsuccessfully stood as a Liberal Imperialist parliamentary candidate in 1900 and was, from November 1903, Director of the LSE. Given, too, the at times lukewarm support within the University, geography's successful establishment at Oxford between 1899 and 1915 and longer-term influence (several of its students were appointed to teaching posts in other newly-established departments) owed greatly to the energies and scholarship of Herbertson (who was appointed to a Chair in 1910), and to the continued financial support of the RGS (whose annual grants came to an end only in 1924).

Herbertson's death in 1915 and the effects of the First World War lessened the School's effectiveness. Although numbers of Diploma students rose in the immediate post-War period, they declined sharply from the later 1920s. This was due in part to illness amongst incumbent staff and to internal differences over the nature of the appointment to succeed Herbertson. Mackinder's view, expressed in a letter to Douglas Freshfield in 1919, was that 'In my opinion no man will make an influential position for himself as a Geographer at Oxford who is not strong on the Humane [sic] side . . . What we need at this moment is a man with personality and broad ability, and the essential geographical standpoint' (RGS Archives, H. J.

Mackinder to D. Freshfield, 7 April 1919). The decline in student numbers was chiefly due, however, to competition with other British departments, most of which had established an Honours programme from the outset. Oxford was slow to do likewise. Not until 1932 were further major changes effected with the appointment to a Chair of (Major) Kenneth Mason and the establishment that October of the Honours School in Geography, the first examinations for which were held in 1933.

In April 1919, in a letter to Arthur Hinks, Douglas Freshfield noted 'Oxford is human where Cambridge is natural scientific & if you want G[eography] to flourish at Oxford you must conform to the local conditions' (RGS Archives, D. Freshfield to A. Hinks, 10 April 1919). The sense in which we might think of geography's different institutional foundation as itself a matter of geographical difference—of 'local conditions' in a constitutive sense—is, to some extent, borne out in the case of Cambridge where the subject was initially established more as a physical science. But the circumstances of geography's foundation at Cambridge also had much in common with those for Oxford.

Geography's modern foundation in Cambridge effectively began, as at Oxford, with the memorandum of the RGS in 1871 in which the Society, through Keltie, its Secretary, emphasised 'the special importance of geography to Englishmen in the present age . . . We speak of geography, not as a barren catalogue of names and facts, but as a science that ought to be taught in a liberal way, with abundant appliances of maps, models, and illustrations . . . there is no country that can less afford to dispense with geographical knowledge than England' (Keltie 1886, 83). (Keltie's use of the term 'England' should be here taken to mean 'Britain': the two are too often used interchangeably, then and now, by people who should know better).

The initial prospects for geography at Cambridge were favourable inasmuch as the special board for geology and biology in the University had recommended in 1883 that university teaching staff be appointed in the subject. But, as in Oxford, established figures in other disciplines were sceptical about the need for geography to be separately recognized, or, indeed that the subject was deserving of examinable attention at all. Several influential authorities in Cambridge refused to countenance jointly funding geography's development with the RGS. Despite opposition that was, then, both personal and disciplinary, the *Report on the Teaching of Geography* that was advanced in March 1887 by advocates for formalising geographical teaching in Cambridge, was approved in June 1887. Lectures in geography designed to further its status within the University were given in February 1888 by General Richard Strachey, President of the RGS. In June 1888, Francis H. H. Guillemard was appointed as the first university lecturer in geography at Cambridge.

Guillemard had trained as a zoologist at Cambridge (where he had attended the lectures of Professor Alfred Newton, geography's leading advocate within the University, and to whom he owed his preferment) and had travelled widely before his appointment. Like Keltie and others, he was critical of the current state of geographical education, notably in schools. In his unpublished autobiography, written years after the events he describes, he remarked 'At the time of which I speak there was no teaching of Geography either at Oxford or Cambridge. The British, who needed it most, a race of wanderers and owners of colonies in all parts of the world, were remarkably ignorant of that science. It appeared to be taught at School with the express aim and object of inspiring loathing in the bosom of the pupil' (Guillemard 1932, 105).

Guillemard's tenure—most of which was spent in Madeira and Cyprus—was brief. He resigned in December 1888, citing ill health, the death of his mother, lack of resources (such as a room and teaching equipment). The University's request that he actually begin teaching in the following year was the last straw. His successor, J. Y. Buchanan, the first person formally to teach in geography at Cambridge, was cut from altogether different cloth.

Buchanan, who established his reputation as a physical scientist as part of Sir John Murray's 1872-1876 *Challenger* voyage, was to become a pioneering glaciologist and oceanographer. His teaching suffered (in the view of his contemporaries) from being overly specialised. Although he later went on to be much honoured by scientific societies, he stood down from the lectureship at Cambridge in 1893. He was succeeded by H. Yule Oldham (Reader from May 1898), who was to teach geography in Cambridge for twenty-eight years.

Oldham's lectures—in physical geography, the history of geographical discovery, and the geography of Europe—were paralleled, from 1903, by special lectures from other scholars (for those wishing to undertake exploration or intending to work overseas) in anthropogeography, geomorphology, and surveying. The growing interest in geography—apparent in numbers of undergraduates—was, from February 1904, reflected in the establishment of the Board of Geographical Studies that drew up regulations over examinations and the syllabus. Internal disputes in 1908-1909, principally over the place of surveying, undermined the

Board, Oldham's position as Reader and briefly threatened the status overall of Cambridge geography. But new appointments, together with the continued support of the RGS, secured its position.

After 1918, earlier proposals to establish a geographical tripos (i. e., final honours examination) were formalised. Philip Lake's appointment as Reader and Head of Department in 1919 (he had been elected to a lectureship in 1908) coincided with the beginnings of tripos teaching. The first Cambridge honours degrees in geography in 1921 marked fifty years of support by the RGS and a new beginning for the subject in Cambridge. The status of the subject in Cambridge was, in the view of the Board of Geographical Studies anyway, further secured by the decision to hold the Congress of the International Geographical Union there in 1928 (Stoddart 1975).

The formalization of geography teaching in Scottish universities in general followed the pattern identified by Stoddart for England and Wales: an initial appointment, of a single person, teaching the subject at Diploma level or as part of the General M.A. or B.Sc. degree, before the establishment of an Honours school (Stoddart 1986, 45). Establishment of the discipline in Scotland was more protracted than in England, despite the fact that geography was taught in each of the country's ancient universities well before the nineteenth century. In Aberdeen, for example, the University's Principal promised in 1909 that the University would have a teaching post in geography. Yet this commitment was not realised until 1919, even though a lecturer in education in Aberdeen, John Clarke, had been giving geography classes and publishing on the subject since 1898 (Bridges 1985, 82).

As early as 1889, the Council of the Royal Scottish Geographical Society prepared a *Memorandum* to the Commissioners appointed by Parliament then investigating the state of Scottish universities, in order 'to press on them very urgently the claims of Geography to be recognised as a department of the Higher Education'. The *Memorandum* employed the same justificatory language as the RGS had in its appeal to Cambridge and Oxford: the need for geography in order to understand and better manage the Empire; the fact that Britain lagged far behind other European nations in the place of geography in higher education (especially in comparison with Germany); and the dull nature of geographical teaching in schools (*Memorandum of the Council of the RSGS*, 1889, 1). Individual universities and other places of learning throughout Britain were contacted in 1901 by Sir Clements Markham, acting on behalf of a delegation of British geographical societies, in which he insisted upon the desirability of recognising geography in their curricula.

In a letter of April 1902, Keltie reported to both the RGS and to the RSGS how 'In the case of the University of Edinburgh, and of Yorkshire College, Leeds, it is stated that steps are being taken to establish Chairs of Geography; but in all cases except Edinburgh University, regret is expressed that nothing can be done for Geography without Financial assistance from outside' (RSGS Archives, Keltie to the Hon. Sec., RSGS, 16 April 1902). Keltie's general conclusion was true. His particular remark about the readiness of Edinburgh University to secure teaching posts in geography was not. Despite earlier financial support from J. G. Bartholomew, head of the map-making firm, from the RSGS and the Carnegie Trust for the Universities of Scotland, no position of any sort was established before 1908. Prolonged disputes between the University, the Bartholomew family and the RSGS over the terms of a legacy left by the family, together with an uncertainty in the University about the benefits of geography as a subject, delayed the appointment of a Chair until 1930 (despite the fact that the Senatus of the University again prompted the RSGS in January 1905 about the need for geography teaching posts and that over £2000 was available by 1906-1907).

In opening the first course of geography at Edinburgh in October 1908, George G. Chisholm remarked that he was 'the first lecturer in geography appointed at any Scottish university' (Chisholm 1908, 561). His self-description is true, strictly speaking, but he made no mention of Hugh Mill's earlier lectures in physiography and commercial geography, on behalf of Heriot Watt College (formerly the Edinburgh Mechanics' Institute). When Mill left Edinburgh to become Librarian at the RGS, his teaching in 'Industrial and Commercial Geography' was continued by his assistant, William Couper, by Herbertson (for the academic year 1896-1897), and in 1899-1900 by William Speirs Bruce, the polar explorer and future leader of the *Scotia* expedition. For this reason, and in terms of the subject's modern origins, the distinction might usefully be made between the first teaching post in geography in the city of Edinburgh (Mill) and the first teaching post in geography at the University of Edinburgh (Chisholm).

In Glasgow, in contrast to Oxford and Cambridge, the establishment of geography was broadly welcomed by geologists. This, I suggest, is because the two advocates of geography's establishment there, Professor J. W. Gregory, Professor of Geology, and, notably, the University's Principal, Sir Donald MacAlister (one of the first recipients of the RGS Medal scheme in geography for promising schoolboys), were much involved in establishing the subject in Cambridge. The first teaching appointment in Glasgow, in 1909, was H. G. Lyons, former Director General of the Egyptian Survey.

By 1913, for MacAlister, 'the Glasgow School of Geography had already proved both its quality and its public usefulness, and efforts were in progress to place it on a footing commensurate with its importance as a branch of liberal education, by procuring for it a suitable endowment'. Other evidence would suggest that wider support for geography was not as forthcoming as MacAlister and other university figures would have wished. 1913 was the centenary of the birth of David Livingstone, the explorer and African missionary. As MacAlister reported in 1921, 'The proposal was made and cordially adopted, that among the local memorials to be established in his honour, as missionary, explorer, physician, and pioneer of empire, a Livingstone Chair of Geography in the University should be included'. This never happened. As MacAlister further noted, 'Perhaps because the effort was dispersed among a multiplicity of objects . . . The War and Peace have diverted men's minds to other needs and other memorials' (MacAlister 1921, 56). Such lack of support is surprising given that geography's academic foundation owed much to the concerns of civic bodies to promote geographical knowledge.

2 The civic impetus: the rôle of geographical societies in geography's academic establishment

When the Royal Geographical Society was founded in 1830 as the Geographical Society of London, it was the third such body in the world (after Paris in 1821 and Berlin in 1828). In part the RGS originated from earlier bodies such as the African Association, begun in 1788. Its establishment in Britain as a formal institution for the promotion of geography was pre-dated by a short-lived scientific journal, *The Edinburgh Journal of Natural and Geographical Science* (1829-1831), which sought to do much the same in published form (Withers 1999).

The RGS's foundation was part of a wider outbreak of associational recognition for the emergent professional sciences in the early nineteenth century: others included the Geological Society, established in 1807, the Zoological Society (1826), the British Association for the Advancement of Science (1831), the Meteorological Society (1836) and the Chemical Society (1841). Charles Dickens's term for the BAAS—'The Mudfog Association for the

Advancement of Everything'—perhaps unjustly satirizes the institutionalising activities amongst science's new professionals at that time. But it is important to the wider context in which geography should be understood that we see the RGS as part of broader trends and as one civic institution among many acting to advance scientific interests.

The aims of the RGS at foundation included the publication of new geographical discoveries, the support of exploration and of Britain's imperial interests and (implicitly) its self-promotion as 'a centre of geographical calculation' within Britain, but to serve the world. It had a fair number of men for whom the Society was more a dining club than an academy (most such bodies then did), but to think of it as not making a contribution to the development of geography in the absence of formal professional provision is misguided. From 1830 until 1880, the RGS published its *Journal*, from 1855 to 1892, its *Proceedings*, and, from 1893, the *Geographical Journal*. It made major contributions to exploration, not least through its famous *Hints to Travellers*, first published in 1854 (Driver 2000).

RGS support from the 1870s for the academic establishment of geography was part of a heightened awareness of the importance of 'thinking geographically' at that time apparent in, for example, African exploration, polar travel and in the BAAS's Section E (Geography and Ethnology) in promoting geographical education (the RGS effectively organised Section E from its foundation in 1851, until 1914). The fact that the RGS gave £7250 to Cambridge between 1883 and 1923 and £11,000 to Oxford from 1887 to 1924 in support of geography is testimony to the Society's commitment to its subject. Similar bodies elsewhere in the United Kingdom did likewise from the later nineteenth century.

In Manchester, the initial expression of a need for a geographical body came in 1879 at a meeting in the town's Chamber of Commerce discussing commercial connections with Africa, and in a failed attempt to establish a Society of Commercial Geography. In 1884, however, the Manchester Geographical Society was begun and that body was instrumental in establishing geography at Owens College and in awakening 'interest in Geography by lecturing to 150,000 in eight years' (or so it claimed in 1901: Brown 1971, 49).

The RSGS with its several branches also promoted commercial interests as a geographical society, principally with Africa. It was no accident that several of its branches were opened with an inaugural lecture from H. M. Stanley, the Africanist then lauded for his heroic endeavours in the 'Dark Continent', and, recently, the subject of more critical attention (Driver 1992b). Yet the RSGS was also strongly professional and academic from the outset, given both the status of its leading proponents—J. G. Bartholomew and Professor Archibald Geikie, to name but two—and the wider social and intellectual 'heritage' of later nineteenth-century Scotland. Elspeth Lochhead has described this 'heritage' in terms of Scotland's democratic intellect, a 'geographical heritage' in relation to traditions of map-making and exploration, and a 'scientific background', notably in geology, oceanography, meteorology and natural history, whose practitioners were disposed to the 'new' subject of geography (Lochhead 1980, 1981, 1984).

Lochhead's argument must now be modified in several important respects. Whilst the contribution to geographical thinking of geologists such as the Geikies, of John Murray in oceanography and of others in the natural sciences, together with the human-environment work of Marion Newbigin, is not in dispute, most of their important work post-dates the foundation of the RSGS. In Aberdeen and in Glasgow, the imperial and commercial interests of the RSGS branches there owed to the prior existence of civic bodies that held lecture programmes in geography: the Aberdeen Philosophical Society, and the Geographical Section of the Glasgow Philosophical Society. The last-named, for example, was the moving force in getting Stanley to Glasgow to talk on 'geographical science as it affects Africa' in 1885. Finally, we must not neglect the contribution of men like Patrick Geddes, who gave lecture programmes in geography in the Outlook Tower in Edinburgh between 1892 and 1915 (having previously held such classes at the Granton Marine Station from 1886), and whose vision for geography did much to promote popular interest in the idea of geography as regional survey and in geography as a practice of citizenship (Matless 1992; Meller 1993).

We should not see the civic geographical societies in Britain as everywhere the same. The RGS was always primus inter pares. It was, from 1871, much the most involved with the foundation of modern academic geography and was followed in this respect by the RSGS and the Manchester society. Several others did not last long: Southampton (1897) and Hull only just survived into the twentieth century; Liverpool (1891) closed in 1933, and the Tyneside Geographical Society, founded in 1887, had effectively closed by 1937. For MacKenzie, these bodies were concerned with 'municipal imperialism', with their own civic identity as well as with geography, and they succeeded rather than preceded the imperial scramble for Africa (MacKenzie 1995).

Yet their place on the map of geography's disciplinary emergence should not be overlooked. As with the foundation of the Geographical Association (which was set up by B. B. Dickinson in Mackinder's rooms in Christ Church College, Oxford, on 20 May 1893 to promote geography teaching), civic geographical societies reflected that wider trend in society of institutionalizing good intentions. In the thirty years from 1884, geographical societies embodied a public geographical and imperial consciousness in Britain. After the 1914-1918 war—in which the RGS was much involved with cartography and military intelligence (Heffernan 1996)—public support was much less evident and the RGS and RSGS became more strictly academic.

3 Disciplinary contexts and curriculum development

One of the reasons why geography was not welcomed by geologists and other natural scientists was because many of geography's first professionals anyway espoused interest in the physical world: in, for example, topography and physiography, in the distribution of vegetation and other natural phenomena, in earth surface processes and in oceanography and climatology. Many scholars shared similar interests, moreover, in the philosophical underpinnings of natural science, notably in evolution and evolutionary thinking (Livingstone 1992, 177-215).

Some of geography's early lecturers also claimed a competence across several subjects, a fact that sometimes made their replacement hard to manage and the exact nature of this 'new' discipline hard to discern in the minds of others. When Herbertson resigned his lectureship in Manchester in 1896 (to move into publishing in Edinburgh), A. W. Ward wrote to Keltie to note that 'The Geographical teaching will be divided between the professors of History and of Geology (Physiography) and their assistants, special provision being made for instruction in cartography. A special course in Trade-routes &c. will be given by Mr. A. W. Flux in Political Economy' (RGS Archives, A. W. Ward to J. S. Keltie, 30 June 1896).

The sense in which geography both embraced a range of disciplinary practices yet, in so doing, seemed not to satisfy any given constituency is apparent in Herbertson's rather despairing tones in a letter to Keltie:

> I believe in using mathematics when you can in Geography but history is far too complex for mathematical notations. I am constantly criticised here

> by one set of people for making geography too mathematical & by another set for not making it mathematical enough! The one set forget that descriptions are worth little if they are without art. You cant [sic] understand land forms unless you can see them as an artist as well as a scientist; & that is true of the more complex geographical whole. I have to fight people who can see nothing in maps & people who can see nothing but maps (RGS Archives, A. J. Herbertson to J. S. Keltie, 1 August 1913).

Further illustration of the ways in which geography's establishment owed much to the vision of its individual practitioners is apparent in H. J. Fleure's letter of application in 1916 to the University of Wales concerning the geography professorship at Aberystwyth (to which he was appointed in 1917). We are told that 'Fleure's printed statement [on his understanding of the nature of geography, a requirement of the appointment process] is somewhat vague & a bit hysterical. His conception of the field of geography lacks precision; some may answer he claims too much . . . The Physical & Mathematical (including the Cartographical) sides of the subject should be included . . . He ought to realise that the physical side is indispensable for the adequate treatment of the human side' (RGS Archives, 'Dr. Fleure', 1). It is surprising that Fleure did not so realise given that, between 1908 and 1910, he was in charge of geology at Aberystwyth, and, from 1910 to 1917, Chair of Zoology there.

The fact that similar interests and capacities are apparent in studying virtually any of the 'founding fathers' and their departments and institutions allows us to make a number of generalizations concerning modern geography's early disciplinary context. Geography was established by people whose intellectual training incorporated one or other (and, often, several) different subjects: history, geology, anthropology, economics, mathematics, zoology, ethnology, botany and natural history, even classics. Geography—understood, for example, as a matter of natural distribution (in botany and zoology), or as the physical 'stage' against which human events were to be examined (history and economics)—was, in several places, incorporated into the curriculum of other subjects.

In these terms, what we might think of as the discursive context of geography's disciplinary emergence was, simultaneously, both a strength and a weakness: the first because geography was taught within *and* outwith its founding departments by people with broad intellectual interests; the second because geography seemed not to have a core, an 'essence', which marked it apart from other subjects. The latter problem, of definition and of disciplinary identity, plagued geography from the outset.

The local 'flavour' imparted to geography departments in their early years in part owed to the particular capacities of incumbent staff: Chisholm on commercial geography in Edinburgh, for example; Mackinder on political, historical and economic geography in Oxford, Buchanan on physical processes in Cambridge, Fleure and anthropogeography at Aberystwyth. Yet there is a sense, too, that a collective effort was made to establish a rounded and general geographical curriculum without there ever being a common sense view of what such a thing was. And on numerous occasions, the curriculum offered by people we might think of as tenured staff was complemented by specialist teaching from others.

Curriculum history sensu stricto is an acquired taste, even amongst educational historians. But let me briefly use one illustration here in support of the above point, to connect this section with what follows and to establish the links between the civic promotion of geography and the teaching content of the first geography programmes.

In its 1890 *Memorial* to the Parliamentary Commissioners appointed under the Universities (Scotland) Act, the Council of the RSGS indicated that '. . . the kind of Geography that the Council wishes to promote in the Scottish Universities, and through them in Scottish Schools of every grade . . . includes:- Mathematical and Astronomical Geography. Physiography, including Meteorology. Topography. Commercial and Industrial Geography. Historical Geography. Ethnology'.

George Chisholm's course in Edinburgh from 1908 came, remarkably, quite close to this ideal general geography programme. His single course in geography included attention to the scope and nature of geography, cartography, the form of the Earth, meteorology and climatology, economic geography, political geography, the history of geographical ideas and the progress of discovery, and, finally, the elementary principles of surveying. In 1909-1910, two half-courses were added, the first covering 'the history of exploration and of geographical ideas, cartography, the distribution of typical landforms and distribution of plants, animals and the races of mankind', the second 'the economic and commercial development of the leading commercial countries during the past 40 to 50 years with special reference to the geographical conditions that have influenced that development'.

The curriculum was taught entirely by Chisholm alone until 1918-1919, when, with the appointment of Alice Lennie as an assistant lecturer and with recognition of geography's status as a Diploma subject, greater attention was given to economic geography, including 'economic ethnography'. From 1923-1924 (in which year Chisholm retired and A. G. Ogilvie was appointed: he took up the Chair in 1930), the curriculum further changed to include 'The physical basis of geography in outline – the leading natural regions of the earth, their relationships to political divisions and their economic importance', and (for a decade), a course on 'Railway Geography', undertaken at the request of the London and North Eastern Railway Company. Further attention was given to physical geography in changes made in 1928-1929, and additional half-courses included historical geography and 'the human geography of the monsoon lands'. The Honours degree was established in Edinburgh in 1930. Honours examinations were based on seven papers, three of which were on regional geography, with a dissertation, usually based upon description of a 100-square-mile part of Britain. Field classes were held in the Scots College in Montpellier, France. In 1935, the Department opened its 'Earth Sculpture' laboratory (essentially, a flume tank) to demonstrate physical and hydrological processes and, in the same year, the students established the student Geographical Society (University of Edinburgh, Department of Geography Archives, MS DG1/31/21).

What was happening in Edinburgh should not straightforwardly be seen as applicable to other departments in the period. Yet several points are worth making. Geography's initially few professional practitioners were, at first, mostly generalists whose energies were devoted to teaching. Only latterly did specialisms emerge within the curriculum, usually, as now, based upon given research interests. The numbers graduating in Honours were low but increased steadily after about 1928-1929. The numbers taking the subject in the first years of their studies increased in all departments established before 1914, with a dip in numbers during 1915-1918 and a surge after 1920. The relative number of students attending human/historical geography was, in the early years, greater than those attending physical geography classes, since, until they were formally recognised for examination purposes, the latter counted for more if taken within geology departments.

How, then, might we summarise the institutionalisation and professionalisation of modern geography in the United Kingdom in the first half-century of its existence? Recognising local differences in emphasis, several points may be made.

Geography's foundation as a modern academic subject owed to a widespread recognition that geography, variously understood as historical understanding, human social order *and* physical processes (including mapping), was useful to Britain as an imperial power. It owed, too, to recognition that other imperial powers, notably Germany, were further advanced in the provision of geography in higher education. Geography's establishment in universities and colleges was intended to be useful at that level and to provide better teachers in order that school-level geographical understanding would be improved. Geography's establishment owed much to the key rôle of civic geographical societies, chiefly of the RGS in relation to Oxford and Cambridge, to the Manchester Geographical Society and, in Scotland, to the RSGS. Indeed, given the initially uncertain support for geography's establishment, in Cambridge, Oxford, Edinburgh and Aberdeen anyway, it is possible to suggest that the discipline's modern foundation was more the result of civic impetus—itself motivated by imperial and educational concerns—than of any concerted support from within institutions of higher education.

Geography's foundation also owed much to the energy and vision of its first professional practitioners. As David Stoddart has noted, distinguishing characteristics shared were pragmatism, the emphasis on pedagogy, their own interdisciplinary intellectual training, and an interest in fieldwork (Stoddart 1986, 51). If not the direction of the discipline as a whole, then its local character in particular places was shaped by the qualities of individual staff. Only from the late 1920s was there a generally discernable move towards sub-disciplinary specialisms as Honours programmes reflected and attracted the growth in student numbers.

Matters of disciplinary credibility since 1933

The first fifty years of the history of modern geography in the United Kingdom might be thought of as a story of achieving international respectability for the subject as a whole. In general terms, the period since 1933 was distinguished by debates within the geographical community as to its own 'credibility', using that term loosely to mean a sense of geography's own institutional status, and, more particularly, to describe the intellectual issues and subject matter with which geographers were concerned.

Recognising that it is not possible to document even in outline the circumstances of individual departments or schools, of individual geographers or of disciplinary trends, this section examines through two examples the credibility of modern British geography in the two senses outlined above. The first considers the rôle and functions of the Institute of British Geographers (IBG) from its foundation in 1933 (including its merger with the RGS in 1995). The second examines something of the evidence for sub-disciplinary specialisation, and, I argue, credibility within geography, by discussing both the changing nature of historical geography from the late 1920s to the present, and the quantitative revolution in British geography as one element of geography's search for scientific credibility.

1 Disciplinary institutionalization and organizational activity: the rôle of the Institute of British Geographers

It has been argued that 'the foundation of the Institute of British Geographers came as the culmination of the movement to establish geography in British universities which began with Keltie's report to the RGS in 1885 and Mackinder's subsequent appointment to Oxford in 1887' (Stoddart 1983, 1). This is broadly true, but we ought not to see the IBG's appearance as an unproblematic matter of academic predestination. Several other circumstances were prompts to its foundation.

The IBG in part owes its origins to a move away from the intellectual traditions of the RGS and, indeed, to tensions within Britain's academic geographical community in the late 1920s between those we might typify as 'explorer-teachers' and those, often younger men, appointed to expanding departments after 1918, who were more concerned with geographical research. A further prompt was the 1928 International Geographical Congress, held in Cambridge. Participants to this meeting not only considered it important in terms of British geography's international prestige: many were also activists in the plans for a consciously academic and research-based association for professional geographers. A further stimulus was the desire for research publication and from a feeling, directed at the RGS's *Geographical Journal*, the Geographical Association's *Geography* and, to some degree, at the *Scottish Geographical Magazine* of the RSGS, that geography's journals and institutions did not devote sufficient attention to academic publication (Steel 1984).

Initial moves to do something about the needs and demands of geography's new professionalism were made in early 1931 in a meeting in London held, as one participant remembered it, 'to discuss some form of organisation to serve the common interests of geographers'. There was, however, no indication given as to quite what those 'common interests' were other than the felt need for greater recognition of the subject's academic credibility. At a meeting held in York in September 1932, a committee working on behalf of what was initially called the 'University Group of Geographers' determined the Group's two 'Objects in View' as the holding of meetings (including a full annual conference held in January, short visits to departments of geography and fieldwork), and publication. The Institute was formally inaugurated on 3 January 1933 in the LSE.

Despite initial declarations of intent, membership levels were slow to increase: only thirty-nine persons were confirmed members by April 1933. The much-cited figure of seventy-three 'original' or 'founder' members listed in the IBG's first publication, *Transactions*, included many individuals who had sent only 'intimation of intended membership'. Membership had increased to eighty-three persons by 1936, but of that total, only thirty-three attended the annual meeting. There were no formal meetings of the IBG between 1939 and 1945 and numbers remained low until the late 1940s, from which period, however, membership increased dramatically (Steel 1984; Stoddart 1983).

The Institute itself was also different after 1945. For Alice Garnett, one founder member, 'a new "IBG Phoenix" arose, vastly increased in size and strength from the ashes of World War II' (Garnett 1983, 34). This 'new' IBG was evident in increased numbers, in a wider societal recognition of geography's standing stemming from geographers' contributions to the war effort and in the increased size of academic departments in institutions in which geography was already established and in geography's inclusion in newer institutions altogether.

One historian of the IBG has commented on these issues thus: 'That the Institute was first resuscitated and then grew steadily in the post-war years—in parallel as it were with the development of geography as a discipline that was increasingly accepted as a proper subject for university study—was largely due to the efforts of a handful of devoted geographers with a real faith in their subject, and to the support they received from colleagues' (Steel 1984, 27). In these terms, geography's history is reduced only to the actions of altruistic individuals, working selflessly for the good of a cause whose essential tenets are, at best, only implicit to historians of the discipline, and, in truth, were probably not much clearer to the contemporaries involved. It is also to explain geography's more recent institutionalisation in more or less the same ways as its foundation: through the efforts of a few individuals.

Men like S. W. Wooldridge, (Sir) Dudley Stamp, Alan Ogilvie and R. O. Buchanan, for example, are often mentioned in this respect. R. W. Steel's account of the IBG is, in many ways, an account of his involvement with the IBG (Steel 1984). Buchanan's 1954 overview of the IBG after twenty-one years offered one man's view of the Institute's history and, in part, a prospective view as to its future. His views are interesting given his own declaration of partiality—'What I have to say will be in no sense a detached, impersonal, critical history'—as well as for the fact that he had been in office for all bar one year between 1933 and 1954 (Buchanan 1954). Alice Garnett's remembrances are notable in being a woman's perspective and for her involvement with Section E of the BAAS, the Geographical Association and the IBG. For the last two, Garnett offers a personal view of their separation as they represented the worlds of school geography and of geography in the academy. Many academic geographers were, of course, members of both, but only within the last ten years or so have the two bodies again made more formal connection in annual meetings, partly under the aegis of COBRIG, the Council for British Geography, and partly, too, given the need for liaison over the nature of geography at different levels in the curriculum.

Yet to see the IBG as simply the result of several active individuals' work and as the concern of only a few is to ignore the sense that that body both represented and stimulated a collective engagement amongst professional geographers with their subject as an intellectual *and* as a social enterprise. Extant histories of the IBG are consciously autobiographical. But understanding institutional histories in those terms alone is to ignore other possible histories. These should include noting the overwhelming male domination of positions of executive office; the structures put in place by the IBG to validate itself as a credible organisation working for geographers' 'common interests' (study groups, the Annual Conference, field meetings, international seminars, guests from other countries and disciplines); the fact that the IBG was strongly London-biased in membership until well after 1945; and, indeed, the geography of its annual meetings, which were based (until 1952) on the 'golden triangle' of Cambridge, London and Oxford.

Writing in 1983, Stoddart noted that 'The Institute of British Geographers on its fiftieth anniversary is thus an altogether more substantial and complex organisation than ever envisaged when it was created. It is accepted as the representative body of British academic geography . . . It has a formidable record of publication . . . and an equally impressive record of activity in the Study Groups' (Stoddart 1983, 10). For one founder member, 'The trouble now is not too few avenues of publication, but too many . . . [but] on the whole, I don't think I would wish it otherwise' (Beaver 1983, 37).

Publications of the Institute initially centred on monographs. The move to establish *Transactions* as a research journal with a 'mixed' collection of human and physical papers only occurred after 1945. There was no formal office of Editor until 1949. *Transactions* appeared twice yearly from 1963, three times a year from 1968 and quarterly from 1976. It has always been a mixed journal, despite moments when its contents would suggest the contrary, and, to some degree in recent years, a fall-off in numbers of papers in physical geography. James Bird's 1983 subjective summary of fifty years of *Transactions* discerned no over-arching 'essential' message but rather, in eleven themes, a diversity to its contents that some might see as formless eclecticism, others as unity in diversity (Bird 1983). In the early 1980s, there was talk of scrapping *Transactions* precisely because it was felt by some that its general nature made it of little value to geographers whose work was reflected in increasingly specialised ways in increasingly specialised journals. The fact that *Transactions* is now regarded as one of the world's leading geographical journals says much about the strength of modern geographical writing.

Yet the rather uncertain beginnings to academic publication in the activities of the IBG and the gradual moves to scholarly diversity within *Transactions* and *Area* means that we can see geography's publishing history as, in outline terms, a measure of the increasing disciplinary credibility of the subject itself. This is true even of student geographical publications where, in one case anyway, undergraduates' articles and commentaries fairly closely followed shifting trends in the discipline in the period 1955-1996 (Philo 1998).

There is a certain irony in the fact that the recent fortunes of the IBG as 'the representative body of British academic geography' have become closely bound with those of the RGS, the very body at one time held not to represent geography's new research professionals. On 1 January 1995, and following proposals outlined in a report of a joint working group that effectively centred upon the greater political authority of a united geographical society, the RGS and the IBG merged to become the 'Royal Geographical Society with the Institute of British Geographers'.

In many respects, it was, and is, a combination of opposites. The RGS has done much for the public status of geography but its members and associates of the RGS (11,625 persons at merger) were, largely, lay geographers, and, for many of the younger professional academics, it continues to be seen 'as a somewhat decrepit descendant of its imperial Victorian ancestor' (Driver 1995, 410). The language used to represent the merger to the two constituencies is revealing. For one past President of the RGS, it was a 'marriage'. The outgoing President of the IBG (with unnecessary unction) described the events as the end of a family squabble: 'It isn't true to say that parents who have had to cope with unruly children welcome them back with quite such courtesy, forethought and negotiation as has happened on this occasion' (Diamond 1994, 363). The fact that many younger academics resigned their membership of the IBG rather than be associated with the 'new' merged family of British geographers is a reflection of the fact that the structures of disciplinary credibility always have fault lines. And, in an age of rising divorce rates, single parents and open marriages, we might anyway want to be cautious abut the rhetorical hazards of using familial and marital metaphors to document disciplinary history.

2 Examples of disciplinary credibility: historical geography, and the quantitative revolution in British geography

Several 'survey texts' cover the most recent (post-1945) general history of British academic geography. Others offer a review of the state of geography at given moments (see, for example, Brown 1980; Freeman 1971; Johnston 1979; Peet 1998; Taylor 1951; Unwin 1992). In these terms, it is possible to assess, following the methodology of the historical geographer, modern British geography's 'process of becoming' over time, and something of the condition of the discipline at given moments.

It would be instructive to consider, for example, the differences between Alan Ogilvie's *Great Britain: essays in regional geography* (1928) with Jean Mitchell's edited *Great Britain: geographical essays*, published in 1967. Ogilvie's edited collection was considered by one leading British (historical) geographer 'as an achievement of the greatest interest in the history of British geography' since it was 'a self-conscious production that showed the subject "as it was" in Britain in the late twenties . . . it had something of the character of a manifesto, a declaration of intent, a promise of the future' (Darby

1983a, 16). In both, British geography's forte was descriptive regionalism: authors began with an account of the physiography of the given area, proceeded through its agricultural and industrial geographies and ended with a discussion of social and demographic issues.

Such a comparison might be meaningless, however, without our recognising that by the date of Mitchell's book and for some time before, regionalism as the distinctive product of geographical writing and, even, of geographical methodology (so far as this was explicit) was in crisis. Thus, George Kimble's famous 1951 essay on the inadequacy of regionalism and his advocacy of systematic applied studies undermines the seeming strength of regionalism apparent in 1967 (Kimble 1951). By this date, anyway, the advocates of a more theoretically-informed quantitative and systematic geography had made their presence felt in important works in human and in physical geography such as Peter Haggett's *Locational Analysis in Geography* (1965) and Haggett and Chorley's co-edited *Models in Geography* (1967). To consider Mitchell's volume British geography 'as it was' in 1967 in the same way as Ogilvie's was in 1928 would, then, be quite misleading.

Seeing geography's recent history as paradigmatic in the strictly Kuhnian sense is not necessarily any more revealing. Stoddart's 1981 interpretation of Haggett's 1965 book as one of four more widely-applicable paradigm 'shifts' in geography places too little attention upon the wider context and the 'processes of becoming' behind that book (Stoddart 1981b). What is certainly true, however, is that by the later 1960s, geography's established academic respectability and, for some, rejection of earlier expressions of disciplinary credibility were apparent in 'an ever-rising and widening flood of geographical publication'. These words were used in the preface to *Progress in Geography*, begun in 1969 by four leading British geographers: Christopher Board (LSE), Richard Chorley (Cambridge), Peter Haggett (Bristol) and David Stoddart (Cambridge). As they further noted:

> The 'explosion of the geographical data matrix', the increasing popularity and relevance of geographical studies, their widening scope and their deepening analytical sophistication have been the causes of this upsurge. It has taken place, furthermore, not only in recognised geographical organs, but in a host of allied journals . . . Indeed it is apparent that much of the most advanced and exciting geographical research is being

> reported in first instance outside the established geographical periodicals (Board et al. 1969, v).

Such has been the continuing range of geographical work that this 'flood' is itself now monitored in two separate publications, *Progress in Human Geography* and *Progress in Physical Geography*. Assessment of these two journals would provide perhaps the best overview of the content of modern British geography in international context.

These remarks carry both methodological and substantive weight in understanding the history of disciplinary trends in modern British geography. Arguably, the history of British academic geography since 1945, certainly since 1969, can never be written as a *general* disciplinary history since that history is one of separate kinds of endeavour, separate kinds of disciplinary credibility and of different connections with practitioners other than geographers. In studying disciplinary trends over time, as in what follows with my few remarks upon historical geography, we are at risk of ignoring contrary trends and of asserting false beginnings. In considering the character of geographical enquiry at a given 'moment' in order to understand the involvement of different people in different places, as in the discussion of the quantitative 'revolution' in Britain, we are in danger of neglecting causal processes. Both methods of examining sub-disciplinary histories need to be seen as part of wider contexts--institutional, personal, discursive, political—themselves never fully recoverable.

My consideration here of historical geography is for several reasons. It reflects my own disciplinary loyalties and interests, the fact that the modern history of academic historical geography in Britain mirrors the history of the discipline as a whole and the fact, too, that for one notable practitioner of the subject, historical geography was 'one of the least touched by the upheaval' (that is, my other illustration—the quantitative revolution) (Darby 1983b, 426)

In the late nineteenth century, attention to historical geography understood as a matter of changing frontiers and the geographical influences upon history were central elements of that historical geography taught in Oxford and in Cambridge by historians and published in works on the historical geography of the British empire, of Europe, and, in Adam Smith's work, on the Holy Land (Butlin 1993, 1995). Historical geography was often taught, where it was taught by geographers at all, in association with the histories of exploration and of geographical discovery (as we have seen for Chisholm's curriculum in Edinburgh). By the later 1920s, however, distinctions were being made between history from the point of view of the geographer, and, much less commonly, the (human) geographies of the past. The latter sense received greater credibility following the First International Congress of Historical Geography, held in Brussels in 1930, albeit in a meeting dominated by non-British historians and not by geographers.

In Britain, the move to formalise modern historical geography may be dated to January 1932 and a joint meeting of the Geographical Association and the Historical Association held to discuss the question 'What is historical geography?' Clifford Darby looked back on that meeting after more than fifty years:

> Thus it was that with the development of academic geography in Britain in the 1920s and 1930s came the rise of historical geography as a *self-conscious* discipline . . . There was a high degree of unanimity among us. We had something of the dogmatic fervour of new converts to a faith, heightened by the fact that the position of geography was not all that well established in our universities (Darby 1983b, 423).

Evangelical fervour soon had its methodological hymnary. In Britain, Darby (and in the United States, Jan Broek) moved to promote and to combine 'cross-sectional' accounts of a certain date or the mapping of a certain source with the 'process studies' of a particular element studied over time. Even so, disciplinary evangelicism has as one of its pitfalls the fact that potential converts might not realise the means to (or the need for) their salvation. When the historian Marc Bloch reviewed Darby's path-breaking *An Historical Geography of England before A.D. 1800* (1936), Bloch began by remarking 'Our vocabulary is so imperfect that to entitle a book "An historical geography" is to risk not giving in advance a very precise idea of its content' (Quoted in Darby 1983b, 423).

British historical geography was, between the 1930s and the later 1960s, the most resolutely scholarly area of human geography, partly because of the training of its practitioners, and, to a lesser extent, because of its attention given to particular historical sources—such as the Lay Subsidies, the Census, the Tithe returns, and, notably, Darby's work on the Domesday Book. Darby's knighthood in 1988 'for services to the study of historical geography', reflects his contribution, of course, but his being the first geographer to be so honoured on the grounds of scholarship says something, too, for the subject he avowed (Williams 1994).

Sub-disciplinary credibility in organisational senses was apparent in the founding of the *Journal of Historical Geography* in 1975 by John Patten and Andrew Clark, and from 1979, in the international CUKANZUS meetings of historical geographers (from *C*anadian, *U*nited *K*ingdom, *A*ustralian, *N*ew *Z*ealand and *US* historical geographers: this is now the more simply-named and less consciously Anglo-American 'International Congress of Historical Geographers'). Such expressions of credibility were, of course, common within the discipline as a whole and, in that sense, British historical geography was not distinct. Where historical geography was less like some of its sub-disciplinary counterparts, however, was in its conscious engagement with matters of theory and with the ideology of the past and in its much less evident engagement with the quantitative revolution.

Reflecting upon the 'momentous' nature of that 'revolution', Darby noted in 1983 (with a characteristic turn of phrase, and, from my memory of hearing it delivered, a noticeable twinkle in the eye to accompany the precise spoken delivery) that 'There was indeed an earthquake, but I doubt whether for every geographer it registered (or needed to register) 8 or 9 on the Richter scale, more likely 5 or 4, or, maybe for many historical geographers only 3 or 2' (Darby 1983b, 425). This is not to claim that historical geography was then (or is now) somehow 'above' quantification. Neither would be true. It is to make the point, however, that what is often portrayed as a shared disciplinary 'moment', a 'revolution' of epistemic proportions, should itself be seen as a matter of difference, with its own missionaries, non-believers, converts and recusants and, moreover, with its own local historical geographies in which certain institutions and individuals embraced the 'new' faith, and others did so less evidently, if at all.

This point about the nature of the quantitative revolution in British geography has been made, if not in quite these terms, by one leading 'quantificatory evangelical'. For Stanley Gregory, early British quantifiers were of three sorts: the 'had-beens', younger scholars trained in the USA in quantitative methodology; the 'elsewhere', established geographers whose experiences outside the UK had converted them; and the 'home-based'. Gregory's distinctions were to make two points: 'first, that the influences underlying the initial phases of the quantitative revolution in British geography were not homogeneous . . . so that the different lines of development since 1960 must be seen, in part, as resulting from varied academic backgrounds; second that in the early 1960s there were still many battles to be fought in a British context' (Gregory 1976, 386; see also Gregory 1983).

Quantification's institutional credibility in Britain came in 1969 with the formation of the Quantitative Methods Study Group of the

IBG: a group to that purpose, in existence since 1964, had initially shunned the IBG, fearing, as Gregory put it, 'that incorporation would be the kiss of death'.

Rather more than noting what actually was being done by the quantifiers, in Britain or elsewhere, it is here interesting to note the language used to describe the 'quantitative revolution', and to know something of how those involved then felt. Gregory's language in considering the issue used, in part, curiously mixed metaphors of sexual transgression and Whiggish progressivism:

> So many classical myths include an element of rape that it is hardly surprising that the quantitative revolution has been seen as the rape of geography, the latter having been wrenched violently from the delights of rustic purity and simplicity by the brutal and licentious quantifiers. But has the translation from the underdeveloped to the developing academic world proved to be a fate worse than death? (Gregory 1976, 389)

For Peter Taylor, the quantification debate in Britain was one of social conflict between and within communities who, simply, had different views as to what was, and was not, academically credible geography. What, for some, was a mathematical language employed to verify statistically-measurable processes was, for others, a form of epistemological camouflage. For its exponents anyway, quantification was seen as the route to geography's credibility as a science: 'In geography this new methodology was nothing less than *the* scientific method . . . emerging professional geographers could be, and undoubtedly were, attracted by a new geography which claimed the whole weight of the prestige of science behind it' (Taylor 1976, 136).

'Traditionalists', as Taylor described them, reacted to the young quantifiers on two counts: they objected to their behaviour and to their arrogance, and to their ignorance of past geography. Such reactions have more to do with attitude and age than with claims about the 'essential' nature of geography, however, and with how, exactly, different geographers imagine their subject's past as a template for its future. For some contemporary leading geographers, then caught up in the 'quantitative revolution' as graduate students or newly-appointed staff, recollections of geography's adoption of statistical methods vary from being, for one, a 'pleasant pain', to another's more widespread engagement with theoretical deliberation that served only to disenfranchise him from his subject, to an almost spiritual devotion from a third (Billinge, Gregory and Martin 1984).

What Ian Burton described as the 'radical transformation of spirit and purpose' that, in Britain, geography underwent between the early 1960s and the mid-1970s, should not be seen as a 'disciplinary revolution' (Burton 1963). In origin, it was a series of local affairs. Only later and more generally was it a social *and* an intellectual movement whose lasting contribution to geography rests, arguably, more in the seriousness with which theory is taken than in the mathematical languages some geographers use to establish the credibility of what they do.

Quantification's impact lessened from the later 1970s as human geographers especially became more concerned with a socially relevant geography, and, in more recent periods still, with matters of representation and of meaning whose interpretation demands recourse to qualitative and ethnographic methods. It is irony indeed, then, that the essential features of modern geography's route to academic credibility—associational activity, research-led publication, teaching excellence, sub-disciplinary specialisation, the adoption of theory and of more precise methodologies and close connections with other disciplines all differently expressed in local institutional settings—are themselves now subject to qualitative judgement.

Ensuring disciplinary quality in contemporary British geography

The theme of quality has dominated British higher education in the 1990s. This is despite the fact that there is no clear consensus as to what 'quality' means. In general terms, five common uses of the term have been identified in respect of teaching and learning: fitness for purpose; the achievement of standards; value for money; value added (the extent to which student performance at completion exceeds the capacities for achievement signalled by qualifications at entry); and customer satisfaction (Chalkley 1994). In addition to judgement of teaching quality through the TQA process (Teaching Quality Assessment), British higher education has periodically assessed research quality through the RAE (Research Assessment Exercise). Whilst, to date, there has been only one TQA in the UK, there have been four RAE: 1986, 1989, 1992, 1996 (that for 2001 is in preparation as I write). In addition, the UK Government has overseen reviews of the entire higher education sector. The most recent was undertaken by Sir Ron Dearing in 1997.

This section briefly considers the quality debate and geography in the UK in the 1990s with reference to teaching, research and to what the Dearing Report (1997) had to say concerning geography. What follows is less concerned, then, with the intellectual nature and content of geography in Britain at the end of the twentieth century, and more with how the subject has fared, in quasi-political terms, in being, like every other discipline, subject to review.

The 1992 RAE, like the preceding two, evaluated disciplinary research on a scale from 1 (the poorest) to 5 (the best). In 1996, the category '3' rating was subdivided, as was the category 5, giving a seven-point scale from 1, 'Research quality that equates to attainable levels of national excellence in none, or virtually none, of the sub-areas of activity', to 5*, where the research quality 'equates to attainable levels of international excellence in a majority of sub-areas of activity and attainable levels of national excellence in all others' (HEFCE 1996, 13). In addition to criteria such as numbers of graduate students, numbers of research-active staff, and research grant income in each department, quality judgements were made by a panel of professional geographers drawn from across the UK in relation to four identified publications nominated by each research-active professional geographer in each submitting department.

Each RAE has generated debate on its outcomes. Whilst there is some evidence to suggest that the exercise is loaded against smaller departments where, it is argued, 'critical mass' of the research community is much less evident, other evidence would suggest that institutional size and the nature of the research ethos and departmental culture play an important rôle in geography's research grading. Most of the 'old established' geography departments in the UK were rated 4 or above in 1996. Yet that pattern of research gradings in geography takes little account of the fact that some former polytechnics and colleges, with institutional cultures geared towards teaching and only formally accredited as Universities under the terms of the 1992 Further and Higher Education Act, only entered the RAE in 1992, and did so from a much lower research 'base'.

In the sense that geography's contemporary research profile reflects differing senses as to the place of geography in an institution's portfolio of academic work, explaining the 'map of geographical activity' in the 1990s is not dissimilar to explaining it for the 1890s: it matters how seriously one's institution takes geography as a subject. Institutional differences make a difference. Geography does better where there are specialist research groups, where the institutional infrastructure permits more attention to research and where teaching programmes are structured so as to free time for research.

In the 1996 RAE, the 5*-rated departments of geography were (with numbers of research-active staff in parentheses), Bristol (21), Cambridge (29), Durham (33), University College, London (31), and Edinburgh (18). The 5-rated departments were Leeds (32), Newcastle upon Tyne (26), Royal Holloway, University of London (20), Sheffield (25), and Southampton (21). Although, in terms of the formal criteria of assessment, these departments represent, in the UK, the sites of internationally- and nationally-important research in geography, it would be quite wrong to see these departments and the cumulative total of research-active geography staff there represented (256 persons, or 24% of the total population of research-active professional geographers in 1996), as the only places where good quality research in geography is undertaken. Many other departments have scholars of international repute, and/or groups with particular research strengths. Taking a long-term view of geography's professionalisation, what is, perhaps, the most remarkable statistic is the difference between the handful of academic geographers in Britain by 1896, and the 1,050 recorded as research active a century later.

There is a direct correlation between the RAE grade and the achievement of excellence in the TQA (in which context, quality is measured in terms of 'excellent', 'satisfactory' and 'unsatisfactory' with, in Scotland only, the additional category, 'highly satisfactory'). In geography, all the departments in England and Northern Ireland with an RAE grade 5 obtained an excellent grading, whereas two-thirds of those with a grade 4, a half of those with a grade 3, one-third of those with a grade 2 and one-eighth of those with a grade 1 achieved excellence.

It would seem, indeed, for the UK as a whole in respect of geography that research activities have a beneficial impact on the quality of student learning. This is true where, for example, a larger number of staff can reduce the staff:student ratio and where the effect of research funds generated in relation to RAE grading have an impact on resources available for teaching. Unlike the TQA, there are significant financial rewards for achieving a high grade in the RAE. It is no surprise, then, that academics and institutional authorities have placed more emphasis on that quality exercise than on quality in teaching, and that there has been something of a 'compliance culture' emerging, with institutions, departments and individual geographers structuring their activities in order to maximize gains (Johnston 1994a, 1994b).

As authorities on quality in British geography have noted, 'the quality era' has not been without criticism (Chalkley 1994, 1996;

Healey 1997). Those agitating for the benefits of quality assessment in teaching stress that quality assessment is about ensuring that all is satisfactory, that action is taken where it is not and that excellence and good practice are rewarded. Others, mainly the academic geographers whose teaching is so assessed, have taken a less charitable view of quality matters, seeing quality agencies as threats to academic autonomy and as unwelcome surveillance, and, through the burdensome preparation of quality documents, threatening the very thing that quality agencies seek to measure.

In a conference organised in 1993 by the Higher Education Study Group of the IBG, Ronald Johnston, a leading British geographer who was also at that time a University Vice-Chancellor, noted that, despite a rhetoric of quality as 'fitness for purpose', quality was actually being assessed on implicit 'quality as standard' criteria that effectively paid no attention to institutional difference. Quality is a matter of politics. As Johnston further noted:

> What is it all for? Why are we doing it? The simple answer is that it has been imposed upon British HE [Higher Education] by an ideologically motivated government, which believes that it is not getting value for money from the country's universities, where many of the staff are not attuned to the enterprise culture and the disciplines of the market place. Of course, the government has no clear view of what quality is (Johnston 1994b, 191).

Undaunted by its own ignorance and committed to a culture of evaluating what is done rather than providing resources to do it better, the UK Government, in July 1997, published the Dearing Report, the first major government-initiated review of higher education since the early 1960s. Amongst other recommendations, the Dearing Report proposed a professional training course for all new teaching staff accredited by a new Institute for Learning and Teaching in Higher Education, increased attention to work experience and to employers' needs, abolition of the current TQA procedures, steps to promote regional collaboration among institutions, the establishment of an Arts and Humanities Research Board and the revision of the RAE to discourage participation by departments unlikely to obtain at least a 3A grading.

The place of geography and Dearing has been considered at length elsewhere (Chalkley 1998a, 1998b). The RGS-IBG official position was broadly in favour of many of the Dearing Report's recommendations: widening student access, improving IT skills, accrediting teaching staff in geography in institutions of higher education, approving the abolition of the TQA and its replacement by panels of external examiners, promoting inter-disciplinary funding, furthering research across all departments. In truth, it would be hard not to be in support of these sorts of things: the trick will lie in making the sensible things work given declining units of resource.

The RGS-IBG response to the Dearing Report was prefaced with a clear view as to what it took geography to be.

> Geography is a holistic and integrative university discipline that bridges the natural and social sciences. It is rare for a subject to make integration a priority and to do so successfully. Geography is strongly focused upon the relationships between people and the environment. Geography is an intellectually challenging subject. It requires an understanding and application of scientific logic, principles and laws, a flexibility and openness of mind to deal with a range of different conceptual paradigms in both physical and human geography, and analytical capabilities to deal with empirical primary and secondary data sets (Gardner 1998, 98).

One cannot help but feel that Mackinder, Keltie, Clements Markham and others a century ago would, in general terms, have agreed however much the evaluative culture that besets British higher education would also dismay them.

Conclusion

In review, how should we understand the professionalization and institutionalization of academic geography in Britain in the last century?

In general terms, the evidence for the self-conscious fashioning by professional practitioners of geography as an intellectual discipline is clearer in relation to geography's foundation than it is today. This is to be expected. The concerns of contemporary geographers in Britain are more to do with the credibility of an accepted discipline, with pushing back disciplinary research frontiers and with ensuring its quality, however understood, than with establishing conditions for the discipline's existence. One hundred years ago, concerns over the lack of geographical knowledge amongst the citizenry, poor teaching in schools and the comparative advantages enjoyed by places of geographical learning overseas were used to justify geography's place as an academic subject.

More specifically, the establishment of departments of geography owed greatly to the activities of 'geographical activists', to the support of civic bodies for whom geography was seen as 'useful knowledge', to individual institutional circumstances, and, in particular in those circumstances, to the nature of the relationship with persons of influence in subjects other than geography. There was only a handful of departments and of professional geographers before 1914, and not many more before 1939. Increased numbers—of institutions offering geography, of students and of staff—were marked after 1945. At first, the principal concern of most of geography's few professionals was teaching. The curriculum was initially delivered by one staff member with subject specialisation only coming later in reflection of increased staff numbers and of particular research interests.

This partial biography has not in detail documented the changing subject content of geography in Britain with, over time, its varying emphases upon (to name only a few) regionalism, systems theory, quantification, geographical information systems, humanism, feminism and methodological innovation. Whether or not one sees the changing fashions in disciplinary attention as revolutions, incursions on the frontier by methodological banditti, matters of epistemological moment or diversions from an 'essential' geography, will, I suspect, be a matter of individual concern and, perhaps, even of indifference to some. But three more general conclusions can be advanced.

Firstly, the professionalization and institutionalization of modern British geography is apparent in the means by which its practitioners have sought over time to make the subject academically credible to others and to themselves. Credibility in the first instance depended upon making a case to others for geography's legitimacy. Once established, geography's credibility has been and is apparent in institutionalised activities—in publications, conferences, and research output—directly related to sub-disciplinary specialisation. For an overview of much of this activity, see Thrift and Walling (2000). Most recently, geography's credibility has been evident in peer review of its quality.

Secondly, understanding why geography was professionalized and institutionalized in the ways it has been must acknowledge individual institutional circumstance. The particular nature of the curriculum, the interests of the staff, the adoption (or not) of new methods have all taken a different shape because of the institutions, and the people there, in which geography has been established. In this respect, understanding geography's disciplinary history must always be a matter of explaining geographical and institutional difference.

Finally, understanding the history of modern British geography will always depend upon how that history is written. My attention to questions of emergence, credibility and quality is not intended to suggest an unproblematic process of disciplinary maturation, a predetermined path to perfectibility. After all, it depends which way you look at the events themselves and at their causation. From a contemporary standpoint, British geography's modern institutionalization and professionalization from the 1880s appears recent indeed. But if those few early modern British geographers active in the 1580s had been able to see the future, they might well have wondered why geography's establishment in these ways took so long.

Acknowledgements

I am grateful to Gary Dunbar for the invitation to contribute to this volume and for facilitating access to his manuscript collections on the RGS in particular, and to David Livingstone for his comments and criticisms. My thanks go to those scholars whose work I have drawn upon here and, perhaps, too, I should offer my apologies to those whose work I have not cited, either by design or through my ignorance of it. Finally, I should like to dedicate this essay to the memory of Peter Gould.

References:

Alexander, R. and D. Gillmor, eds. (1989) *Geography in Education in the Republic of Ireland*, Smith, Dublin.

Baker, J. N. L. (1963) *The History of Geography*, Blackwell, Oxford.

Beaver, S. H. (1983) Recollections of a founder member, *Transactions of the Institute of British Geographers* 8, 36-37.

Billinge, M. D., D. J. Gregory, and R. Martin, eds. (1984) *Recollections of a Revolution: geography as spatial science*, Macmillan, London.

Blouet, B. W. (1987) *Halford Mackinder: a biography*, Texas A&M University Press, College Station.

Board, C., R. J. Chorley, P. Haggett, and D. R. Stoddart, eds. (1969) *Progress in Geography: international reviews of current research* 1, Arnold, London.

Bridges, R. (1985) The foundation and early years of the Aberdeen centre of the Royal Scottish Geographical Society, *Scottish Geographical Magazine*

101, 77-84.
Brown, E. H., ed. (1980) *Geography Yesterday and Tomorrow*, Oxford University Press, Oxford.
Brown, T. N. L. (1971) *The History of the Manchester Geographical Society 1884-1950*, Manchester University Press, Manchester.
Buchanan, R. O. (1954) The I. B. G.: retrospect and prospect, *Transactions of the Institute of British Geographers* 20, 1-14.
Burton, I. (1963) The quantitative revolution and theoretical geography, *Canadian Geographer* 7, 151-162.
Butlin, R. A. (1993) *Historical Geography through the Gates of Space and Time*, Arnold, London.
Butlin, R. A. (1995) Historical geographies of the British empire, *c*.1887-1925, in M. Bell, M. Heffernan and R. A. Butlin, eds., *Geography and Imperialism 1820-1940*, Manchester University Press, Manchester and New York, 151-188.
Chalkley, B. (1994) The quality era, *Journal of Geography in Higher Education* 18, 163-167.
Chalkley, B. (1996) Editorial 1. Geography and teaching quality assessment: how well did we do?, *Journal of Geography in Higher Education* 20, 149-158.
Chalkley, B. (1998a) Editorial. Arena Symposium. Dearing and Geography, *Journal of Geography in Higher Education* 22, 55-101.
Chalkley, B. (1998b) Setting a post-Dearing agenda for geography, *Journal of Geography in Higher Education* 22, 92-98.
Chisholm, G. G. (1908) The meaning and scope of geography, *Scottish Geographical Magazine* 24, 561-575.
Cormack, L. (1991) Twisting the lion's tail: practice and theory at the court of Henry Prince of Wales, in B. Moran, ed., *Patronage and Institutions: Science, Technology, and Medicine at the European Court*, Boydell Press, Woodbridge, pp. 67-84.
Cormack, L. (1997) *Charting an Empire: Geography at the English Universities, 1580-1620*, University of Chicago Press, Chicago.
Darby, H. C. (1983a) Academic geography in Britain, 1918-1946, *Transactions of the Institute of British Geographers* 8, 14-26.
Darby, H. C. (1983b) Historical geography in Britain, 1920-1980: continuity and change, *Transactions of the Institute of British Geographers* 8, 421-428.
Diamond, D. (1994) [Discussion of the RGS-IBG proposed merger], *Geographical Journal* 160, 365.
Driver, F. (1992a) Geography's empire: histories of geographical knowledge, *Environment and Planning D: Society and Space* 10, 23-40.
Driver, F. (1992b) Henry Morton Stanley and his critics: geography, exploration and empire, *Past and Present* 133, 134-166.
Driver, F. (2000) *Geography Militant: Cultures of Exploration in an Age of Empire*, Blackwell, Oxford.
Driver, F. and G. Rose, eds. (1992) *Nature and Science: Essays in the History of Geographical Knowledge*, Historical Geography Research Group Publications Series, vol. 28, Cheltenham.
Driver, F., D. Matless, G. Rose, C. Barnett, and D. N. Livingstone (1995), Geographical traditions: rethinking the history of geography, *Transactions of the Institute of British Geographers* 20, 403-422.
Fahy, G. (1981) Geography and geographic education in Ireland from early Christian times to 1960, *Geographical Viewpoint* 10, 5-30.
Freeman, T. W. (1971) *A Hundred Years of Geography*, Duckworth, London.
Garnett, A. (1983) I. B. G.: the formative years - some reflections,

Transactions of the Institute of British Geographers 8, 27-35.

Gardner, R. (1998) A response to the Dearing Report from the Royal Geographical Society with the Institute of British Geographers, *Journal of Geography in Higher Education* 22, 98-101.

Gregory, S. (1976) On geographical myths and statistical fables, *Transactions of the Institute of British Geographers* 1, 385-400.

Gregory, S. (1983) Quantitative geography: the British experience and the role of the Institute, *Transactions of the Institute of British Geographers* 8, 80-89.

Guillemard, F. H. H. (1927-1932) The Years that the Locusts Have Eaten, 7 vols., unpublished typescript in the Rare Books Room of the Cambridge University Library.

Healey, M. (1997) Geography and education: perspectives on quality in UK higher education, *Progress in Human Geography* 21, 97-108.

HEFCE (Higher Education Funding Council for England) (1996) *1996 Research Assessment Exercise - the Outcome*, HEFCE, London.

Heffernan, M. (1996) Geography, cartography and military intelligence: the Royal Geographical Society and the First World War, *Transactions of the Institute of British Geographers* 21, 504-533.

Johnston, R. J. (1979) *Geography and Geographers: Anglo-American Human Geography since 1945*, Arnold, London.

Johnston, R.J. (1994) Quality assessment of teaching: inputs, processes and outputs, *Journal of Geography in Higher Education* 22, 184-193.

Johnston, R. J. and E. V. Brack (1983) Appointment and promotion in the academic labour market: a preliminary survey of British University Departments of Geography, 1933-1982, *Transactions of the Institute of British Geographers* 8, 100-111.

Keltie, J. S. (1886) *Report of the Proceedings of the Royal Geographical Society in Reference to the Improvement of Geographical Education*, John Murray, London.

Kimble, G. H. T. (1951) The inadequacy of the regional concept, in L. D. Stamp and S. W. Wooldridge, eds., *London Essays in Geography*, Longmans, Green and Company, London, 151-174.

Livingstone, D. N. (1992) *The Geographical Tradition: Episodes in the History of a Contested Enterprise*, Blackwell, Oxford.

Lochhead, E. N. (1980) The Emergence of Academic Geography in Britain in Its Historical Context, Unpublished Ph.D. thesis, University of California, Berkeley.

Lochhead, E. N. (1981) Scotland as the cradle of modern academic geography in Britain, *Scottish Geographical Magazine* 97, 98-109.

Lochhead, E. N. (1984) The Royal Scottish Geographical Society: the setting and sources of its success, *Scottish Geographical Magazine* 100, 69-80.

MacAlister, D. (1921) Geography at the University of Glasgow, *Scottish Geographical Magazine* 37, 53-57.

Mackinder, H. J. (1887) On the scope and methods of geography, *Proceedings of the Royal Geographical Society* 9, 141-160.

MacKenzie, J. M. (1995) The provincial geographical societies in Britain, 1884-1914, in M. Bell, R. A. Butlin, and M. Heffernan, eds., *Geography and Imperialism, 1820-1940*, Manchester University Press, Manchester and New York, 93-124.

Matless, D. (1992) Regional surveys and local knowledges: the geographical imagination in Britain, 1918-1939, *Transactions of the Institute of British Geographers* 17, 464-480.

Mayhew, R. J. (1998a) Geography in eighteenth-century British education,

Paedagogica Historica 34, 731-769.
Mayhew, R.J. (1998b) The character of English geography *c.* 1660-1800: a textual approach, *Journal of Historical Geography* 24, 385-412.
Meller, H. (1993) *Patrick Geddes: Social Evolutionist and City Planner*, Routledge, London.
Memorial of the Council of the Royal Scottish Geographical Society to the Commissioners Appointed by Parliament in the Universities (Scotland) Act, 1889, Royal Scottish Geographical Society, Edinburgh.
Parker, W. H. (1992) *Mackinder: Geography as an Aid to Statecraft*, Clarendon Press, Oxford.
Peet, R. (1998) *Modern Geographical Thought*, Blackwell, Oxford and New York.
Philo, C. (1998) Reading *Drumlin*: academic geography and a student geographical magazine, *Progress in Human Geography* 22, 344-367.
Royal Geographical Society (RGS) Archives: A. J. Herbertson to J. S. Keltie, 1 August 1913; A. W. Ward to J. S. Keltie, 30 June 1896; D. Freshfield to A. Hinks, 10 April 1919; D. MacAlister to anonymous correspondent, 11 March 1897; 'Dr. Fleure' (one-page ms. by 'J. S. K.' [John Scott Keltie], n. d. [1916]); H. J. Mackinder to D. Freshfield, 7 April 1919; Supplementary Papers.
Royal Scottish Geographical Society (RSGS) Archives: J. S. Keltie to Secretary of the RSGS, 16 April 1902.
Scargill, D. I. (1976) The RGS and the foundations of geography at Oxford, *Geographical Journal* 142, 438-461.
Steel, R. W. (1984) *The Institute of British Geographers: The First Fifty Years*, Institute of British Geographers, London.
Stoddart, D. R. (1967) The growth and structure of geography, *Transactions of the Institute of British Geography* 41, 1-19.
Stoddart, D. R. (1975) The RGS and the foundations of geography at Cambridge, *Geographical Journal* 141, 216-239.
Stoddart, D. R. (1981a) Geography, education and research, *Geographical Journal* 147, 287-297.
Stoddart, D. R. (1981b) The paradigm concept and the history of geography, in D. R. Stoddart, ed., *Geography, Ideology and Social Concern*, Blackwell, Oxford, 70-80.
Stoddart, D. R. (1983) Progress in geography: the record of the I. B. G., *Transactions of the Institute of British Geographers* 8, 1-13.
Stoddart, D. R. (1986) *On Geography*, Blackwell, Oxford.
Taylor, G., ed. (1951) *Geography in the Twentieth Century*, Methuen, London.
Taylor, P. (1976) An interpretation of the quantification debate in British geography, *Transactions of the Institute of British Geographers* 1, 129-142.
Thrift, N. and D. Walling (2000), Geography in the United Kingdom 1996-2000, *Geographical Journal* 166, 96-124.
University of Edinburgh, Department of Geography Archives, MS DG1/31/21.
Unwin, T. (1992) *The Place of Geography*, Longman, London.
Williams, M. (1994) Henry Clifford Darby (1909-1992), *Proceedings of the British Academy* 87, 289-306.
Withers, C. W. J. (1996) Notes towards an historical geography of geography in early modern Scotland, *Scotlands* 3, 111-124.
Withers, C. W. J. (1999) Towards a history of geography in the public sphere, *History of Science* 37, 45-78.
Withers, C. W. J. (2000) Towards a historical geography of the Enlightenment in Scotland, in P. Wood, ed., *The Scottish Enlightenment: Essays in Re-Interpretation*, University of Rochester Press, Rochester, NY, 63-97.

Charles W.J. Withers
Department of Geography
University of Edinburgh
Drummond Street
Edinburgh EH8 9XP
Scotland

5 Modern Geography in Italy: From the Archives to Environmental Management

Ilaria Luzzana Caraci

Introduction

As in other European countries, the history of geography in Italy can be traced from two different points of view: either focusing on the historical roots of the geographical knowledge of our time, or tracing the history of geographical learning from its most distant origins of knowledge organised in a rational system. In the first case, we favour the didactic function of the history of geography, as we use it to better define the epistemological and methodological values of current geographical research. In the second case, we choose the historical function, in the true sense of the word, bringing to light the many deep transformations of contents and methods of that complex knowledge that was collected over the course of time and put together under the label of "geography".

If we wanted to plot a history of Italian geography from the second point of view, we would have to go back to an extremely distant past, perhaps to a point only two centuries after the Milesian philosophers made their first geographical reflections on a logical basis and not simply by analogy. It would be necessary in fact to retrace the intellectual flourishing of the cities of Magna Graecia, with Pythagoreans like Philolaos of Croton (Calabria), who lived in the 5th century B.C. While observing the movements of the planets, he was apparently the first to sense that the Earth was not the centre of the universe. The Greek scholars who established a colony in Elea in Campania, south of Paestum, in that same period were also interested in the relationship between the Earth and celestial phenomena.

The Pythagorean and Eleatic interest in geography (and that of the other thinkers of that time who lived in the Greek colonies of southern Italy) is closely linked to the Greek culture of their homeland. This culture was spread around the shores of the Mediterranean by the great Greek migrations of the 6th and 5th

G.A. Dunbar (ed.), Geography: Discipline, Profession and Subject since 1870, 121–151.

centuries B.C. that led to the founding of numerous flourishing urban centres. Geography also played an essential support rôle for territorial expansion and rule by the Romans, who basically used the patrimony of knowledge accumulated by the Greeks.

Excluding these forerunners and also the thinkers of the early Middle Ages, exponents of a culture that even in the limits of its epistemological narrow-mindedness appears rather uniformly diffused all over Christian Europe, an autochthonous Italian geographical tradition can be discerned in the late Middle Ages and Renaissance in the maritime and mercantile centres of the Italian peninsula. On this subject we have only to mention the valuable production of Italian nautical charts in the 14th and 15th centuries, which are evidence of a legacy of the geographical knowledge that was being assembled at that time, and the reports of contemporary travelers such as Odoric of Pordenone and Marco Polo, which together with the classics of Greek and Roman geographical literature were rediscovered and diffused in Italy in their original versions in the middle of the 15th century. These supplied the materials for the drawing up of a scheme of geographical reference for the planning and realisation of the first important transoceanic exploratory expeditions.

The discovery and the beginning of the exploration of America were decisive for the history of geography, as for all the knowledge that Western Europe had inherited from classical antiquity and the Middle Ages. The existence of a "new world" that was unexpected and unknown, full of plants, animals and people totally different from those classified and catalogued by European culture in the course of the previous centuries, marked the end of medieval dogmatism and prepared the way for the Galilean revolution.

More than just an event, the discovery of America was the result of a long process of the widening of the geographical horizons of the European people. The theoretical knowledge and the practical experience of first the Italians, Spaniards, and Portuguese and then of the people of Northwestern Europe came together in this enterprise. But it was certainly not a coincidence that it was precisely some Italians, from Christopher Columbus to John Cabot, from Amerigo Vespucci to Giovanni da Verrazzano, who made decisive contributions to that course of events. These and other less well known contemporary explorers were not lone examples of exceptional genius, but they were products of a fortunate time in Italian history, in which the rediscovery of classical culture went hand in hand with its critical interpretation and with a new thirst for knowledge that urgently demanded theoretical knowledge just as much as practical experience. As is well known, the fervour for study and research that had characterised Italy in the Renaissance period had its active centres in the elegant courts and in the cities, where the mercantile, educated, and elitist middle classes had established themselves. They initially encouraged and also financially supported the process of the great geographical explorations. Paradoxically, however, these expeditions later brought about Italy's economic isolation when new and wider horizons for commercial traffic were opened up. Excluded from the transoceanic routes to which it had contributed so much to map out, and divided into too many small states in eternal conflict with each other, Italy soon began a process of political decadence together with a cultural provincialism, destined to last even after the eventual national unification. Naturally, this did not mean that there was suddenly a complete lack of intellectual strength and men of great intelligence. Many outstanding individuals could be mentioned. As regards geography in particular, we have to remember at least Galileo Galilei and his mechanistic conception of reality, which led to a new formulation of the problem of knowledge according to a vision of a harmonious world, where everything can and must be studied in a rational way, with consequences especially for mathematical geography. But, most certainly, the central rôle that Italian geographical culture had played from Humanism to Renaissance in Europe was now assumed by other countries.

The dawn of modern geography

In the 17th and 18th centuries there was a rapid evolution of scientific thought and the establishment of a great number of branches of learning. Even though they came under the general title of "Natural History", each asserted a claim to autonomy in the ambit of the sciences of the earth. They started to draw on the substratum of empirical knowledge traditionally considered the subject matter of geography, using it to create the basis for new sciences, with a rigorously experimental method and with the use of techniques that became increasingly more perfected.

The cosmographical and astronomical knowledge handed down from antiquity gave way to a documented knowledge of the morphology and dynamics of the universe. The measurement of the dimensions of the earth allowed the basis of geodesy to be laid down. The study of rocks and fossils developed into geology. The great amount of knowledge obtained through the exploration of the seas and other continents brought about new branches of learning, such as oceanography, and enormously enriched others, such as botany and zoology. The boundaries between geography with its traditional meaning of "science of the earth" and the other sciences were becoming more and more uncertain.

On the other hand, the human or social sciences, which geography had not had so much contact with in the past, but which certainly grew in strength after the era of the great geographical discoveries, showed a remarkable vitality and the need to go beyond the traditional boundaries of geography.

At the beginning of the 19th century, geography was fundamentally left with only two well defined and closely connected tasks. The first, which can be defined as technical, was to supply a description that was as exact as possible of the earth's surface, a task that at that time was of great importance, since the continuation of the explorations was gradually perfecting this knowledge. The second, which broadly speaking can be considered didactic, was to assemble, catalogue, and update all these elements, which together could help to outline the characteristics of the various regions of the earth.

At that time, geography stood out primarily for its usefulness: for governors, who were supplied with essential information for both political and military purposes; for administrators, to whom it gave a precise, detailed mirror of the natural and human conditions in the places entrusted to them; for travellers and merchants; and, more generally, all those who were interested in finding out about various parts of the earth. Its usefulness was also acknowledged from a strictly cultural point of view, as a subsidiary of history, as the Herodotean concept of geography as "a description of places where human actions take place" was still valid.

Reduced to a position within these restricted boundaries, geography had little to offer compared to the previous century. This was especially the case in Italy, because the influence of the two German "Fathers of Modern Geography", Alexander von Humboldt and Carl Ritter, which was already noticeable in other countries by the 1830s, was not to be felt in Italy until much later, after geography had taken on the rôle of an autonomous science with the institution of university chairs.

In the first decades of the 19th century in Italy, as in other European countries, "geographer" was synonymous with "cartographer", and geography was, therefore, primarily the profession of those who drew maps. Cartography was at that time the most lively expression of the vitality of geography. At the beginning of the century, geodesic cartography spread to almost all the Italian states. On the whole it was a production of an optimum level, considering the times and the means available, destined to provide a basis for work both for subsequent maps and for detailed geography.

The other dominant aspect of Italian geography at that time was the interest in data and detailed information. Many "universal treatises" of geography were produced in the first decades of the century. It was the final great attempt to assemble all geographical knowledge in exhaustive, monumental encyclopedic works. The effort of learning was directly proportional to the need to compensate for the lack of first-hand, updated knowledge with second-hand sources.

Only a few of these works left their mark. Among these is certainly the *Abrégé de géographie* by Adriano Balbi (1782-1848), a Venetian by birth who lived in Paris for many years. The *Abrégé* brought universal fame to its author and was extremely successful in Italy. It was considered a model for a long time, taking its place in the tradition of geo-political treatises, a tradition that in the course of the 19th century was to produce what was called "the geography of statesmen", or rather a descriptive geography for the use of governments and administrators, full of demographic and other kinds of data. Giovanni Botero (1534-1617) may be considered a forerunner of this tradition, and his *Relazioni universali* have recently been revalued for their modern geographical approach.

At the same time, however, two other trends characteristic of the 19th century were emerging—the trend towards greater emphasis on topical or "general" geography, above all physical geography, and a development of greater interest in the history of exploration. The latter was to become a particular specialty of Italian geographers. In other European countries, after an initial phase when the history of exploration was dabbled in by scholars in many disciplines, including geographers, it became the exclusive domain of historians. In Italy, for reasons that will shortly be explained, it was the geographers who took a special interest in this field.

As for general geography, the most outstanding exponent around the middle of the century was Francesco Cesare Marmocchi (1805-1858), a naturalist and author of various works, including the *Corso di geografia universale*, which was very well known in its time. On the whole, the *Corso* is a treatise in general geography, with few concessions made to description and with a consistent methodological basis. Although not so original as Balbi's *Abrégé*, Marmocchi's work shows more intuition. In addition, the considerable weight attributed to historical values made it closer to the needs and expectations of the Italian culture of the time.

If the above are the forms and functions that were attributed to geography during the first decades of the 19th century, as in the previous century, scholars of different vocation and culture dealt with many other topics with which modern geography acknowledges more solid links. The subjects that are studied by geographers today were to a great extent already present in the Italian culture at that time, only they were not dealt with by geographers, or, to be more precise, those who studied them did not call themselves geographers. They were often politically committed people with different aims but determined to improve the social conditions of the country and to make up for the cultural and economic delays caused by long-standing political divisions. Culture for them was still a coherent value, in which—at the most—one could distinguish between the natural and the human sciences (but which were always closely linked). Modern Italian geography willingly grants the title of forerunner to these scholars.

Melchiorre Gioia and Carlo Cattaneo were examples of this trend, even though neither ever considered himself a geographer. However, geographical concepts and methods are undoubtedly present in Gioia's *Filosofia della statistica*, where the mutual relations that exist between the phenomena of the Earth's surface and Man are evaluated. He began the study of the State, seen as a union of living people with the same social constraints, with analyses of topographic, climatic, and demographic conditions.

Closer to the modern conception of geography, and maybe for this reason more often remembered by our contemporaries, are the writings of Cattaneo. They began with his work as a journalist in 1831 and continued with ever more compelling essays in the "Annali di statistica" and "Il Politecnico". Published between 1839 and 1844, "Il Politecnico" played an important rôle in stimulating the study of social and economic problems in Italy. Among these, of particular geographical interest were town planning, agriculture, rail transport, and emigration. Cattaneo's historical background allowed him to face territorial or regional problems with a much wider perspective. His best work was perhaps *Notizie naturali e civili sulla Lombardia*, which was written in a short space of time for presentation at the Congress of Scientists in 1844.

Many other names could be mentioned to confirm how strongly the themes of economic geography and territorial organisation were felt in Italy. As the premises were those of the Enlightenment, far from presenting obstacles, geography was considered an instrument of work, a means of explanation, and an aid for modification. The idea of region was proposed, but its bases were still uncertain. As the concept of landscape was not defined, attention continued to be placed on single phenomena, separate from each other and considered easily modifiable by Man.

The institutionalisation of geography in Italy

A crucial phase of Italian geography began in the second half of the 19th century. It is important to remember that between 1850 and 1870 Italy finally became a unified State with the gradual expansion of the Kingdom of Sardinia. The Kingdom of Italy was officially established in Turin on 14 March 1861. Serious problems regarding internal politics were immediately raised. The huge economic imbalance between the North and the South, the conditions of cultural backwardness, and the necessity of rapidly creating the conditions for a real process of national unification required an enormous effort of planning and aggregation.

Italian culture, both at the initial phase of the Right-wing governments and above all after the coming to power of the Left in the 1874 elections, became aware of the responsibility of its own rôle. This responsibility was also felt by the geographers and other scholars, often coming from very different cultural experiences, who shared an interest in the different cultural ambits in which geography then found itself.

At the same time, innovative ideas from German geographers began to circulate. Geography was beginning to be thought of as an autonomous science—that is, as a subject in its own right, with its own methods, whose definition was a subject of debate. Above all, two events were significant regarding this new situation: the establishment of the Italian Geographical Society and the admission of geography into the universities.

The establishment of university chairs in geography began in 1859, with the gradual application in all areas that came under the jurisdiction of the Italian Kingdom of the Casati Laws, which established geography as a compulsory part of some university curricula. Initially, there was some difficulty in finding professors. Therefore, the chairs were assigned to statisticians, who limited themselves to a teaching based on superficial factual information, in line with the conception of geography as encyclopedic knowledge directed towards the learning of history or towards the preparation of future managers. However, as time went on, the university teaching staffs became made up of specialists, who, with the enthusiasm of beginners, did their utmost to define the work and objectives of geography in Italy.

The Società Geografica Italiana was established in 1867. It had been preceded by attempts to create geographical societies, such as the Società Toscana di Geografia e Storia Patria, founded in Florence in 1826 by Father Inghirami. Its members were some of the greatest intellectuals and political figures of the time, such as Ottavio and Antonio Targioni Tozzetti and Carlo Cattaneo. The Ufficio di Corrispondenza Geografica was established in Bologna in 1840 by Annibale Ranuzzi, who drew in Carlo Cattaneo, Francesco Cesare Marmocchi, Ferdinando de Luca, and the hydrologist Elia Lombardini.

The organiser and inspirer of the Italian Geographical Society was Cesare Correnti (1815-1888), an important political figure who was part of Cattaneo's circle. The latter was probably connected with the project, but his name never appeared on the list of members. Perhaps this was because the Society, although it was conceived as an institution with multiple functions, one of which was to favour a better knowledge of Italian geographical problems, soon moved noticeably away from its initial objectives. Among the first members were numerous politicians, experts on "allied sciences", military men, and aristocrats, but very few actual geographers, who were still few in number in Italy. Thus, although the Geographical Society pledged itself, as its first statute states, to promote "the science of geography, whatever its branch", from the very beginning the most favoured branch of geography was that which went along with the interests of the ruling class—that is, the so-called "exploratory geography", with particular reference to Africa. More than half of the articles in the *Bollettino della Società Geografica Italiana* dealt with African themes. As Cesare Correnti declared after he became president of the Society in 1873, Africa was "a sort of vocation for our Society".

In 1881 the French began the occupation of Tunisia, which initially had been the target of Italian colonialism, and a rather considerable and influential Italian community already existed there. Italy then began to look eastward to the Horn of Africa. The first step was the acquisition of the Bay of Aseb on the Red Sea, from the Rubattino Company, who had obtained it from local sovereigns through the missionary father Giuseppe Sapeto. In 1884, when Italy took part in the Berlin Conference (1884-1885), a large part of public opinion, which initially had shown no interest whatsoever in colonial expansion, ended up convincing itself that it was necessary in order to be able to count for something on the international chessboard.

Massawa was occupied in February 1885, and the Italian colony of Eritrea was established in 1890.

The rôle played by the Italian Geographical Society in these affairs and also later in the conquest of Libya in 1911-1912 has been strongly criticised by modern geographers, who are more concerned with getting rid of the colonialist label than in reconstructing the historical context. In reality, that rôle was no different from that of other European geographical societies during those years. As in the other geographical societies, the Italian Geographical Society had the aim of keeping up the attention of the cultural-political élite of the country regarding explorations or field trips in areas of interest to the government. The Society honoured its explorers with medals, and their reports were given much space in the "Bollettino". Like the other European societies, the Italian Geographical Society formed ever closer ties with the government, acting as a close, if not totally convinced, ally.

Apart from political motivations, there were other reasons for the interest of the Italian Geographical Society in "exploratory geography". The principal one could be broadly defined as "didactic". The Society was clearly aware of its responsibility to keep the general public informed. After having achieved Unification, Italy still appeared isolated and rather closed in its geographical interests. Not only people of average education, but also many prestigious intellectuals had very limited geographical knowledge. Moreover, this knowledge was almost always secondhand, obtained from books and atlases. Very few Italians had travelled, and even fewer could boast of having made a European itinerary that was more than just a tour of the main capitals.

It was therefore necessary to broaden the field of vision as far as possible to faraway places: Africa, Asia, America and Oceania. Following the encyclopaedic model on which its most authoritative members had formed their intellectual culture, the officers of the Italian Geographical Society believed that the first aim of geography was to describe the world in order to learn about it. Travel reports from anywhere in the world—at least, outside Old Europe which was considered to be already too well known to arouse interest—could be useful in capturing the attention and broadening the geographical horizons of a people who still had to recognise themselves as such. In this way, the confrontation with the habits and customs of people in distant places was important for drawing attention to the characteristics that would favour the unification and help the Piemontese, the Neapolitans, the Milanese and the Romans to feel part of one society. "Exploratory geography" would also allow the essential historical values of the Italian culture to be retrieved through the history of exploration. In the late 19th and early 20th centuries, a school of geography for the study of the history of exploration was established in Italy. It allowed various Italian scholars to maintain a certain intellectual autonomy even through the years of the Fascist dictatorship.

Italian geography at the end of the 19th century

In 1881 the Minister for Education, Guido Baccelli, appointed Roberto Ardigò to the chair of the history of philosophy at Padua University. This date marks the official entry of positivism into Italian universities and the establishment of a philosophical movement that in previous years had been strongly contested, even though intellectuals and the ruling classes had already had some knowledge of positivism from the time of the founding of the Kingdom of Italy. Indeed, in 1865, Pasquale Villari had started his lessons at the Istituto di Studi Superiori in Florence with a lecture on "La filosofia positiva e il metodo storico", which was published the following year in the journal "Il Politecnico", with wide-ranging effect, extending a debate that had been timidly begun by Cesare Correnti in 1840.

According to Wilhelm Buttermeyer, the late diffusion of positivism in Italy was due to the belated affirmation of the lay and liberalist middle classes. While it had already taken place in the first decades of the century in other European countries, in Italy this affirmation had to wait for the establishment of the Kingdom and the advent of the Left. The delay was even greater as far as geography was concerned, because of its nature as a science. With strong systematic features and a firm sciolistic basis, the geography of the time had the tendency to isolate itself and refused a confrontation with philosophy. However, the first geographers in united Italy were convinced of the social and political rôle of their studies. They considered geography to be a didactic instrument, a means of broadening the cognitive horizons of the masses, of overcoming the profound differences between the North and the South, and of speeding up the process of national unification on a cultural level.

Giuseppe Dalla Vedova (1834-1919) was the most famous Italian geographer of the period. A pupil of Friedrich Simony at the University of Vienna and therefore brought up on Ritter's ideas, he basically had to remain faithful even in later years to a harmonious and idealistic vision of geography. Reconciling his university teaching (from 1867 in Padua and from 1880 in Rome) with his appointment first as secretary (from 1877) and later president (1900-1906) of the Italian Geographical Society, Dalla Vedova soon became a point of reference for all his colleagues, especially those teaching in peripheral universities. He exercised great power. Even though he did not agree with the teleological principle, he was attracted to Ritter's geography, above all by its claim to be an autonomous science. Later he accepted Peschel's ideas, combining the Ritterian method, which established the correlation between the human and physical factors, with the study of the geographical distribution of phenomena. Later on he moved closer to the positivist positions but without modifying the contents of his geography, which to the end remained firmly fixed to the Ritterian principles. His relatively low level of production—his best works were the speeches given on various occasions where he obviously limited himself to expounding general ideas without ever giving a practical application—contributed to making his position weaker over the years.

The era of positivist geographers started with Bartolomeo Malfatti (1828-1892). He, like Dalla Vedova, also came from the Austrian school at a time when Ritter's ideas were fashionable, and his initial position was Ritterian, until he was asked to fill the chair of geography at the Istituto di Studi Superiori in Florence (1878). There he met Pasquale Villari and the Florentine positivist school. From then on, his production, which remained modest, had a different philosophical basis and became methodologically more thought out. This resulted in the concrete expression of studies in toponymy and historical geography. But it was above all in his teaching that Malfatti committed himself to providing a new vision of geography, trying to reconcile the Ritterian study of the relationship between Man and environment with the prospects opened up by positivism and evolutionary theory. This attempt, which proved traumatic for him personally and which he never managed to believe in wholeheartedly, was taken up with better results by his successor at the Florentine school, Giovanni Marinelli (1846-1900).

Marinelli's vocation as a geographer developed in an unusual way, not through studies, but with the practise of scientific alpinism. The lack of a solid preparatory basis, above all in the natural sciences, led him to be quite careful when facing particular subjects; on the other hand, the need to reach an epistemological clarification encouraged him to study questions of theoretical geography in depth. His most important works are in fact dedicated to such questions.

On the whole, Marinelli's scientific production was quite vast, much greater than that of his predecessors and contemporaries and completely justifying the title of "The Father of Italian Geography" that was to be attached to him later on. His works included a great treatise, *La Terra*, the first of its kind in Italy. Marinelli planned the work and wrote many sections himself: the chapters on general geography, biological geography and those on Europe and Italy. Marinelli's fame as Master was justified also by his leadership qualities. Above all, they found expression in his two inaugural lectures in the universities of Padua and Florence, and in particular in the second where he faced the crux of the definition of geography: *Concetto e limiti della geografia* was the title of his inaugural lecture at the Istituto di Studi Superiori in Florence on 20 November 1882. In this lecture geography is defined as the science of how and where, as it investigates the phenomena that take place on the Earth, be they natural, social or human, in their external form and in their localisation and distribution. The problem of the relationship between geography and the other sciences is resolved by accepting that they may have the same subject matter in common but claiming a distinctiveness of method. In the same lecture, Marinelli also faced another problem that was then beginning to be felt by geographers, even though it was still in an embryonic form: the dualistic nature of geography. But the positivist premise of the essential unity of knowledge allowed him to affirm that there is no conflict between physical and human geography.

Arcangelo Ghisleri and the Italian geographical journals of the late 19th and early 20th centuries

Giuseppe Dalla Vedova and Giovanni Marinelli both had very strong personalities. In addition, both of them taught in the most prestigious Italian universities of their time and had many students. This meant that even after their death they were considered the Masters of Italian geography. The school or, rather, the two schools founded by them were firmly established, to the point of monopolizing the university teaching of geography in its methods and content and also in university politics. Alberto Magnali, a geographer in the first half of the 20th century who paid dearly for not aligning himself with either of those schools, noted that in 1916 more than half of the thirteen university professors of geography came from Friuli, where Giovanni Marinelli had been born, or from Florence, where he had taught for a long time. Ten years later the number of professors was much larger, but the predominance of the two schools had not decreased, partly because Giovanni Marinelli's son Olinto had established himself as the new Master of Italian geography.

Olinto Marinelli (1874-1926) was destined to leave perhaps even more of a mark than his father, despite his premature death. There is a line of continuity in Italian geography from Giuseppe Dalla Vedova to Olinto Marinelli, outside of which there are only people of secondary importance, who had little impact on the development of geography in the following period. This was the case with Arcangelo Ghisleri (1855-1938), a great thinker who has recently been reappraised, but in his time he was totally isolated. In a historical perspective, Ghisleri's geographical merits did not go beyond the battle that he fought in favour of the diffusion of geography in schools and the promotion of the geography of the homeland ("geografia di casa nostra"). Even though he was several years younger than Giovanni Marinelli, Arcangelo Ghisleri was a source of inspiration for him and his contemporaries, providing a stimulus towards a geography that was much more aware of Italy's internal affairs and the relationship between geography and society. However, this relationship sounded like blasphemy to most Italian geographers, who regarded any contact between politics and science as a dangerous mixture. This was probably the reason why Ghisleri was excluded from a university career.

The increasing interest in geography experienced in Italy on all cultural levels at the end of the 19th century and the beginning of the 20th was also reflected in the unusual proliferation of geographical journals. Before 1890 only the *Bollettino della Società Geografica Italiana* had existed for more than twenty years and had gained an excellent standing abroad. Dalla Vedova's great care had given it a very dignified form and assured a constant scientific support, even though, as an organ of the Society, it continued to have primarily an informative function. The aristocratic journal *Cosmos* was founded by Guido Cora in 1873. It was an expression of the cultural élite of which Cora was a member. It dealt with voyages, exploration and the history of geography. The typography was excellent, and the journal included a wealth of maps of a rare quality. It ceased publication in 1896 and then appeared again in a single issue in 1913. In 1888 there was an unsuccessful attempt by Giovanni Marinelli and others to produce a new publication, *Rassegna Geografica Italiana.*

This was the situation in 1891 when, on the initiative of Ghisleri, *La Geografia per tutti* was established. It was a revolutionary periodical for its time. It was aimed at a much wider public than the existing journals and proposed to spread the knowledge of geography to more and more levels of society. The contribution of *La Geografia per tutti* was more methodological than epistemological, even though it included a column that proposed a theoretical debate on the concrete problems of the country. An objective that the new periodical targeted perfectly was to meet the need for information of the numerous groups of geography teachers in Italian schools, with whom it established a direct and extremely lively debate. In the four years that Ghisleri was the editor of *La Geografia per tutti*, it dealt with many themes of great relevance at the time, such as those that were discussed in the section entitled "Geografia di casa nostra". This section soon became a point of reference for all types of geographical topics concerning Italy, favouring information about places that were not well known and doing its best to establish the exact spellings of place names.

In 1893 Giovanni Marinelli came up with the idea of the *Rivista Geografica Italiana*. From 1894 to 1900 it was edited by Marinelli himself. In this short period of time it became the most authoritative voice of Italian geography. Compared to the *Bollettino della Società Geografica Italiana*, to which it immediately assumed a cautiously critical position, the *Rivista* stood out because of its greater specialisation and the greater space given to young researchers. Compared to *La Geografia per tutti*, it maintained a more severe tone and greater scientific precision. After the death of Giovanni Marinelli in 1900, the *Rivista* was edited by his son Olinto and Attilio Mori (who had practically edited it in the previous two years, during the Master's long illness), without changing any of its features.

Geo-historical studies

By the end of the 19th century, geography has assumed a well defined rôle in schools and universities in Italy. The teaching of Giuseppe Dalla Vedova and Giovanni Marinelli dominated almost everywhere over the statistical-descriptive method of the first years of the Italian Kingdom. This new geography claimed the right to call itself a science and wanted to establish general principles beyond the contingent facts, but it nurtured the illusion of being useful to civil progress in the country without having to face up to its real problems. Outside the university, the "geography of non-geographers" such as Arcangelo Ghisleri tried to form a closer relationship with Italian social life. Despite the different positions, everyone was affected by the same enthusiasm, desire for change and will to work towards making Italy more civil, modern and united, thus overcoming the imbalance among the different regions.

The need to establish a national identity was also at the basis of the geo-historical studies, ranging from historical geography to the history of map-making and the history of exploration. The trend towards erudite research in these fields was widely developed in the various Italian states in the first half of the 19th century. The unification of the Kingdom provided scholars with an enormous archive patrimony of historical-geographical interest. The Italian archives that everyone could now visit contained important documents that testified to the relations between the Italian states and other European countries. Not only did they possess numerous maps that made it possible to reconstruct the history of the territory and governmental intervention (reclaiming and cultivating land, irrigation, agricultural and urban planning, etc.), but they contained precious collections of nautical charts and rich documentation concerning the activities of Italian merchants and traders in the late Middle Ages and Renaissance. This patrimony encouraged and favoured geo-historical research.

Many geographers dedicated themselves to historical research. Curiously, they did not all come from the Arts faculties. For example, Luigi Hugues, who was an engineer, and Guglielmo Berchet, who had a Law degree, were among the authors of the *Raccolta Colombiana*, a monument of geo-historical studies, which marked the IV Centenary of the discovery of America. Gustavo Uzielli, another engineer, and Pietro Arnat di S. Filippo, who was not even a university graduate, edited the *Studi biografici e bibliografici sulla Storia della Geografia in Italia* (1882), a work that has retained much of its value down to the present day. Yet these men were all extremely accomplished scholars, capable of reading, interpreting and putting to use all types of archival documentation. Besides being geographers, they all had a good classical background, both in history and in philological-literary learning. This third trend in Italian geography (beside the official academic geography and the "geography of non-geographers") was made possible by the activities and encouragement of the Italian Geographical Society. The geo-historical studies, and in particular those regarding the history of exploration, appear connected from the start to "exploratory geography", to which in many cases they became a complement and a support.

In October 1892 the first volume of the *Raccolta Colombiana* appeared, following the wishes of the Italian Government to honour the memory of its most famous explorer. The publication was carried out in collaboration with the Italian Geographical Society and the Italian Historical Institute. The rôle of the Society on this occasion was decisive, because it brought together all those who were interested in the history of exploration and map-making, so that while the schools of Giuseppe Dalla Vedova and Giovanni Marinelli both grew stronger and the Ghislerian idea of "geografia di casa nostra" became more popular, the study of the history of exploration found its cradle right in the Society.

There was also another reason for the growth of geo-historical studies. In the years before the IV Centenary of Columbus' discovery, there was intense activity in research on the themes of discovery and conquest everywhere, in Europe and in the Americas. The level of debate was at its highest. Never before had so much energy been spent on affirming the merits of this or that people, of this or that explorer. The stakes were high. The United States then appeared at the international forefront as a rich, free nation with immense resources and a future of unending development. It was natural that everyone in old Europe wanted in some way to take part of the credit for America's "discovery". In the general histories, as in the works dedicated to individual explorers or single episodes in the conquest of the New World, the Italian contribution was often underestimated. In part, this was due to the political divisions of the country, which had favoured sterile, limited and parochial arguments, such as those between the supporters of Columbus (usually from Genoa) and Vespucci (usually from Florence) in the 18th and 19th centuries. Italian culture, as soon as it could identify itself, tried to react to this alienation from history. It did so with inadequate means, with a rather vague, undiscriminating preparation and with a strong inferiority complex, therefore with impetuosity and prejudice. This explains the hagiographical tones and the polemics in the works of that time. Today these tones are amusing and appear out of place, but they were perfectly in line with what was being written outside Italy. The claim of "the Italian genius of the past", which some have interpreted as an inferior expression of Italian nationalism at the end of the 19th century, was the reply of a somewhat provincial culture to an atmosphere of extreme tension and to the events of European colonial politics.

In the first decades of the 20th century, the relationship between geography and the history of exploration was further strengthened. Later, above all in the period between the two world wars but also afterwards, Italian geographers continued to dedicate themselves to such studies. In the curriculum of any geographer who aspired to a university career, at least one historical research project became essential. Until the 1960s, it was normal for every young Italian geographer to have done some archival research, usually on a cartographical topic. Nevertheless, beside this sporadic production often carried out with not much competence and little enthusiasm, which often gave rise to a great number of publications about individual antique maps or about some of the more obscure explorers, there was a vast amount of more demanding work. We have only to think of the *Monumenta Cartographica* of Italy and the Vatican edited by Roberto Almagià (1884-1962), the studies of the history of cartography and of the work of Amerigo Vespucci by Giuseppe Caraci (1893-1971), and more recently of the collaboration of Italian geographers in the compilation of the *Nuova Raccolta Colombiana*.

The contribution that Italian geographers made to geo-historical studies, and in particular to the history of map-making and the history of exploration in the course of the last century, was outstanding not only in quantity but above all in quality. It was critical and methodological, and it showed that the geo-historical studies require specific preparation, both historical and geographical, and that it is not possible to turn oneself into an historian of exploration without knowing how to use both the methodology of the historian and that of the geographer. The most recent research has largely superseded the conclusions reached by geographers of previous generations, so that their research could seem to have been useless or unproductive. In reality, however, the breakthrough that they made in that area of study during the last century, contesting authorities who had never been questioned before and doubting or rejecting deeply-rooted traditions, was a serious and constructive contribution, making historiography revise its own assumptions and verify them one by one.

Olinto Marinelli

The second generation of positivist geographers was much more firmly grounded than the first. It soon became evident that the lively intelligence of Olinto Marinelli would make him stand out among his contemporaries. He was Giovanni's son and his successor to the chair of geography in the Istituto di Studi Superiori in Florence. Like his father, Olinto had a short life, but, having been interested in geography from a very early age, he had an influence on the field for a relatively long period of time.

For his contemporaries, Olinto Marinelli represented the ideal of the perfect geographer, just as his father had imagined it: with a solid natural science background, a good classical culture, a steadfast commitment to the direct observation of geographical phenomena and a sincere love for the "geografia di casa nostra" (although conceived more as a means than an end for research). To these qualities of a scholar, the young Marinelli added the rare human quality, which allowed him to unite colleagues and followers around him and to make himself their Master. In many ways, he followed in his father's footsteps, but, in reality, he reached a higher epistemological and methodological level. With Olinto Marinelli Italian geography found the individuality it had so long been looking for, and there was no longer any worry about suffering in comparison with the standards of other European countries.

Olinto Marinelli aligned himself early on with the theories of Friedrich Ratzel. In his inaugural address, *Alcune questioni relative al moderno indirizzo della geografia* (Istituto di Studi Superiori, 1902), he expounded these theories for the first time. It was a critical and well reasoned exposition, presenting a plan for geography as a dualistic but coherent science whose essential function had to be the study of the distribution of phenomena on the Earth's surface. The problem of the relationship with other sciences was resolved by postulating that all phenomena could be studied from the geographical point of view. Geography was no longer considered as the summa of all the sciences, even though physical geography had always been proposed as the basis for all the biological and human sciences. Another great merit that Marinelli's contemporaries always appreciated was his ability to put such premises into practise and to give concrete models of research into the themes considered to be the subject matter of geography. He realised, however, that such models could not be valid forever, and that "local exploration", even though it might be considered "the most urgent work", could only be "a provisional work". According to Marinelli, local observation did allow for the acquisition of basic data on individual regional units ("elementary regions"), but the final objective had to be the transition from the study of detail to the classification of geographical phenomena and the formulation of general laws. As a consequence of this conception, which subordinated regional geography to topical or thematic geography but which nevertheless considered the region to be the geographer's experimental laboratory, the starting point for most of Marinelli's surveys was Friuli, his native country, which he had begun to study when he was very young. He devoted himself in particular to the lakes, glaciation, glaciated terrain and altitudinal limits—the perfect links between physical and human geography—and to the study of houses and settlements. Glaciation and limnology, together with karstification, remained some of Marinelli's favourite subjects in his more mature years, but he was then interested in more complex topics, among which were population density, cartography (including its relations with geomorphology) and toponymy.

Italian geography between the two world wars

Seemingly, the situation of Italian geography in the interwar period was unproblematic. Following a tradition that went back to the time of Dalla Vedova, Italian geographers had always been good servants of the State, with little or no political commitment. With their political neutrality and their growing cultural isolation, Italian geographers witnessed the advent of Fascism with hardly any reaction. With few exceptions, they were unaware of its dangers. Indeed, they were even led to appreciate certain aspects of Fascism, such as the exultation of national values and the respect for hierarchical order. Thus, allegiance to the State turned almost imperceptibly into allegiance to the regime, which was opening up prospects of new and stimulating studies.

The foundation of the National Geographical Committee in 1921 was the first step towards consent, even though it was originally only a way of bringing together the interests of geographers towards common objectives, to organise research and to establish a link with the International Geographical Union that was set up at that time. In reality, in the years that immediately followed, the work of the president of the Committee, Nicola Vacchelli, was also to be of use in preparing the ground for a series of initiatives that the Fascist regime was interested in. But as these were scientific initiatives, and because geographical skills were sought for the first time in the service of the State, geographers were flattered and pleased to adhere to them. They gave their total approval to the land reclamation projects and to colonial policies.

The subject of land reclamation contributed to modifying the way Italian geographers viewed the rural environment. Up to that time, they had above all tried to identify the types of settlement and to analyse in detail the basic techniques of human adaptation to the physical environment. The greater interest in the natural environment, however, now led to the paradox that in a deeply humanised country such as Italy, the physical, more than the cultural, aspects were emphasised.

Italian geographers felt only the positive and innovative aspects of the land reclamation project and saw it only in terms of landscape modification. The concept of landscape (It. *paesaggio*) was beginning to be discussed. In fact, from 1914, Antonio Renato Toniolo, in accordance with the theories expounded by Albert Oppel at the end of the 19th century, had defined landscape as a geographical topic. A small dispute on the subject arose between Toniolo and Olinto Marinelli, thus serving to arouse wider interest among Italian geographers.

The land reclamation projects, particularly in the Pontine region, were, however, more described by geographers than studied. Similarly, research on the depopulation of mountain areas, which involved geographers together with economists and agronomists, was limited to presenting only a static picture of a situation that was actually undergoing rapid evolution, and the causes of change were greatly simplified in the analyses that were carried out. These investigations revealed the basic fault that also limited the validity or usefulness of other initiatives of the National Geographical Committee—i.e., the illusion of being able to provide homogeneous study models for research on the whole national territory or on well-defined parts or aspects of it, such as mountains or rural settlements. These models were so rigidly laid out that they left no room for individual initiative, and even the diversities of the regions and the distinctive characteristics of their various phenomena were neglected in favour of stereotyped and generalised analyses.

Research on rural housing was promoted by Renato Biasutti in 1926. This research began with a volume on rural housing in Tuscany and was continued for the next two decades. It broke away from rigid schematisation, thanks to its coordinator's broadminded ideas. Although not refusing to consider natural environmental factors, right from the beginning Biasutti intended to place emphasis on the economic and social elements that regulate the living conditions of the agricultural community and the relationship between the traditional features and the innovations determined by more recent economic factors.

At the end of the 1920s, some Italian geographers gave a more explicit consent to the Fascist regime through their studies of geopolitics. As is well known, geopolitics originated from Friedrich Ratzel's concept of *Lebensraum*, which was to achieve its greatest notoriety when it was taken up by some German geographers to support Adolf Hitler's expansionist policies. In Italy, the concept was used as a reminder of the historical greatness of Rome and as a justification for Italy's function as a colonising power. It reflected the aspirations of nationalists, who, after claiming the territories in the north that had once been part of the Austro-Hungarian Empire, then turned their attention to other Mediterranean regions, Corsica and Malta in particular, where an Italian minority lived or which in the past had belonged to ancient Italian states. Studies of these topics appeared in the journal *Geopolitica*, which was modeled after the German *Zeitschrift für Geopolitik* and was published from 1938 to 1942.

The extent of the involvement of Italian geographers in geopolitics and especially with the aberrant theses on racial superiority in the Fascist period is still being debated today. In a cultural environment that not only justified but also exalted imperialism, it was logical that geographers were attracted by the prospects of geopolitics. But actually they had a much more moderate and tolerant attitude in such matters than many other Italian scientists, economists, historians and anthropologists.

Roberto Almagià and Italian geography between the two world wars

Although Italian geography was set in an increasingly rigid and monolithic system, it showed a considerable vitality and variability. The important rôle that the political powers assigned to the universities and that, as in all dictatorial regimes, the Fascists attributed to geography gave its teachers great authority and prestige. As a consequence, they enjoyed many privileges, and their research was encouraged and supported. Some of the geographers wrote large works, which appeared in magnificent formats and were distributed not only at home but also outside Italy to promote the image of a renewed and modern nation, as it wanted to be perceived.

Roberto Almagià and Renato Biasutti were certainly the most representative of the Italian geographers of this period, as well as the best known outside Italy. Of the two, it was Roberto Almagià (1884-1962) who most influenced the geography of the time. He was a student of Dalla Vedova and a friend of Olinto Marinelli. He had a solid classical background, a lively curiosity and an uncommonly versatile mind. He began his university teaching career at an early age (in 1911) and in 1920 became part of the editorship of the *Rivista Geografica Italiana*. It was natural, then, that by the 1930s (Marinelli had died in 1926) Almagià appeared to be the most suitable heir of the masters of the previous generation and the logical person to assume the rôle of directing Italian geography. He took on this rôle by editing the geographical section of the great *Enciclopedia Italiana*, for which he compiled numerous items between 1929 and 1938. "Geografia" was one of the longest articles that Almagià wrote for the *Enciclopedia*. Beginning with an exhaustive and well-articulated historical section, the article then established the nature and aims of modern geography. This was defined in a series of preliminary statements:

1. Geography is a unitary science. The dualism between physical and human geography no longer has a reason to exist, because the two "branches" are united by *the unity of method* and the *final object*.
2. The final object of geography is the description, explanation and classification of the various aspects of the Earth's surface, "as they result from the work of physical agents, from the presence and the differentiation of biological associations, and from the presence and activity of Man".
3. Geography is a complex subject, "as it concerns interdependent facts, and reciprocal relations and connections".
4. "Forms and aspects that are continually modified" stem from such correlations, making up the "geographical landscapes".

The problem of the relationship between geography and the so-called auxiliary sciences, which for many years had been the most difficult epistemological problem, together with that of dualism and its logical consequence, was overcome by clarifying the principles on which geographical research was to be based:

1. The principle of *expansion*, which is necessary for the study of the distribution of phenomena;
2. The principle of *coordination*, which allows the comparison of phenomena in various parts of the Earth and therefore their classification;
3. The principle of *causality*, which makes it possible to go back to the causes that determine the distribution of phenomena;
4. The principle of *correlation*, which is necessary in order to show the interdependence of the various phenomena in the same region.

The description is completed by the division of geography into three branches: *physical*, *biological* and *anthropic*.

This description, which had already been proposed in previous works and which Almagià was to remain loyal to in subsequent works, seemed, at least in principle, to resolve all the theoretical and methodological questions of Italian geography for at least a century. However, the description was absolute and was not supposed to be debatable. There was no comparison with any previous or contemporary definitions; it simply incorporated and took on as its own the theories of old and new geographers with whom Almagià felt an affinity. The principle of expansion was the traditional geographical approach, that of causality came from Alexander von Humboldt and that of coordination from Emmanuel de Martonne. Alfred Hettner's method provided the definition of geography, and the emphasis on regional geography came from the school of Paul Vidal de la Blache. The respect for Italian tradition from Dalla Vedova and the Marinellis made the model acceptable to the majority of Italian geographers. At the same time, the model was innovative, as it left room for regional geography, the main direction of future research. The definition of geography was based on the functions that it had to carry out, and these were seen as stable in time, despite the fact that even in this model they appeared very clearly linked to an extremely precise phase of its development. This was the inevitable conclusion of a history of geography established according to an evolutionary scheme: from a pure and simple description of the world, geography became established as a science with the great German geographers of the 19th century. It grew and matured with the explorations at the end of the century, and when these were over, the field was pretty much set. As the experience of the following years was to show, the limits of Almagià's model were not so much theoretical as practical, because of the difficulties faced by geographers who found it impossible to cover many different fields of study with equal authority, while still adhering to the notion of the unity of method. Therefore, the dualism that had been overcome in theory was to remain a practical problem.

Even though he exhibited interest in all fields of geography, Almagià himself showed his preference for certain areas of research—in physical geography for geomorphology, and in human geography for the study of the distribution of population and settlements. The study of landslides in Italy, which he began in 1905, was Almagià's first and largest concession to "scientific geography". Around the same time, he began his research on the history of geography and exploration, to which he was to dedicate most of his energy for the rest of his life.

In the years between the two wars, Renato Biasutti (1878-1965) was also rising in prominence. He was a more versatile scholar than Almagià and more open to requests that came from abroad. It was perhaps for this reason that he was destined to have a more lasting, if belated, influence on Italian geography in the postwar period. His scientific production includes various titles on ethnology and anthropology. As for the geographical subjects, it is easy to see an evolution from his first works, where there is a strong link with the school of the two Marinellis, to those in his more mature years, which are much more original and personal. It was perhaps his long-standing friendship with Olinto Marinelli that caused him to place little importance on epistemological problems. Indeed, Biasutti was only occasionally interested in the nature and aims of geography. His activities mainly involved human (and even humanistic) geography, a field in which he moved with a sensitivity uncommon in that era. This was the result of a rich and varied cultural formation, which was also open to the social and economic consequences of geographical matters and their correlations with the environment. In 1948 his book *Il Paesaggio terrestre* brought an end to the most active and creative phase of Biasutti's experience as a scholar. In the study of landscape Biasutti found a way to channel and bring together his geographical and ethnological interests. At the same time the concept of landscape allowed him to revive the traditional Italian anthropogeography, still based in the best examples on an ecological study with the new French reflections on the relationship between Man and the environment. Biasutti's *Paesaggio* immediately aroused great interest in Italy. There was an increasing number of studies on landscape, as well as debate on the concept, the value of its physical components and its central position in geographical research. The debate provided one of the first motives for confrontation between Italian geographers in the immediate postwar years, but it also reflected the difficulties of the times and the need for theoretical reflection and confrontation that geographers had done without for too long.

Reconstruction after the second world war

After the war, geography also promptly got ready to reorganise its structure of research. The decentralisation of the universities increased the numbers of Institutes of Geography. The institutes operated as autonomous centres of learning, publishing memoirs and periodicals. But the fundamental structure of research was still the Consiglio Nazionale delle Ricerche, of which the Comitato Nazionale per la Geografia was a part. The committee was directed by Roberto Almagià and was united with the commitees for geology and oceanography. The CNG had been very active in the immediate prewar years and in the early part of the war in publishing numerous monographs on such topics as land reclamation, depopulation of the mountain regions, beaches, historical variations in climate, the altitudinal distribution of vegetation, cities, ports and rural housing. Starting up again after the war, the CNG was especially concerned with completing the previous researches and even more in coordinating the work of four study centres that had been set up: centres for physical geography (directed by Antonio Renato Toniolo), for anthropogeography (Roberto Almagià), for ethnological geography (Renato Biasutti) and for economic geography (Carmelo Colamonico). The centres were very active, encouraging the publication of maps, monographs and bibliographies.

In 1963, following the reorganisation of the CNR, the geographical centres were closed down (the centre for physical geography was closed in 1968), and the CGN was divided into three parts. A difficult situation was created for the coordination of research, which has only been solved in recent years with the formation of the Associazione dei Geografi Italiani (AGeI). During the years of reconstruction, the publication of the main geographical periodicals also started up again at a normal pace. The most important of these were still the *Bollettino della Società Geografica Italiana* and the *Rivista Geografica Italiana.*

In the 1950s, Italian geography was still to a great extent tied to the models and problems of the prewar period. Liberated from the weight of political but not cultural conditioning, the geographers applied themselves to new horizons of study, above all in the field of human geography, yet the interpretative models with which they faced geographical reality only changed on the surface.

Studies of regional geography became fashionable. The intention was to achieve for Italian regions something similar to what had been done in France by the school of Vidal de la Blache. Many scholars even thought that the aim of geographical research was essentially to describe what existed on the land surface and to complete with all the details the descriptive knowledge of the Earth. Paradoxically, in other parts of the world, many geographers were beginning to criticise this approach for its superficiality.

A turning point in the direction and contents of Italian geography only came about in the 1960s. Several factors contributed to this change. First of all, there was the profound renewal of Italian culture, conditioned by the rapid and substantial transformations of the social-economic structure. Seemingly, the debate regarding the rôle, organisation and nature of culture, which was very lively in

Italy in those years, did not involve geographers, but, although the majority of them did not declare their political position, they were indirectly influenced by the debate. The problems that Italian intellectuals discussed brought about a revision of the epistemological values of Italian geographers, although they were not geographical problems in a strict sense.

Secondly, in their search for more solid and more suitable roots, Italian geographers established new contacts with their foreign colleagues. These contacts were becoming easier, thanks to the new international political climate and to the increasing opportunities for travel. The government's economic support through research funds allocated by the minister in charge or by the CNR allowed longer and more frequent field trips. The rather provincial and suspicious view with which foreign theoretical or practical innovations were at least initially received in the past gave way not only to some enthusiastic consent but also to a more mature deliberation. The frequency of the contacts established with French, English, American and German geographers was one of the main features of Italian geography in that period.

Another feature, more decisive for the subsequent evolution of geography, was the rapid growth of economic geography. The establishment of new universities and the foundation of the Political Science Faculties required new teachers in this field. When the number of economic geographers increased, some of them took on a leadership rôle. That was the case with Umberto Toschi (1877-1966), author of the well known *Trattato di Geografia economica,* published in 1959.

A fourth important feature of Italian geography in the 1960s was the adherence to possibilism by a large segment of the young university professors. The generation born in the 1920s and '30s was now coming to the forefront. At a very early age they had experienced Fascism and the Second World War without having been directly involved. They felt that they were a strong part of the new Italian reality, which they perceived as problematic. They were convinced that geography could and should say something about current affairs. These geographers were far removed from the clear certainties of Olinto Marinelli, and, apart from rare exceptions, they could not boast of the deep, refined culture of previous generations. However, they had grown up on the historicism of Benedetto Croce (1866-1952) and therefore were more able to perceive the importance of possibilism than their masters. Their adherence to the theories of Vidal de la Blache and his school (which to a certain extent also the old masters such as Almagià now politely took into consideration) was the fruit of a profound epistemological and methodological maturity, conscious of the evolution that the same Vidalian thought had undergone elsewhere, taking care to interpret rather than merely follow, and above all adapting theories and methods to Italian reality.

The adherence to possibilism was a triumph for human geography in Italy. On the level of theoretical premises it allowed for the proposal of a new link with history, no longer in the traditional position of dependence and subordination, but complementary to it. Human geography no longer aimed at merely establishing a pedestal on which the field of history could rest, but it sought to study the dialectical relationship of humans with their environment and the reasons for their distribution over the Earth. Such a perspective also opened the way for geography to return to the heart of Italian cultural tradition, from which it had been moved far away by positivism. Possibilism brought to the forefront of research the rôle of human societies and their history in the formation of geographical landscapes.

One of the first consequences of this change of outlook was that regional research was favoured. A second was that geographers no longer limited themselves to description and classification but devoted themselves to interpretation. The concepts of landscape, geographical regions and genres de vie were revived and analysed in a way that privileges the historical dimension of the relationship between Man and environment.

The greater intellectual activity that characterised those years made the continuing predominance of the prewar masters even more difficult and less bearable. They still monopolised the politics of the open competitions (which are used to allocate teaching posts in Italian universities), but from an epistemological point of view, their influence declined. Most of them still supported a unified geography and tried to resolve the problem of dualism with erudite logical pronouncements that were increasingly less convincing. This highly conservative position, despite a passive acceptance and superficial support of the more modern epistemological positions, led to disastrous university policies. While other disciplines in those years saw a proliferation of specialised courses with a corresponding increase in the number of specialised teachers, in geography the tendency to specialise was seen as an attack on its unity. Consequently, while, for example, the teaching of history, or Italian or history of art increased noticeably, geography hardly increased at all. This caused a considerable proportional decrease in the weight of geography not only in the universities but also in the way it was perceived by the general public.

In 1964 Lucio Gambi published a little volume entitled *Questioni di geografia*, putting together some of his previously published essays. One was a ruthless analysis of the situation of Italian geography. Applying to academic geography a term that was then very fashionable in regional research, Gambi called it a "depressed region". He strongly criticised the conservative positions of the Masters, which he said were anchored in an extreme view of geography that had been dead for at least half a century. Gambi's accusation, which was initially brushed aside with a very negative review, was to have an incredible effect on Italian geography a few years later. It even became the creed of the new geographers after the deaths of the old Masters in the 1960s and '70s. Even today, many of that generation identify themselves with the theoretical premises of Gambi, claiming the distant roots of the Enlightenment for Italian geography and underlining the reformative rôle of the "geography of the non-geographers" of the second half of the 19th century and the first years of the 20th century.

The most recent tendencies in Italian geography

In the following years, quite a few geographers, mainly those from the classical faculties who had only a faint gloss of physical geography, understood the impossibility of cultivating enough competence in a field of study that was so vast and distant from their *forma mentis*. For them the obvious answer was not the recognition of two fields of study, different not only in content but in method, within the same subject, but the subordination and alienation of physical geography. Quite a few claimed that human geography was the only truly modern geography. The problem of dualism became more complicated with the plurality of fields: not only physical and human, but biological, ecological, economic, political, historical . . . geography (or geographies?). A definition of geography on the basis of content became increasingly more difficult, and one on the basis of method was no longer acceptable.

Acknowledgement of the specialisations and the multiplicability of the various aspects and methods of geography was slowly reached. The widening of the horizons of human geography necessitated further internal subdivisions, each of which required particular skills. It was at this point that new interpretative models of geography were beginning to be accepted in

Italy. These models assumed the existence of various levels of study, not subordinate to each other but reciprocally intersecting.

The theoretical debate started to become very lively in the early 1970s. In 1977 a new professional organisation of university geographers, the Associazione dei Geografi Italiani, was formed, supplanting the Comitato Nazionale that had operated in an elitist manner in the past. In the spring of 1980 the Institute of Human Geography of the University of Milan organised a meeting in Varese on behalf of the AGeI to survey the state of geographical research in the last twenty years. This meeting resulted in the publication of a book, *La ricerca geografica in Italia, 1960-1980,* which superseded a volume published in 1964 by the Italian Geographical Society that covered the previous sixty years (*Un sessantennio di ricerca geografica italiana*). The latter book was carefully subdivided by the branches of physical and human geography, with no critical commentary whatsoever, but the 1980 volume was much larger and was divided not only by subjects but by problems. The second section, *Settori e Problemi,* contained chapters by more than forty contributors, many more than had contributed to the entire earlier book. Even more significant were the three sections of the 1980 volume that contained critical analyses of the current situation of Italian geography. The first of these sections, *Strutture della ricerca,* pointed out how both inside and, above all, outside the university, Italian geographers had learned to come together in different types of structures according to their interests. Another section, *La geografia italiana e la società moderna,* focused on the contribution that geographical knowledge had given, or could have given, to the development of the country. The last of these analytical sections, *Metodologia ed epistemologia della ricerca geografica in Italia,* was of special interest to the "young geographers", bringing up not only the problem of geographical paradigms but also a series of subjects and methodologies that had come up at recent debates on both the national and the international level. This was the case, for example, with the problem of underdevelopment, which Italian geographers had started to face at least thirty years earlier but with very limited results and often, as Francesco Adamo pointed out, hiding the dominant ideology behind a claim of "the neutrality of science". The Varese papers also illustrate the new perspectives in regional and applied geography, involving interdisciplinary relationships among geographers, economists, town planners and historians. For the first time at a meeting of geographers, there were discussions of the geography of perception and the Marxist trend in geography.

In the following years, Italian geographers have, to a large extent, put these enthusiasms into perspective, showing the limits of a theoretical deliberation that was detached from the practise of research, but they also allowed for the broadening of horizons that previously would have been unthinkable. Time has also shown that the variety of fields of study, interpretative paradigms and methodologies of Italian geography was not a sign of crisis but, rather, the demonstration of a great disciplinary vitality. This variety has increased in recent years, along with a greater specificity in the study programmes followed by geographers, given that geography is taught in Science faculties in Italy (such as Geology and Natural Science), in Arts faculties (such as General Arts and Education) and in Economic-Political faculties (such as Economics and Political Science). There is a close relationship between geography and Italian universities. Indeed, there are very few examples of geographers who are active in research outside the university. For this reason, it will be very interesting to see what happens in the next twenty years, following the great educational reforms that the universities are now facing.

References

For general essays covering about a century beginning around 1870, see especially L. Gambi (1973) Uno schizzo di storia della geografia in Italia, *Una geografia per la storia*, Turin, 3-37, and Ilaria Luzzana Caraci (1987) Storia della geografia in Italia dal secolo scorso ad oggi, in G. Corna Pellegrini (ed.) *Aspetti e problemi della geografia*, Milan, 47-94. Regarding the institutionalisation phase of Italian geography, see Ilaria Luzzana Caraci (1982) *La geografia italiana tra '800 e '900 (dall'Unità a Olinto Marinelli)*, Istituto di Scienze Geografiche della Facoltà di Magistero, Università di Genova.

For the Italian Geographical Society and its involvement in colonial politics, the following works are still fundamental: M. Carazzi (1972) *La Società Geografica Italiana e l'esplorazione coloniale in Africa (1867-1900)*, Florence; L. Gambi (1992) *Geografia e imperialismo*, Bologna; and C. Cerreti, ed. (1995) *Colonie africane e cultura italiana fra Ottocento e Novecento* , Rome.

The most recent information about Adriano Balbi, Melchiorre Gioia and Francesco Costantino Marmocchi can be found in the *Enciclopedia Italiana* and in I. Luzzana Caraci, *op. cit.* (1982). For Carlo Cattaneo and Arcangelo Ghisleri, a larger bibliography exists, and we can cite L. Gambi, *op. cit.* (1973), M. Quaini (1977) *La costruzione della geografia umana*, Florence, and, above all, N. Bobbio (1977) *Una filosofia militante: Studi su Carlo Cattaneo*, Turin, and G. Mangini, ed. (1989) *Arcangelo Ghisleri: Mente e carattere*, Bergamo (see especially the chapter by W. Buttermeyer, "Il Positivismo nella società italiana," pp. 19-34).

Regarding Italian geography in the first half of the 20th century, in addition to the works already mentioned, one should also see C. Caldo (1982) *Il territorio come dominio: La geografia italiana durante il Fascismo*, Naples, and M. Quaini (1992) *Tra geografia e storia: Un itinerario nella geografia umana*, Bari. The scientific production of Italian geographers for the first sixty years of the 20th century can be found in R. Riccardi, ed. (1964) *Un sessantennio di ricerca geografica in Italia*, Rome. The next twenty years are covered in G. Corna Pellegrini and C. Brusa, eds. (1980) *La ricerca geografica in Italia, 1960-1980*, Varese. For the subsequent years, it is useful to consult the Acts of the Italian geographical congresses.

Finally, the following list provides some additional bibliographical references (in order of publication) on individual authors or on particular subjects dealt with in the text:

Marinelli, G. (1908) Concetto e limiti della geografia, *Scritti minori* 1, Florence, 143-179.

Magnaghi, E. (1916) *Geographi Italici Maiores*, Florence.

Almagià, R. (1946) Giovanni Marinelli (nel centenario della nascita), *Rivista Geografica Italiana* 53, 57-59.

Almagià, R. (1947) Concetto e indirizzi della geografica attraverso i tempi, *Introduzione allo studio della geografia*, Milan, 7-47.

Migliorini, E. (1963) Roberto Almagià, *Rivista Geografica Italiana* 70, 2-24.

Nice, B. (1965) Renato Biasutti, *Rivista Geografica Italiana* 72, 333-337.

Sestini, A. (1974) La figura e l'opera di Olinto Marinelli, *Rivista Geografica Italiana* 81, 523-544.

Caraci, I. (1978) *A sessant'anni dalla morte di Giuseppe Dalla Vedova*, Genoa.

Taviani, P. E. et al. (1993) Giuseppe Caraci, geografo storico, umanista, in Grillotti Di Giacomo, ed., *Momenti e problemi della geografia contemporanea: Atti del Covegno internazionale in onore di Giuseppe Caraci*, Rome.

Ilaria Luzzana Caraci
Università di Roma Tre
Via Ostiense 159
00154 Roma
Italia (Italy)

6 Geography and Geographers in the Netherlands since the 1870s: Serving Colonialism, Education, and the Welfare State

Ben de Pater

More than the natural sciences, geography was and is a field of scholarship that is bound to a country and a linguistic area. Geography's dependence on diverse national contexts explains why nothing much is ever written about *the* history of *the* field: geography is "a form of situated knowledge" (Livingstone 2000, 7) and "has meant different things to different people in different places" (Livingstone 1992, 28; cf. Heslinga 1978a and Taylor 1996).

This national dimension comes to the fore in the traditional histories of the discipline. They all cherish their own celebrities and standpoints. In contrast to this diversity, there is a similarity in the way the national histories of the field are portrayed. Attention is usually drawn to the work and thought of individual "great men" (and a few women) as well as to the rôle of institutions, particularly of geographical associations. Much less interest is shown in the societal contexts in which the geographical sciences develop, other than in journeys of exploration.

Mainly under the influence of geographers engaged in social critique, a growing criticism of this personal, internalist approach to history has emerged since the end of the 1970s. It is as if the history of a nation's geography had been separate from the social history of that country or linguistic area. There were calls for a more externalist or contextual approach. One voice was that of David Harvey speaking out in his "historical materialist manifesto": "The history of our discipline cannot be understood independently of the history of the society in which the practices of geography are embedded" (Harvey 1984, 1). Referring more specifically to the period around the turn of the century (c.1900), Brian Hudson was one of the first to point out the relation between "the new geography and the new imperialism"; it is no coincidence that his contribution appeared in the radical journal *Antipode*, the vanguard of critical geography. He observed that scientific geography "was vigorously

G.A. Dunbar (ed.), Geography: Discipline, Profession and Subject since 1870, 153–190.

promoted at that time largely, if not mainly, to serve the interests of imperialism in its various aspects including territorial acquisition, economic exploitation, militarism and the practice of class and race domination" (Hudson 1977, 12).

In the meantime, numerous case studies supporting the statements of Harvey and Hudson have become available. Geography's historians see close links between the emergence of fin-de-siècle geography at universities in Europe and the dominant social developments of that time: colonialism, imperialism, the formation of states, militarism, and nation-building. The collections of articles compiled by Hooson (1994), Bell et al. (1995), and Godlewska and Smith (1996) provide dozens of examples.

Of course the contributions of individuals remain important in the contextual approach—at least in its more subtle forms—even though they have lost some of their autonomy (and thereby some of their luster and glory). Geographers, who work in the sphere of influence of social and scientific forces, are continually giving shape to their profession. They may conform to the dominant forces of liberalism, capitalism, and nationalism (and that is what the majority of geographers have done). They could try to keep from getting involved (claiming scientific independence). Or they could try to practice their profession from a counter ideology (such as pacifism, anarchism, and socalism). At the end of the 19th century, such geographical dissidents were rare in Europe.

It is self-evident that historiographical studies should find a balance between the individual actors and the structures and developments of society. Paraphrasing a statement made by Karl Marx, we might say that geographers make their own professional history, but not under conditions that they have been able to choose themselves (cf. Harvey 1984). How they make the history of their profession, and under which conditions, will be examined in this chapter with regard to the practice of geography in the Netherlands since the 1870s. That decade was chosen because it marks the point when the contours of human and physical geography as institutionalized disciplines became visible.

An expedition to Sumatra, 1877-1879

The process of institutionalization was given strong impetus when the Koninklijk Nederlands Aardrijkskundig Genootschap (KNAG—The Royal Netherlands Geographical Society) was founded in Amsterdam in 1873. The Society was only a year old when a plan was hatched to outfit an expedition to Sumatra. That island belonged to the colonial empire of the Netherlands in the East Indies. Officially, Sumatra became a Dutch colony in 1824, when the British gave up their claims to the island in exchange for Dutch possessions elsewhere in Asia. Sumatra was one of the many islands that were formally ruled by the Netherlands but where the Dutch had hardly exerted any real authority. In North Sumatra, the Sultanate of Atjeh, the Dutch had even engaged in a colonial war against the Atjehnese, who refused to recognize Dutch sovereignty (and who again rebelled in 1999, this time against Jakarta and Javanese dominance).

None of these events prevented the KNAG from hammering out plans for an expedition to Central Sumatra; there in the interior lay the Sultanate of Djambi. That area, still unknown to Europeans, would prove to hold great promise for scholars, industrialists, and merchants alike. This combination of potentials was important, since the Society was not a purely scientific organization. The interests of trade and industry were also supposed to be promoted (Velde 1995).

Indeed, the Dutch expected to find a network of rivers in the Sultanate, something that was certain to interest the geographical scientists. But their explorations "also opened up the perspective that suitable natural transport routes would be designated for the products of the highlands beyond, renowned to be so fertile and so rich." The "chance to create multifarious oppportunities, to discover new sources of welfare for Colonies and Motherland" was decisive. This motivation led the KNAG to choose the Sultanate as the destination for the expedition and not other parts of Sumatra, which were equally interesting from a scientific point of view. These were the considerations expressed at the Society meeting held on 20 June 1874 (Hasselt and Snelleman 1881, Part 1, Chapter 1).

It took the autumn of 1874 and all of 1875 for the Ministry of Colonial Affairs in The Hague and the Government of the East Indies in Batavia—the capital of the colonial empire, today called Jakarta—to obtain permission and arrange for cooperation. The reactions coming from the Dutch East Indies were for the most part optimistic: the colonial administration expected to reap many benefits and encounter few political problems with the sultan, as long as the members of the expedition kept a low profile in the Sultanate. Meanwhile, funds for the expedition had been obtained from wealthy entrepreneurs in the Netherlands. In lectures—like the ones given by C. M. Kan, who was a member of the board of the Society and involved in the organization of the expedition—the audience was presented with a picture of paradise. In view of Sumatra's presumed wealth of minerals, fertile land, and potential markets, there was money—much money, in fact—to be made. Moreover, Kan cleverly played on the nationalistic feelings of his audience—the honor of the Fatherland was at stake—as well as on their humane feelings. If the expedition were to make it easier to expand the colonial administration, then this would be a blessing for the population of the interior, since they were suffering from poverty and inept government under local headmen (Kan 1876).

The expedition itself took place in the years 1877-1879. It was presented as a private enterprise, so as not to alarm the indigenous leaders unnecessarily, and, strictly speaking, it was. But the relations with the colonial government were intensive, and it should not come as a surprise that the Sultan of Djambi in particular felt his relative autonomy was in jeopardy. In fact, the expansion of the Dutch authority and of the colonial economy did play a rôle (Velde 1995). Annexation of autonomous areas in the colonial administration and the colonial economy was hardly a secret aim. The members of the expedition saw it that way themselves. Thus, one of them, D. D. Veth, investigated the possibilities for transporting coal from the Ombihin region in the Padang Highlands to the East Coast. To carry it out, a railroad would have to be built. Veth concluded that his exploratory study demonstrated that no problems could be expected (Hasselt and Snelleman 1881, Part 1, 199-200). Whether or not the sultan would agree with the idea of putting in a railroad did not concern Veth. From his point of view, that was an irrelevant question: the Dutch governed all of Sumatra, thus also the Sultanate of Djambi.

Another member of the expedition, A. L. van Hasselt, also referred to annexation of the interior in his reports. He observed contentedly that in the coastal areas under Dutch administration, affluence comes with the new roads—built by the residents who are bound to provide services as vassals. The local population derives "advantage and pleasure" from the roads. As Van Hasselt concluded:

> Where the road network is the most extensive, the affluence is greatest; by building roads, we will see enjoyment increase, and if they are extended through to the borders of independent regions [such as the Sultanate of Djambi], then a peaceful annexation thereof will not be long in coming. Beyond the borders there will be controllers who know the people and their language and feel sympathetic towards that people, and good transport roads up to the border post—these are the means that, faster than the bayonets, will spread Dutch authority over all of Central Sumatra (Hasselt 1882, 194).

There is little doubt that the members of the expedition—as children of their time—saw their journey in part as a means to extend the sphere of influence of the Netherlands. And they were not alone: the Sultan of Djambi and the native headmen saw things that way too. Because of their animosity toward the Dutch, the success of the two-year expedition was ultimately only very partial. The original goals and destinatons were attained only partly, if at all. Nonetheless, the members of the party—just like Alexander von Humboldt eighty years earlier when he returned from his travels through Central and South America—brought back countless previously unknown objects (plants, animals, rocks, agricultural and forest products, and ethnographic materials) to the fatherland, where these items were donated to public collections, such as museums.

In addition, there is a long series of books reporting on the research that was carried out by members of the expedition. The title of the series, which was of course written in Dutch, may be translated as "Central Sumatra: Travels and Investigations of the Sumatra Expedition, Outfitted by the Geographical Society, 1877-1879, Described by Members of the Expdition, under the Supervision of Prof. P. J. Veth." The first two volumes in the series appeared in 1881-1882 and offer a chronological account of the travels. An evaluation of the expedition is included at the end of the second part, where among other issues the authors look back on the political problems that the travelers had encountered. They distinguished three kinds of regions. There are no problems in regions of the first kind, because those places are completely under Dutch authority. There, travelers are safe and can count on the full cooperation of the administrators. Regions of the second kind belong to the Dutch Empire in name only. They are still administered by the local headmen, who see no harbingers of effective colonial authority in an expedition. There, a small party, if aptly led, can conduct research without many problems.

It is the third type of regions—like the Sultanate of Djambi—that pose "ponderous problems". These are territories "of which the population has been half subjected but not wholly robbed of self-government, or where, though in all but name still independent, yet to such an extent confined by Dutch territories or enclosed therein in enclaves, that the headmen and people thereby as of their own accord are led to fear an impending annexation" (Cornelissen et al. 1882, 250). The advice of the members of the expedition is straightforward: to organize no more expeditions to such regions, lying in the broad zone of contact between areas that formally and in actual fact fall under Dutch authority, on the one hand, and areas where the precolonial administration still continues its relatively undisturbed and unthreatened existence.

Eighty years of expeditions to the Dutch tropics

In the Netherlands, just as elsewhere in Western Europe, the institutionalization of the geographical sciences cannot be separated from the colonial ambitions of the country. Far into the 20th century, the territorial expansion of societies always went hand-in-hand with the practice of and interest in geography, and the Netherlands is no exception. The KNAG organized numerous expeditions to regions that fell under the Dutch colonial administration, both in the East Indies (present-day Indonesia) and in the West Indies, particularly to Surinam (Schrader 1974, Brink and Moree 1996, and Ormeling 1998). For instance, besides the various expeditions to Sumatra, there were three to Borneo (in the period 1893-1925) and one to Celebes (1909-1910). Furthermore, in the first decade of the 20th century, no less than seven expeditions were organized into the interior of Surinam to chart the course of the rivers and map the location of mountains. All of these projects were mounted and financed chiefly by the KNAG. Even as late as 1926, a KNAG expedition left for the Wilhelmina Mountains in Surinam.

The last expedition in which the KNAG was involved was the one to New Guinea. By the beginning of the 20th century, only the coastal areas had been mapped. In the interior, the snow-covered peaks beckoned, peaks visible from afar. The KNAG organized two expeditions to this mountain range. The first, in 1904-1905, did not reach its goal—to scale the Snowy Mountains. The party did discover a new snow-covered mountain, the 4750-meter Wilhelmina Peak. In 1939, the KNAG came back to New Guinea; because of the Second World War, it was not until 1948, when the KNAG had been in existence for 75 years, that the results were presented in a book, *De Bergpapoea's van Nieuw-Guinea en hun woongebied*, I ("Mountain Papuans of New Guinea"), written by the expedition's leader, C. C. F. M. le Roux. Volume II and an atlas were published in 1950.

The last expedition that the KNAG helped organize took place in 1959. Once again, the destination was New Guinea, but this time to the Star Mountains. The members of the expedition were carried to their research area by helicopter. The interest shown by the media back home was huge, and so were the financial worries. Thanks to a generous gift by the shipbuilder Verolme, the journey could proceed. Three years after this expedition, the Netherlands was forced to give

New Guinea, the last Dutch colony "in the East", to Indonesia. At that point, the curtain fell for the last time on more than eighty years of exploration by the KNAG, expeditions that had been faithfully reported in the journal of the Society. At the same time, it marked the definitive end of an era; the KNAG was no longer an association oriented towards the Dutch colonies.

The colonial past was revived just one more time, in 1998 when the KNAG celebrated the 125th anniversary of its founding. At that event, fragments of a previously lost film of the 1959 expedition were shown. As the Dutch voice-over commented on the images: "never before had man set foot in these regions." (In 1959, "Man" would still have been understood to mean "*European* Man.") The Star Mountain range in New Guinea was no unknown territory to the Papuan people. It was this Eurocentric attitude—so characteristic of the entire colonial period—that was ridiculed, albeit in a mild manner, at the 1998 anniversary celebration by Jetty Mathurin, a cabaret artist who herself comes from a former colony, namely Surinam. In the postcolonial era, she briefly personified generations of colonized peoples, about whom members of expeditions had so frequently reported with a certain condescending air.

Geography at the fin de siècle

At the end of the 19th century, geography was not actually practiced at any of the Dutch universities, with a single exception (see below). Geography flourished largely outside the walls of the university as a colonial science, among its other concerns. Besides that, a domestic geography had also been cultivated as a science of the fatherland. Its mission was to extend and disseminate the geographical knowledge of the homeland, especially through secondary education. In that context, the emphasis was on the study of the regional diversity of the cultural landscapes of the Netherlands: their origin and development, their historical context, and their present-day features. Most attention was devoted to the struggle of the Dutch people against the "national archenemy"—water. Polders and drainage, land reclamation and loss of land: these themes were subject to much discussion.

The fatherland's three great non-academic geographers were—and here I defer to Max Heslinga, who knows the history of geography in the Netherlands better than anyone else—Anton Albert Beekman (1854-1947), Hendrik Blink (1852-1931), and Roelof Schuiling (1854-1936). Blink and Schuiling were both certified to teach geography in secondary education, having earned an "MO-akte" or state teaching certificate. After getting that diploma, Blink went on to study geography at university in Paris and Strassburg. In 1892, he received an advanced degree from Strassburg, defending a dissertation on a study of prevailing winds and ocean currents around the Lesser Sunda Islands in the Dutch East Indies, although he had never visited those islands himself. Schuiling achieved notoriety for his book on the geography of the Netherlands, *Aardrijkskunde van Nederland*, which was first published in 1884 and went through six editions, the last appearing in 1934. In view of its colossal size, it would have served more as a reference work than a textbook, but it was used for generations.

Beekman obtained his geographical knowledge as an army officer, not in an educational setting. Under his name on the title pages of his books proudly stands "former officer in the Engineering Corps." He was educated at the Royal Military Academy in Breda, where geography had traditionally been part of the curriculum. After graduating, he practiced his profession for six years (Heslinga 1977). In that period, he picked up a great deal of practical knowledge about the use of land and water in the defense of the fatherland. The Nieuwe Hollandse Waterlinie (its initial layout dating back to the first half of the 19th century) was further fortified in the 1870s and 1880s. This line of defense was supposed to protect the heart of the Netherlands, where the big cities are clustered (the present-day Randstad Holland), from enemy incursions. The defensive power of that line—now on the UNESCO list of world heritage monuments—was based on the capacity to flood low-lying areas along a strip of ninety kilometers running from the Zuiderzee into the Biesbosch. Vulnerable or strategic points in the line were protected by fortifications, above all by forts. A great deal of geographical knowledge about the area was needed for the construction of the line. That knowledge did not come from books but was obtained in practice by military engineers, one of whom was Beekman.

After he left military service, in the rank of lieutenant-engineer, Beekman wrote a book entitled *Nederland als polderland* (1884). There, he devoted attention to the Low Netherlands, thus to those areas where natural drainage was impossible and artificial pumping (by windmills and steam-powered pumping stations) was supposed to provide relief from high water. Like the book by Schuiling, Beekman's *Nederland als polderland* did service for about a half century; the third (and last) edition appeared in 1932 (Heslinga 1977).

Besides the colonial and domestic geography practiced outside the university in the decades around 1900, geography was also able to create a niche for itself in the academic domain. The KNAG did what it could to promote geography's admission to the university. In 1874, the association presented a case to the government to establish independent chairs in geography at the state universities (in the same manner as the Royal Geographical Society in Great Britain supported the idea of establishing chairs in geography at Cambridge and Oxford; see Stoddart 1986, chapter 5). The argument presented by the KNAG did not fall on fertile soil. Mainly in view of the cost, the government decided (in the 1876 Act on Higher Education) to tie human and physical geography to existing chairs—respectively, in the Faculty of Letters and the Faculty of Mathematics and Physics. This saved on salaries for professors as well as on the expense of building up a geographical library and map collection. It should be noted that, in practice, the legislation did not mean that both subjects were actually taught at the state universities.

In 1877, however, the KNAG successfully lobbied the municipal council of Amsterdam, the city most renowned for trade in colonial goods. The capital city's Athenaeum was then expanded to become a university, the Netherlands' only non-state university at that time. Amsterdam's university—which, unlike the state universities, enjoyed a large degree of autonomy—deserved, as the KNAG saw it, to be given a chair in geography. One would suffice, as the Society saw no reason to split geography into human and physical geography. On behalf of the board of the Society, the chairman, P. J. Veth (the father of D. D. Veth, mentioned earlier), and the secretary, C. M. Kan, wrote a memorandum to the municipal council of Amsterdam about establishing a chair in geography in that city. One of the points they addressed was the following:

> Nowhere will a professor in geography be better placed than in Amsterdam, the first trading city of the Netherlands, in earlier times the point of departure of many a traveler and explorer who have distinguished themselves in the field of geography, where even today the relations with other countries and especially with other continents are greater than anywhere else in our fatherland (Veth and Kan 1877, 369).

In the memorandum, the authors played on the chauvinistic feelings of the members of the municipal council. The appointment of a professor of geography would uphold "Amsterdam's repute in the field of geography and cartography". Their main argument was the benefit that such an appointment would offer to secondary education. Geography (in Dutch, Aardrijkskunde) was a relatively important subject at the HBS (Hogere Burgerschool), which was, after the gymnasium, the highest level of secondary education in the Netherlands. But the geography teachers came from disparate educational backgrounds. On the one hand, geography might be taught by graduates in mathematics and physics or in Dutch literature. Even if they never had to take a single examination in geography at the university, on the grounds of their study they were considered to be capable to teach either physical or human geography. In practice, they were supposed to pick up the subject they had missed as they went along. On the other hand, there was a possibility for primary school teachers to acquire sufficient geographical knowledge through individual study to pass the examination for the teaching certificate in geography. Individual study—whether or not coached by one or more tutors—was the only path; teacher training programs did not exist for the field of geography. The new professor would be able to offer such a program. According to Veth and Kan, that would provide capable teachers, and the schools in Amsterdam would be among the beneficiaries.

That argument found a willing audience in the municipal council. On 15 October 1877, the council appointed Cornelius Marius Kan (1837-1919)—the same individual who earlier that year had co-authored the memorandum on behalf of the Society—as professor of human and physical geography, including "geography and ethnography of the East Indies Archipelago." Until his retirement in 1907, Kan provided an education at the University of Amsterdam for persons who wanted to obtain a teaching certificate in geography. For the most part, these were auditors, not regular students; they were primary school teachers who wanted to become secondary school teachers. They studied geography in the spare time that their teaching job left them. The fact that Kan trained geography teachers but not geographical scientists did not seem to bother him much. He too was mainly an educator at heart, having been educated in the classics and thus being first and foremost a humanities scholar.

Two key years: 1907 and 1908

When Kan left the university in 1907, his advice was to appoint two men to succeed him. The degree of specialization in geography at that time was, from his point of view, so far advanced that it was impossible to expect just one man (it would then still be ten years before the first woman was to be appointed professor in the Netherlands) to be able to do research in both physical and human geography.

Kan's advice was not ignored. Several months after Kan retired, the paleontologist Eugène Dubois (1858-1940) was appointed professor of physical geography, whereas the jurist, ethnologist, and sociologist Sebald Rudolf Steinmetz (1862-1940) took over the social side of Kan's post. Steinmetz was named professor in the field of "political geography, ethnography, and the geography and ethnology of the East Indies Archipelago" (Heslinga 1999).

Under Steinmetz, human geography developed into a form of descriptive sociology, which he called sociography. Just as ethnology and ethnography were concerned with non-Western peoples, the fields of sociology and sociography dealt with Western populations. Ethnology and sociology were essentially theoretical, while ethnography and sociography were mainly descriptive. Concretely, sociographers sketch portraits of population groups with a territorial (be it urban or rural) framework, giving a great deal of attention to their social relations, mentality and culture and little attention to their relations with the physical environment (which explains why, according to many geographers at Utrecht University, the field of sociography was not really geography). Thus, the development of theory was not a priority in sociography; it was more concerned with description than explanation. The only time explanations were sought and an effort was made to formulate theoretical insights, there was clear evidence—in line with the spirit of the times—of biological determinism and a racist undertone (Biervliet et al. 1978, Mok 1999). Steinmetz himself, incidentally, wrote few sociographical monographs; he was mainly interested in producing compendiums that cover a broad field, such as the one he published in 1920 about the issue of nationalities in Europe. That work brought him renown as a social-science scholar of European stature.

Dubois, educated as a physician (specializing in anatomy), was famous for his work in Java, where in 1891 he excavated *Pithecanthropus erectus*, which he presented as the missing link between ape and man. In 1898, he was given an endowed chair as professor of geology and paleontology. The scope of his chair was extended in 1907 to include physical geography and was turned into a regular professorship. Dubois, much more than his colleague Oestreich in Utrecht, was involved in the study of the physical geography and geology of the Netherlands itself: groundwater, dunes, and glacial deposits. He never attracted many students.

It was not until Jan Pieter Bakker (1906-1969) became professor in 1946 that physical geography really got off the ground in Amsterdam. There, for a while, the field even overtook Utrecht, the cradle of physical geography in the Netherlands. Bakker introduced the University of Amsterdam to modern forms of research in physical geography (complete with laboratories), whereas Utrecht, at the time under Jacoba Hol (1886-1964), held onto the classic field of geomorphology. Change only came with her successor, the quaternary geologist from Leiden Jan Zonneveld (1918-1995), and that change was gradual.

The two appointments in 1907 at the University of Amsterdam—a municipal institution—were writing on the wall that the state universities could no longer rest on their laurels. Even before Kan's departure, initiatives had already been taken at the state university in Utrecht by the Faculty of Letters and especially by the Faculty of Mathematics and Physics to appoint two geographers in Utrecht: one for human geography and one for the physical components of the field. Although the two other state universities, Leiden and Groningen, would have been glad to appoint geographers, the initiative taken by Utrecht was rewarded by the minister. Besides the extensive memoranda in which Utrecht motivated the move, demonstrating the broad support in Utrecht for geography, there was also a practical argument in play. The Royal Dutch Meteorological Intitute (KNMI) was located close to Utrecht, in the village of De Bilt, and its director, as an endowed professor, could easily teach classes in meteorology, climatology, and oceanography to the geography students. It saved having to pay the salary of a meteorologist. Indeed, until long after the Second World War, it was the rule that a director of the KNMI was also a professor at the university in Utrecht. Ultimately, Utrecht got geography in 1908, Leiden "Indology", and Groningen modern languages (Heslinga 1978b, 1983, 1999).

In September 1907, the month in which Dubois and Steinmetz were appointed in Amsterdam, the minister presented his budget for 1908 to the Dutch parliament. An amount was reserved in that budget to appoint two professors in Utrecht as well as to purchase books and maps. In fact, the minister thereby made it possible to establish a geographical institute at the state university. After parliament approved the budget, the search for suitable candidates got started; by then it was early spring of 1908. In the end, the German Karl Oestreich (1873-1947) was appointed as professor of physical geography, and the chair in human geography was given to the Dutchman Jan Frederik Niermeyer (1866-1923).

Oestreich, who was educated in the natural sciences at German-speaking universities, perceived physical geography as geomorphology. His fame was based on his geomorphological research in mountainous areas of Europe and Asia. No wonder Oestreich soon took Utrecht students to the Eifel Mountains in 1909; his subsequent field trips to the Alps were renowned. Regarding the substance of the field, Oestreich remained oriented toward his German colleagues, particularly Albrecht Penck. Yet Oestreich also kept in close touch with others, such as the American William Morris Davis. Together with Niermeyer—with whom Oestreich did not get along, due to controversies about the field as well as about politics: Oestreich was oriented toward Germany, Niermeyer toward France (Heslinga 1999)—Oestreich formed the Dutch representation on the transcontinental field trip that European geographers took in the United States in 1912, invited by Davis. During the 35 years that Oestreich was professor in Utrecht, he educated the first generation of physical geographers in the Netherlands, including his successor, Jacoba Hol, and J. P. Bakker, who was to bring physical geography to fruition at the University of Amsterdam.

For an international discipline like physical geography, the minister, who made the decisions on appointments at state unversities (at that time, state universities had much less autonomy than they do at present), was still willing in 1908 to appoint a foreigner such as Oestreich. For the field of human geography, which is more explicitly linked to the country and its colonies, the situation was different. Although Utrecht University would rather have appointed a geographer with scientific credentials, even if he were a foreigner (Otto Schlüter was the most likely candidate), the minister, who had to take the mood in parliament into account, deliberately chose to appoint a Dutchman. In doing so, he could count on the support of the board of the KNAG. Many names were going around, including those of Blink, Schuiling, and Beekman. But in the end the post went to Niermeyer, a geography teacher from Rotterdam who had once been taught by Kan (Heslinga 1999).

Of course, Niermeyer was no ordinary teacher. He was extremely active in KNAG circles and had made a name for himself by updating (in collaboraton with Joh. F. Snelleman, one of the members of the expedition to Sumatra in 1877-1879) the four-part standard work on Java by Pieter Johannes Veth (1814-1895). Besides being the first chairman of the KNAG, Veth had also been a respected professor in Leiden covering various fields, including the geography and ethnography of the Dutch East Indies, a region he had never even visited. (This shows the extent to which travel was more the exception than the rule at that time.) With some regret, Veth described his standard work on Java, *Java, geographisch, ethnologisch, historisch*, as "the fruit of book learning, not of personal observation." Niermeyer augmented Part 3 of that work, the part dealing with the geography of Java, with his own observations, which he had made during a half-year trip around Java and Madura. This second edition was published in 1903, 28 years after the first edition. It was partly due to Niermeyer's knowledge of the Dutch colonial areas, demonstrated in this work, that he could be appointed to the post of professor in Utrecht in 1908. As professor, he remained for many a representative of geography as taught in schools rather than an exponent of geography as a branch of science.

Geography and the decline of colonialism

The appointment of Niermeyer is characteristic of the problems prevailing around 1900 in the Netherlands with respect to appointing professors of geography. In view of the fact that the country had no university programs in geography, the universities could not find any Dutch academic geographers for the job. They had to make do with foreigners, predominantly from Germany, where university programs in geography already existed. Oestreich was one such foreigner. Either they could choose a Dutch person, but then would have to accept that the only candidates came from circles of secondary education (Niermeyer), or they could choose among the candidates who had completed a non-geographical course of study at the university, such as Steinmetz and Dubois (Heslinga 1999). Individuals with a military background, such as Beekman, might be acceptable as well.

That situation persisted until well into the 1920s. After the early death of Niermeyer in 1923, Louis van Vuuren (1873-1951) was named as his successor. Originally, Van Vuuren was a professional military man who had fought in the Dutch East Indies in the colonial war in Atjeh. Having suffered a saber gash, and no longer having the full use of one of his arms, Lieutenant Van Vuuren became a colonial administrator in the civil service and then moved up to head the Encyclopedic Bureau in Batavia. In that position, he wrote monographs on regions in the Dutch East Indies in the service of the colonial administration. In 1922, after this bureau was closed, he became a lecturer in colonial geography at the University of Amsterdam; in 1927, he became professor of human geography at Utrecht University. His career was truly remarkable: having had no university education whatsoever, never having been a geography teacher, and yet he was appointed professor in human geography. After that time, such an appointment would be unthinkable.

The apppointment of Van Vuuren illustrates not only the take-off phase in which geography, as a university science, found itself until well into the 1920s, but it also clearly demonstrates the extent to which there was still a connection between geographical practice and the colonies (Heslinga 1996).

It was only after the Second World War that colonial geography disappeared; its demise coincided with the independence of Indonesia. Like most Dutch people at that time, geographers were disappointed about the decolonization: "Indies lost, disaster born" was the saying. The expectation was that independence would break up the Netherlands, as well as the East Indies. The Dutch had done a lot of good work and would continue doing so, if they got the chance. This was the prevailing opinion in the Netherlands. In 1949, the Utrecht geographer Jan Broek (1904-1974), who became chairman of the Department of Geography at the University of Minnesota a year earlier, defended the Dutch viewpoint on American radio. Broek was a panelist on the New York radio talk show *Town Meeting* on 11 January 1949, where the main question of the debate was "What's behind the Crisis in Indonesia?" The question was timely, because the Dutch were engaged in what they euphemistically called the "Second Police Action" but what in reality was an ordinary war being waged to draw Java back into the colonial fold. In the view of Broek and the majority of Dutch geographers, the Dutch had brought European civilization to the Dutch East Indies. Although that view might have been shared by the United States (itself made up in great part of former European colonies), it was not translated into political support. The United States and the United Nations Security Council took the side of the Republic of Indonesia, which became a sovereign state in December 1949. The last Dutch geographers left Indonesia in 1957. One of them was the physical geographer Herman Verstappen; another was the human geographer and cartographer Fer Ormeling. Both were to become internationally known for their administrative work, among other things (Verstappen as president of the International Geographical Union and Ormeling as chairman of the International Cartographic Association).

At the universities, interest in the ex-colonies waned, but at the end of the 1960s attention was again focused on the study of the former colonial world. The new designation was "geography of developing countries" or "geography of the Third World". This trend toward specialization was manifest in the credentials of the newly appointed professors: Jan Hinderink (a specialist on Africa and Indonesia) at Utrecht University, Jan Kleinpenning (a Latin America specialist) at the University of Nijmegen, and Ad de Bruijne (India and Central America) at the Free University and since 1985 at the University of Amsterdam. All three had studied human geography at Utrecht in the 1950s. Hinderink and Kleinpenning have recently retired, and De Bruijne will retire in the autumn of 2001.

Blink's economic geography and "public betterment"

The effort to turn human and physical geography in the Netherlands into established fields of scientific endeavor at the universities got strong impetus shortly after the First World War. In 1921, the government recognized both disciplines as full-fledged and independent studies. The dualism that had characterized Dutch geography for a long time was thus forcefully prolonged and further institutionalized by this decision. From then on, students in Amsterdam and Utrecht could take either human or physical geography in their first year at university. Formally, the full program took at least five years, but in practice it often lasted quite a bit longer. Geography was no longer a veiled teacher-training program or an elective—though seldom taken as a minor—for university students majoring in the arts and letters (who might take human geography) or in mathematics and physics (who might take physical geography).

At the start of the 20th century, eonomic geography was promoted as a separate subject for men "in practical pursuits." As a specialization within human and physical geography, it drifted into the doldrums in 1921. It was not recognized as an independent course of study. It did, however, become a required course in the economics programs of the universities. That is why many a geographer who had an interest in economics ended up working there; in some instances (in Rotterdam and Amsterdam) it even prompted the creation of economic-geographical institutes in the faculties of economics. Such institutes were in existence from the 1920s to the beginning of the 1990s. They have since fallen prey to reorganizations and budget cuts, either to be abolished or to be taken up in other frameworks.

Hendrik Blink, the father of economic geography in the Netherlands, always remained on the margins of the university environment. He felt more at home in the world of administrators and business people. For him, economic geography was primarily a practical profession. Blink was the son of a peasant-cum-blacksmith from the peripheral province of Drenthe, and his unpretentious origins probably explain his devotion to the cause of "public betterment". In 1886, he started up a popular-science magazine, *Vragen van den Dag* ("Questions of the Day"), of which he was the editor-in-chief (and principal author) until his death in 1931. The ideal of adult education, of public betterment, was also propounded by Niermeyer and the Amsterdam sociographers. For instance, before his appointment as professor, Steinmetz held sessions at his home on Sundays for workers who were eager to learn from him. In 1913, he helped found the Volksuniversiteit in Amsterdam, an institution offering adult education for the general public. Between 1913 and 1937 he was the chairman of this organization (Vries 1963). One of Steinmetz' students, Hans de Vries Reilingh (1908-2001), believed strongly in the ideals of the movement for adult education. He established several schools for public education and obtained a Ph.D. in 1945 with a dissertation on the history of adult education in various European countries.

"There the state builds a new society"

More things changed in the 1920s than just the position of the discipline in the universities. There were signs that support for the field in society at large was expanding. Whereas geography in Dutch universities had once rested on two pillars, namely the colonies and teacher certification, a new pillar was introduced: regional socio-economic policy and spatial planning.

A glimpse of that third pillar was offered in the dissertation that the Amsterdam geographer Henri Nicolaas ter Veen (1883-1949) defended in 1925 on the Haarlemmermeer polder as a colonization area. In that work, Ter Veen described the daily life of the population that came to live in the heart of Holland right after the Haarlemmermeer was reclaimed. The Haarlemmermeer was an inland sea of 18,000 hectares and seemed to expand with every storm. In 1839, it was decided to pump the water out of this lake, and in 1848 the work actually started. First, steam-powered pumping stations had to be designed and built; before that time, lakes had always been pumped dry with the help of windmills. By 1852, it was dry, and the state could sell the reclaimed land at auction. The state did not get involved in cultivating the land or putting in infrastructure to service the area. Most of the land fell into the hands of large landowners who lived outside the

Haarlemmermeer polder. Those who actually came to live there were tenant farmers, hired hands, and people on the margins of Dutch society. They found themselves in a swampy wilderness. Life was extremely harsh: poverty was widespread, child mortality high, and alcoholism ubiquitous. And yet, according to Ter Veen, in an argument couched in Social Darwinism, even though this environment was hostile in several respects, it had a chastening effect. Only the strong held out. For the weak, "the struggle was too hard: they gave in and fled" (or died, one might add). Because of this "natural selection", those with the strongest initiative, vitality, and work ethic remained. In Ter Veen's eyes the energetic and progressive population of the Haarlemmermeer polder looked like the individualistic pioneers whom Frederick Jackson Turner had observed on the American frontier. Through their efforts, the pioneers—at any rate, their children—built up the Haarlemmermeer polder; by Ter Veen's day, it was a modern and prosperous agricultural area. (Today Schiphol Airport is located on this territory, while Hoofdorp, the settlement at the center of the reclaimed area, is developing as an edge city of Amsterdam.)

Although the end result was quite an achievement, the wretched social conditions prevailing during the first decades after the reclamation were shameful. The gist of Ter Veen's assessment was "once but never again". And that was a relevant message because work was begun in pumping water out of the Zuiderzee in 1927, two years after Ter Veen's book was published. He hoped that his dissertation would prove to have a practical effect on the process of drawing up the plans for the colonization of the future Zuiderzee province (now known as the province of Flevoland).

The decision reached by the government in 1918 to pump the Zuiderzee dry was based for the most part on a discussion between those in favor of the plan and those against it. The opponents pointed out the high costs and the damages to fisheries and inland shipping. The proponents cited the increase in agricultural land in a densely populated country and emphasized the great security. Other than that, they used emotional arguments. Those arguments were openly nationalistic: a bold and glorious people who are working on their future by bringing their archenemy, the sea, to a humiliating defeat.

The geographer Anton Albert Beekman was one of the most outspoken proponents. By his own count, he had given more than a hundred lectures throughout the country over the years proclaiming the necessity of reclaiming the Zuiderzee. In 1912, for instance, he made the following statement in a speech:

> That country [the Netherlands] will see its international position improve by the peaceful conquest of a territory 11 to 12 times the size of the Haarlemmermeer with a prosperous population. Through their efforts, sacrifice, and struggle, the people will have increased their power by carrying out a great deed (Beekman 1912, 1).

And in 1914, Beekman wrote that the Netherlands—the great and powerful country of the 17th-century Golden Age, "the homeland of Rembrandt" for which many a modern person had so passionately longed—once again could take an important place "on the list of civilized nations" by "wresting [from the sea] an area as large as never before." Other peoples would gaze in amazement at this great deed and appreciate that the "old spirit of enterprise of the Dutch" was back. In short, according to Beekman, after 65 years of talk about closing off and reclaiming the Zuiderzee, the time had come to make a definitive decision.

That decision came four years later, in 1918. The government, which had kept its distance from the colonization of the Haarlemmermeer polder, realized that it had a rôle to play in the Zuiderzee project. But how big did that rôle have to be? Beekman was a liberal and wanted to avoid too much government intervention. At first, Ter Veen, a social democrat, did not want too much government involvement either. In his dissertation, he observed:

> No one wants to see a repetition of the uninterested carelessness that the State showed at the time of the reclamation of the Haarlemmermeer; yet an excess of State care is equally damaging. When it comes to resolve the big issue of colonization, our Parliament will have a decisive voice; in that case may the Members of Parliament let themselves be led by the results of scientific research. The field of human geography has an important role to perform here (Veen 1925, 249).

The argument presented by Ter Veen in favor of managing the colonization process in the Zuiderzee polders, increasingly referred to as the IJsselmeer polders, did not fall on deaf ears. This was a period in which liberalism, the dominant political philosophy, gradually made way for a welfare state, which was based on Christian and social-democratic convictions. The government became increasingly involved in the well-being of the population and in the organizaton of space. Ter Veen was named the secetary of a committee that was installed in 1926 to arrange for the allocation of agrarian land in the future Zuiderzee polders (Heinemeyer 1996).

The first polder in the Zuiderzee became dry in 1930. This was the Wieringermeer polder, 20,000 hectares in size. Unlike the situation in the Haarlemmermeer polder, the state remained actively involved in the area. Ter Veen was by then convinced of the need for more intervention than he had thought necessary in 1925. The colonization of the Wieringermeer polder was centrally led and planned. As Ter Veen approvingly observed, "The State is building up the new society here." The composition of the population was not determined by any kind of natural selection but by a government-led selection. Exactly who might become a "pioneer" or "colonist" and settle on the "frontier" was decided by government-appointed selectors, some of whom were Amsterdam-educated social geographers (usually called sociographers). Thanks to the efforts of the state, "the colonists were saved from unmentionable suffering," without having undermined the pioneer mentality—that of "self-confident, progressive trailblazers, social pioneers" (Veen 1950).

In 1937, work was started on the reclamation of land for the Noordoostpolder (40,000 hectares). After the war, two more new polders followed: Oostelijk Flevoland (54,000 hectares) and Zuidelijk Flevoland (43,000 hectares). In this "new land" the spatial planning effort was even more intensive than in the Wieringermeer polder. How many service centers were needed, how large would they have to be in order to be viable, and what was a desirable location for them? In making various designs, inspiration was drawn from the central place theory of the German geographer Walter Christaller (Dutt and Costa 1985, Wal 1997).

Thus it was that two disciplines—human geography (understood as sociography) on the one hand and socio-spatial planning on the other—became the warp and weft of a single fabric at the University of Amsterdam. Ter Veen took the initiative to weave them together in 1933, when he took on the position vacated by Steinmetz. He may not have been as brilliant or charismatic as Steinmetz, but he was more practical, with a talent for organization and bureaucratic practice (Heek 1978). Unlike Steinmetz, an "armchair" scholar who preferred to work at his desk rather than in the field, Ter Veen organized fieldwork for his students, complete with surveys that they conducted among various social groups. He was the founder of the government-affiliated organizations that carried out practical research on social and socio-spatial issues, both in the "new land" (i.e., the Foundation for Population Research in the Reclaimed Zuiderzee Polders, established in 1936) and in the "old land" (the Institute for Research on the Dutch People, from 1940). These organizations provided many human geographers, sociographers, and eventually sociologists with a living. Various municipal sociographic bureaus were also established in that period to keep the social (and socio-spatial) profile of the local population up to date. Some of these bureaus have survived all the reorganizations that have swept over the municipal bureaucracy and are still in operation to this very day.

Applied regional geography in the "old land"

Spatial planning carved out a niche for itself not just in the Zuiderzee polders but also in the previously inhabited parts of the country, the "old land". The start of this development was marked by the passage of the Housing Act in 1901. The main objective of this legislation was to improve the dwelling stock: it enabled the municipalities to demolish slums. But the act also required large and rapidly growing municipalities to make plans for the layout of the streets. From then on, any plans to extend the built-up area would have to include a rudimentary street plan. When the Housing Act was revised (in 1921 and 1931), the local authorities were given more responsibilities. The municipal street plans gradually became plans in which *all* areas within the extension were assigned a land use.

The oldest spatial plans referred to the local situation. Afterwards, the regional, the national, and even the international level came into view, in that order. During the 1920s, the first trial runs were carried out for regional plans; this time it was the provincial authorities who took the initiative. In the 1930s, there were increasing calls for a national spatial plan—that is, a plan for the entire territory of the Netherlands. In 1941, with the blessing of the German occupying force, which looked favorably on the prospect of "strong centralized control" over the "allocation and use" of land in the Netherlands, the time came to establish the State Agency for the National Plan (Bosma 1993). This was the predecessor of the present-day Rijks Planologische Dienst (National Spatial Planning Agency). Incidentally, the first national spatial plan would not be published until 1960.

In the 1930s, some planners harbored even greater ambitions. For instance, J. M. de Casseres (1902-1991), at the time one of the most prominent planners (he was then working in the city region of Eindhoven), introduced the notion of "planology" to the Netherlands. This term is still used to denote what is known in English-speaking countries as "urban and regional planning". In a speech presented in

1933, De Casseres made a case for drawing up an international spatial plan:

> A close look at the map of the world reveals the very great extent to which a carefully considered ordering of the spatial structure is lacking. Everywhere we see that individualism driven to its extremes has been the cause of an arbitrary dispersal of roads, railroads, and population agglomerations. The national plan exists to bring order to the national chaos; the international, the world plan will help order the world chaos (Casseres 1933, 11).

For De Casseres, who himself had an international background, such a plan was also a means to promote mutual understanding between countries. It was a panacea that would help control the international tensions rising in Europe. (Hitler had been the Chancellor of Germany for only ten days when De Casseres presented this lecture.) To support his vision, De Casseres cited the closing lines from the book *A Nation Plan* (1926) by the American Cyrus Kehr: "The World Plan is therefore urged as one of the most promising means of establishing a logical working basis for world harmony and world peace."

De Casseres presented his lecture at the Geographical Institute of the State University of Utrecht, where he had been invited to speak by Louis van Vuuren, the professor of human geography. Van Vuuren had already in the 1930s started to practice an applied regional geography, which he saw as a kind of descriptive survey. It offered an overview of the characteristics, problems, and potentials of a rural or urban area. Such an inventory would have to precede the process of making a spatial plan for the area. From that survey, recommendations were supposed to be distilled, forming grounds on which the responsible government authorities could base a regional welfare policy as well as a spatial policy.

In fact, Van Vuuren and De Casseres followed the classic motto of one of the founders of spatial planning, the Scottish biologist Patrick Geddes: "survey before plan". In the triad *survey-analysis-plan*, the geographer was in charge of the first phase. In 1935, at the first Dutch Geographers Meeting, Van Vuuren pointed out the new rôle that human geographers were to play:

> [They] will in particular be able to perform a good service with respect to the social research that has to precede any plan for urban extension or any regional plan. There is now a general consensus that a plan for an urban expansion area is not and never has been a technical

> engineering problem. It is much more of a social problem. In the future, whether people like it or not, it will no longer be possible for an expansion plan or a regional plan to be made until in-depth human-geographical research has been done (Vuuren 1935, 189).

On Van Vuuren's initiative, the result was an impressive series of regional "welfare studies" in the course of the 1930s, carried out by Utrecht students and graduates in human geography. Their work was supervised by Van Vuuren and financed by various tiers of government. Those studies were intended to lay the foundation for a policy that would combat poverty and systematically organize space.

The impetus for government intervention on the "old land" came from the worldwide economic crisis of the 1930s. In the Netherlands, the Depression led to rising unemployment and increasing poverty. The idea that the government had to take measures in the areas hardest hit gained more and more support. In rural areas the unemployed were put to work to make arable land out of the country's remaining wastelands, such as the heath and the moorland that had been stripped of its peat. The idea was that the land thereby brought under cultivation would increase the welfare of the regions. The Chambers of Commerce, in cooperation with municipal and provincial authorities, looked for ways to improve the economic prospects for industry in their districts.

Van Vuuren's reports were intended as signposts for this policy. For instance, the *Rapport Salland* (1941) ends with "guidelines for the measures to be taken in the future" with regard to agriculture, industry, traffic, and natural beauty in the region of Salland. With respect to industry, an inventory was made of the problems that the factories had to deal with: difficulty in the supply of raw materials and little profit as yet from the new markets in Germany. Calls were heard for "a fair arrangement, which rests on the new ties of solidarity, by which the Netherlands is bound."

At the beginning of the war, people were fairly complacent, willing to believe that from then on the Netherlands would be part of an empire ruled by Germany. People saw this as an unavoidable state of affairs and went back as quickly as possible to their daily lives. This was also true in geographers' circles. For instance, at the end of September 1941, Gerrit-Jan van den Berg and Chris van Paassen (1917-1996), who had both just graduated in human geography, left for the city of Vlissingen in Zeeland, as directed by Van Vuuren, to do research there. The task that Van Vuuren assigned was to investigate the necessity of extending the city limits of Vlissingen at the expense of the surrounding municipalities. The request to carry out this study came from the municipality of Vlissingen, which wanted to expand the city. While the city, which was occupied by the Germans, was being attacked almost daily by Britain's Royal Air Force, Van den Berg and Van Paassen were collecting data for their *Rapport Vlissingen* (1942). On the basis of that report, Van Vuuren advised the Director of the State Agency for the National Plan that there was no reason for annexation. But ultimately the violence of the war would come to overshadow any and all concerns about moving the boundaries.

Besides the applied geography reports that were commissioned, the dissertations of the time also bear witness to a practical mentality. A striking example is the dissertation that Willem Steigenga (1913-1974) wrote in 1939 on employment in the agrarian sector. One of the topics he investigated was the differentiation among rural areas in the Netherlands with respect to the degree to which the agrarian population could make a living there. Steigenga ended his study with extensive suggestions for an agrarian policy that would reduce unemployment. Such engagement with social issues has been typical of Dutch geography.

Although the majority of the geographers from Amsterdam and Utrecht who graduated in the 1930s ended up as teachers in secondary schools, in that decade more and more graduates took jobs as socio-spatial planners with (semi-)governmental agencies. The geographer increasingly became part of the welfare state's bureaucratic élite that was developing at the level of the central government, the province, and the municipality. After the Second World War, this tendency was to pick up momentum.

After 1945: serving the welfare state

The belief in a "makeable" society, and the "makeability" of its spatial organization as well, was widespread in the Netherlands after the war and continued well into the 1970s. For instance, the Central Planning Office for the national economy was established as early as 1945; each year the office prepared a national plan for the reconstruction of the economy of the Netherlands. The ambition of the State Agency for the National Plan was to achieve a more balanced distribution of population throughout the country. The agency was afraid that people and businesses would become too strongly concentrated in the Randstad, which would mean that the "economic motor of the Netherlands would become overheated." Economic geographers were called in to determine the possibilities to spread out manufacturing (nobody had even thought of the service sector yet) over the regions at the perimeter of the Randstad in order to curb migration to the Randstad by people from the adjoining regions who were looking for work.

Johan Winsemius (1910-1964), an economic geographer and a civil servant involved in industrial planning, made a pioneering study of location factors in the Dutch manufacturing sector. He concluded that not even ten percent of the companies were moveable (Winsemius 1949). A large-scale effort to move existing industries out of the Randstad and into the regions on its perimeter was thus not the way forward. The government hoped that by providing subsidies and taking other measures they could make the adjoining regions attractive, particularly to new manufacturing companies. (In Winsemius' work, geography's transition from a colonial field of scholarship to one in the service of the welfare state is nicely visible. He defended his dissertation in 1936 on a study of the opportunities that New Guinea, which was a Dutch colony at the time, offered for European colonists. To his "bitter disappointment," those opportunities were very limited. In his foreword to his dissertation, he wrote, "I devoted myself to this subject in the hopeful expectation that I might be able to discover a New Netherlands, and I found . . . New Guinea: swampy low plains from which steep mountains rise up.")

In the 1950s, '60s, and '70s, the welfare state came into its own. Its grand finale came in the years 1973-1977, when the Netherlands had a socialist prime minister, Joop den Uyl. His goal was to redistribute power, knowledge, and income. On that platform, Den Uyl wanted to spread these resources over groups in different positions on the social ladder. The spatial component of that effort involved a dispersal over the regions. In those days, regional policy was grounded in the same ideals as those formulated by the British geographer David Smith in his book *Human Geogaphy: A Welfare Approach* (1977). And spatial planning policy had its own Bible: "The Third Report on Physical Planning." It was published between 1973 and 1981 in fifteen volumes (nearly 5,000 pages). Its size alone is an indication of the level of detail in the government plan for the future spatial development of the country (Borchert 1986).

The political ideals that shaped the regional policy and the spatial planning of the welfare state were those of social democracy, mixed with Christian convictions. No wonder that human geography, which found more and more support in this welfare state, was also marked by these ideals. Geographers who attained prominent positions at the universities after the Second World War were sympathetic to socialism. An early exponent of that group was Ter Veen. In 1940, he wrote about the "great constructive value" of socialism with respect to determining "the route for the march toward a better society." In judging the societal value of socialism, one should not be blinded by its "destructive slogans" but rather should recognize its "great positive value." That is how Ter Veen saw it, at least.

Ter Veen's sympathy for socialism was shared by Adriaan Cornelis de Vooys (1907-1993). In 1926, he began his study of human geography in Utrecht and in 1949 was appointed professor there. In Utrecht, he was ultimately the successor to Van Vuuren, after an interim period of two years from 1946 to 1948 when Jan Broek was professor. De Vooys, who dominated human geography at Utrecht until 1973, conceived of human geography as the study of the way in which socio-economic groups subsisted in their geographical territory. That territory could vary from the impoverished interior of Spain and Greece, where De Vooys carried out fieldwork in the 1950s with a small group of handpicked students, to the small towns in the Netherlands, where students collected data for their theses.

De Vooys was a fervent defender of what he saw as the field and the boundaries of human geography. From his point of view, danger was lurking all around. First of all, danger was coming from Amsterdam in the form of sociography. The Utrecht geographers had been fighting a bitter war of paradigms with the Amsterdam sociographers since the 1910s, the era of Niermeyer and Steinmetz. The controversy flared up once again in 1961 when Hans Dirk de Vries Reilingh, who had succeeded Ter Veen in Amsterdam in 1950, published a programmatic article on "human geography as one of the humanities." In that article, De Vries Reilingh accused the Utrecht geographers of an overly economic approach; supposedly, they were too much concerned with the means of subsistence and the pursuit of affluence by regional groups, ignoring anything related to culture and politics—in short, anything that was at all social. De Vooys' response was vehement. He was angered by "the expulsion of economic geography, which in fact comes down to an almost complete desertion of the field of geography." For De Vooys, economic geography was and always would be the core of the field. In reply, De Vries Reilingh retorted that he had certainly not excluded economic geography, "but had merely challenged its monopoly on human geography." He ended his reply on a tone of reconciliation: "Let us then agree on one thing: even though our opinions differ . . . we still have to acknowledge, not renounce, one another, as colleagues working in the same field. There is no need to denounce each other!" But De Vooys was not a conciliatory man (Vries Reilingh 1961, Vooys 1961).

Cooperation with the physical geographers with whom De Vooys shared the building in Utrecht was equally unfeasible. Human and physical geographers inhabited different worlds in Utrecht—as elsewhere in the Netherlands—despite occasional attempts of a few individuals, such as Jan Zonneveld, who was professor of physical geography in Utrecht from 1958 to 1983, to break down, at least partially, the high walls between the two fields. Little came of the agreements over joint research and teaching. For instance, in the mid-1950s, De Vooys had reached an agreement with Jacoba Hol, Zonneveld's predecessor, to jointly teach a class on "the geography of Great Britain." This plan ended in failure. De Vooys discussed the industrialization of the Midlands, while Hol treated the glaciations of Mount Snowdon.

Only after the retirement of De Vooys in 1973 did the younger generation of pragmatic human geographers at Utrecht put an end to these skirmishes. A new self-confidence emerged; no longer was the outside world fraught with danger. The man who took the lead in this generation was Marc de Smidt, who became professor of economic geography in 1977. In 1992, he died after a long illness at age 51.

New professions, new geographical institutes

Two new and expanding fields in the Dutch universities threatened human geography around 1960. The first was planology, the second sociology.

Planology, at least its social-science components (in the present discussion I disregard the field of planning as it evolved from architecture and urban design), developed in Dutch universities thanks to the efforts of the human geographers at Utrecht. In the '30s, under the supervision of Van Vuuren, they had been among the first to experiment with regional research and then later learned the fine points of the profession in practice. Among that group were the above-mentioned founder of planology at the University of Amsterdam, Willem Steigenga, and his counterparts Chris van Paassen (at Utrecht University) and Gerrit-Jan van den Berg (at the University of Groningen) (Hoekveld 2001, Berg 1983). Some of the geographers who studied at Utrecht in the '50s were still among the first generation of professors in planology: in the '60s, Gerrit Wissink was appointed at the Catholic University of Nijmegen and Jaap Buit at the Free University in Amsterdam. By the year 2000, all of the above were either retired or deceased.

In contrast, the first generation of sociology professors in the Netherlands were educated at the University of Amsterdam. Formally, they graduated in human geography; in practice, they were sociographers from the school of Steinmetz and Ter Veen. They did not look back fondly on their training in sociography, which they dismissed as descriptive journalism. Sociology, as the successor to sociography (and perhaps also to human geography on the Utrecht model), was to engage in theory-building and modeling—in short, in "real science".

Competition for the established programs in Amsterdam and Utrecht came not only from planology and sociology but also from new human geography programs elsewhere in the Netherlands. The first new program was established in Groningen at the state university there. The founder was Hendrik Jacob Keuning (1904-1988), who took his doctorate under Van Vuuren in 1933 (Jong 1974, Keuning 1977). His dissertation was a highly esteemed regional monograph—in the French tradition—of the fen communities of Groningen. After the Second World War, he published some excellent regional textbooks on the Netherlands—e. g., *Het Nederlandse volk in zijn woongebied* ("The Dutch People in Their Territory," 1947). His appointment in 1949 to the new faculty of economics in Groningen could take place because of the decision dating from 1921 that economics students were required to take a course in economic geography. On the grounds of that decision, the universities in Rotterdam and Amsterdam had long had their economic-geographical institutes, and in 1949, when an economics program was started in Groningen, a chair in economic geography was created. In 1950 Keuning was given permission to establish a curriculum in human geography. The program did not cost much, because Keuning was already on the payroll and his salary as a professor was the main expense at the time. Initially, the number of students was modest; in 1950 one student enrolled in human geography, and five started in 1951, but by the time Keuning retired in 1974, there were roughly 350 students in all. Most of the graduates became geography teachers in secondary schools; up to the beginning of the '70s, the geographical institute in Groningen served primarily as a teacher-training program and as an educational institution. Research and training in "professional geography" were given greater emphasis only during the '70s (Pellenbarg 1999).

The same pattern of modest growth was also characteristic of the other two new institutes: one in the Catholic University in Nijmegen, and the other at the protestant Free University in Amsterdam. The impetus to establish these institutes came from the desire to produce a supply of geography teachers for Catholic and Protestant schools. In 1958, the human geography program got started in Nijmegen under R. H. A. Cools (1908-1987). Like Keuning, he too had studied in Utrecht under Van Vuuren. In 1942, he defended his dissertation on the thought and work of the French geographer Jean Brunhes. (In Utrecht, French geography was greatly admired, at least up to the arrival of spatial analysis at the end of the 1960s.) After receiving his doctorate, Cools went to work as a teacher in The Hague. As professor in Nijmegen, he was mainly engaged in management and teaching. Research got off the ground when younger men were appointed to the institute of geography and planning (e. g., Gerrit Wissink, Jan Buursink, and Jan Kleinpenning). The student population rose from over 100 in 1965 to roughly 500 in 1974 (Kleinpenning 1975). Afterwards, the growth trend did not continue, making the geography and planning program in Nijmegen vulnerable to budget cuts and reorganizations. Today it is part of the Faculty of Policy Studies.

The Free University followed in 1961, initiating the country's fifth program in human geography, and later, in 1970, also began a program in planology. Max Heslinga paved the way for this program. He had received his doctorate in 1962 under De Vooys on a study of *The Irish Border as a Cultural Divide* and was immediately appointed professor in the Free University. In 1967, Heslinga received reinforcement from, among others, Gerard Hoekveld, who was to bring urban geography to fruition there. In terms of the number of students, the Institute of Geography and Planning of the Free University was one of the smallest human geography programs and thereby one of the most vulnerable when the central government was short on money. That time came in 1983, when the minister of education cut the budget for university education by 258 million guilders. The strategy was to reduce the number of places where the programs were offered. The human geographers and planners at the Free University were the victims of that strategy. In September 1983, no new students were enrolled in these fields; the 55 first-year students who had already registered had to choose a different field of study or move to one of the four remaining programs (at the universities of Amsterdam, Utrecht, Groningen, and Nijmegen). Most of the staff members moved to the University of Amsterdam, a few to Utrecht.

A department of physical geography had been established in 1960 at the Free University with a student of Bakker, A. J. Wiggers, (1921-1990) as professor. Although it was smaller than the department of human geography, it was not terminated in 1983 but remains as an independent program, though more along the lines of a geo-science than as a geographical discipline. At present, besides the department at the Free University, there are still physical geography programs at the universities of Amsterdam and Utrecht, the latter, with over 300 students, having by far the biggest physical geography program in the Netherlands. In Groningen and Nijmegen, the field of physical geography never attained the status of a program of study. There, it has remained limited to a required minor course for human geography students.

Five developments

When we look back over the postwar period, roughly spanning the second half of the 20th century, we can spot numerous developments in the practice of geography in the Netherlands. Those developments are summed up under the following five points.

First of all, geography was drawn into a new network of symbiotic relations. No longer was the field in the service of colonial interests. With the rise of the welfare state, geographers devoted increasing attention to their own country. From then on, human geographers were supposed to serve the interests of spatial planning, regional policy, urban renewal, and so forth. With the rising concern about the conservation and restoration of the environment and landscape, the physical geographers also became involved with the welfare state.

Incidentally, when geographers tied their fortunes to the fate of the welfare state, they also became vulnerable. Indeed, since the 1980s, when the welfare state was forced to retreat, it has been an uphill struggle for geography. But the market-oriented mentality also created new opportunities: the private sector absorbed more and more geographers. For instance, they found jobs in private research and consulting firms specializing in spatial and environmental planning. Figures derived from labor-market surveys show that of every ten geographers who graduated from Utrecht prior to 1980, one is working in the commercial sector, compared to four out of ten of those who graduated after 1980. The picture is no different for the other geography programs.

Secondly, the links with school geography in secondary education have worn thin. Until well into the '70s, many professors had worked for some years in secondary education. There were some striking exceptions—e. g., Keuning and De Vooys, who had both been employed at the Central Bureau of Statistics before their appointments as professors in 1948 and 1949, respectively. Niermeyer and Ter Veen had taught school before the war, and Cools and Hoekveld had done the same after the war. Others, like Rob Tamsma, professor in Groningen, were closely involved in the development of new geography textbooks for secondary education. Today, there are no longer any professors who had ever been geography teachers in secondary schools.

The fact that the number of geography graduates who take teaching jobs has been steadily declining is related to this development. Of the geographers who graduated from Utrecht before 1970, about 40 percent ended up in teaching jobs, compared to just five percent of those who graduated after 1980. The percentages for the other geography programs are similar. To some extent, this development may be ascribed to the establishment of non-university teacher-training colleges in 1971. These new institutions replaced the old teacher-certification programs for secondary education, mentioned earlier. Whoever wants to become a geography teacher nowadays is more likely to study in one of the nine places in the Netherlands where they can train for a job as a geography teacher than to enroll at a university to take a degree in geography.

Third, partly because of the enlargement of scale and democratization, the individual character of the local geography programs is fading, or, rather, we might say that each program has taken on more characteristics. More programs, staff members, and students have appeared; the small and easily accessible geography institutes of old have turned into professional organizations. Democratization has dismantled the old hierarchies, so that it is no longer easy to silence a dissident voice. An individual professor has less say than previously about what is going to happen at his institute with respect to research and teaching.

Objectively, the differences among the programs have faded, yet they have not disappeared entirely. At the University of Amsterdam, for example, there is still much more room for political and cultural geography—studies on nationalism and ethnic groups (Herman van der Wusten, who retired in 2001) and on migrants and minorities (Hans van Amersfoort)—than there is in Utrecht, where geography is more economically oriented. In terms of efficiency, this might look like evidence of complementarity. But in terms of their emotional impact, the differences might stand out instead. For instance, this is true in the experience of Willem Heinemeyer (1922-1999), who had been on the staff of De Vries Reilingh for several years before he himself was appointed professor at the University of Amsterdam in 1968. In 1994, Heinemeyer, by then already retired for seven years, was interviewed by Rogier Noyon. With regard to the relations between Amsterdam and Utrecht, Heinemeyer, "the last sociographer", made the following comment:

> My students don't call themselves sociographers anymore. They are social geographers and on top of that preferably have some specialization. . . . But I do hope that something will remain of the Amsterdam tradition, because the difference from Utrecht geography is still clearly evident. I don't consider human geographers from Utrecht to be my intimate colleagues. They are still just as one-sided as before; the sociological side of the issues remains underexposed. And what's worse: the way they often use spatial metaphors is dubious, while the factor of space is itself nothing (Noyon 1994).

Fourth, Dutch geography has come to resemble the Anglo-American model more and more. Physical geography in the Netherlands had been traditionally modeled on the German disciplines of physical geography and geomorphology, which at the beginning of the 20th century were the trendsetters worldwide, as was the case with so many other fields of science in Germany. For human geography in the Netherlands, a distinction must be made between Utrecht, Groningen, and Nijmegen, on the one hand, and Amsterdam on the other. The former group was strongly inspired by the French *géographie humaine*, even though they diverged from the French model by emphasizing the economic base and by seeking ways to apply knowledge. Amsterdam's sociography was a striking example of a local school.

Since the end of the '60s, English-language geography has made its breakthrough in the Netherlands. The programs replaced the French textbooks with British and American books. Spatial analysis and behavioral geography displaced the regional and ecological approaches. For research, publishing in English-language professional journals became the norm. German and French geographical studies were read sporadically if at all. Language barriers were also created; many students and researchers now have hardly any command of French or German.

The fifth and last point is the fact that over the past ten years, a strong tendency has arisen toward interdisciplinary and multidisciplinary collaboration. The '50s and '60s, when De Vooys in Utrecht steadfastly defended the boundaries of human geography and De Vries Reilingh in Amsterdam protected sociography from attacks by sociologists, are over. The '90s saw the ascendancy of large research schools in which several universities and disciplines participate. For example, urban geographers work together with researchers from other fields in the Netherlands Graduate School of Housing and Urban Research (NETHUR); geographers of developing countries collaborate in the Center for Resource Studies (CERES); and physical geographers work in the Netherlands Center for Geo-ecological Research (ICG) and in the Graduate School for Atmospheric and Marine Research (SAMO).

The advantages of cooperation across the boundaries of the disciplines are recognized. At the same time, such collaboration leads some to fear for the future of geography. Isn't there a danger that the field will dissolve in a substrate of new disciplines? Rural geographers, for instance, have started to work together more and more with other rural specialists. In this way, the contours of the new field of rural studies have been drawn. And just as for rural studies, the same process is underway for developmental studies, urban studies, gender studies, etc. What are the implications of this trend for a classic field? Does a "fragmented discipline" still have any legitimacy? Doesn't this lead to a "discipline without a heart"? Both the British geographer Ron Johnston and the Dutch geographers Gerard and Gerda Hoekveld are looking for a binding agent, a new heart, in places, in regions (Johnston 1991, Hoekveld 1995). Whether that remedy will work only a new century will tell.

Geography and society

In the course of the past few decades, historians of the profession have demonstrated that the practice of geography in many countries has reflected the dominant ideologies of the "ruling class". Is that true of geography in the Netherlands too? The answer to this question is a conditional (but not a straightforward) "yes".

Geography as a colonial science was practiced from a Eurocentric perspective and on behalf of the imperialistic interests of the administrative and commercial élite. As the colonial era was drawing to a close, the welfare state was on the rise, bringing social and spatial planning along with it. Geographers found employment as experts in that field in the service of a paternalistic government, which only wanted the best for the people but did not grant its subjects an effective voice in their affairs. Social and spatial planning, as it developed from the 1920s until well into the '70s, had an authoritarian side but was also grounded in a sincere desire to tackle the problems of society, such as poverty and unemployment. More democratic forms of spatial planning could only emerge once the government had shed light on the limits to the "makeable" society.

Many geographers, inspired by socialist (as well as Christian and humanist) values, have played their rôle in the policy of the welfare state. In addition, they have worked diligently to improve the quality of geographic education for children as well as for adults (as part of the movement toward "betterment of the people"). Among these efforts, the contribution that geographers have made towards an emancipated society, consisting of informed citizens, comes most clearly to the fore.

Acknowledgements. I would like to thank Nancy Smyth van Weesep for her translation of the text and M. W. Heslinga for his thoughtful comments on an earlier draft. Inspiration for the title came from Ron Johnston's book *Geography and geographers: Anglo-American human geography since 1945* (1979).

Bibliography

Beekman, A. A. (1912) Droogmaking der Zuiderzee, *Pro en Contra betreffende vraagstukken van algemeen belang*, Serie VII, no. 7, Hollandia Drukkerij, Baarn.

Bell, M., R. Butlin, and M. Heffernan, eds. (1995) *Geography and imperialism 1820-1940*, Manchester University Press, Manchester and New York.

Berg, G. J. van den (1983) Some personal reflections on geography, in A. Buttimer (ed.), *The practice of geography*, Longman, London and New York, 209-224.

Biervliet, H. et al. (1978) Biologisme, racisme en eugenetiek in de antropologie en de sociologie van de jaren dertig, F. Bovenkerk et al. (eds.), in *Toen en thans: De sociale wetenschappen in de jaren dertig en nu*, Ambo, Baarn, 208-235.

Borchert, J. G. (1986) Urban systems in transition and government actions, in J. G. Borchert et al. (eds.), *Urban systems in transitions*, KNAG/Utrecht University, Amsterdam, 222-230.

Bosma, K. (1993) *Ruimte voor een nieuwe tijd: vormgeving van de Nederlandse regio 1900-1945*, NAi, Rotterdam.
Brink, P. P. W. J. van den and P. Moree (1996) *Met kapmes en kompas: Vier eeuwen Nederlandse ontdekkingsreizen en reisverslagen*, Koninklijke Bibliotheek, Den Haag.
Casseres, J. M. de (1933) *Grondslagen der planologie*, Oosthoek, Utrecht.
Cornelissen, C. H., A. L. van Hasselt and Joh. F. Snelleman (1882) *Reizen in Sumatra, 1877-1879, door de leden der Sumatra-Expeditie, uitgerust door het Aardrijkskundig Genootschap* (Part 2), E. J. Brill, Leiden.
Dutt, A. K. and F. J. Costa, eds. (1985) *Public planning in the Netherlands: Perspectives and change since the Second World War*, Oxford University Press, Oxford.
Godlewska, A. and N. Smith, eds. (1995) *Geography and empire*, Blackwell, Oxford.
Harvey, D. W. (1984) On the history and present condition of geography: an historical materialist manifesto, *The Professional Geographer* 36, 1-11.
Hasselt, A. L. van (1882) *Volksbeschrijving van Midden-Sumatra*, E. J. Brill, Leiden.
Hasselt, A. L. van and Joh. F. Snelleman (1881) *Reizen in Sumatra . . .* (Part 1), E. J. Brill, Leiden.
Heek, F. van (1978) Omzien naar het sociografisch intermezzo (1925-1953), in F. Bovenkerk et al. (eds.), *Toen en thans . . .*, Ambo, Baarn, 15-26.
Heinemeyer, W. (1996) Contribution of applied geography to the colonisation policy of the IJsselmeerpolders, in V. Berdoulay and J. A. van Ginkel (eds.), *Geography and professional practice*, KNAG/Universiteit Utrecht, Utrecht, 205-208.
Heslinga, M. W. (1977) Anton Albert Beekman en de vaderlandse aardrijkskunde in het laatst van de negentiende eeuw, *Geografisch Tijdschrift*, new series, 11, 324-338.
Heslinga, M. W. (1978a) *Geography and nationality*, Queen's University, Belfast.
Heslinga, M. W. (1978b) Sociografie versus sociale geografie, in F. Bovenkerk et al. (eds.)*Toen en thans . . .*, Ambo, Baarn, 53-67.
Heslinga, M. W. (1983) Between German and French geography: In search of the origins of the Utrecht school, *Tijdschrift voor Economische en Sociale Geogafie* 74, 317-334.
Heslinga, M. W. (1996) Colonial geography in university education in the Netherlands, in V. Berdoulay and J. A. van Ginkel (eds.), *Geography and professional practice*, KNAG/Universiteit Utrecht, Utrecht, 173-193.
Heslinga, M. W. (1999) Grondleggers, in B. de Pater, *Een tempel der kaarten*, Universiteit Utrecht, 34-73.
Hoekveld. G. A. and G. Hoekveld-Meijer (1995) Het beeld van de geografie een Rotterdam à la Zadkine?, *Geografie* 4, no. 6, 31-35.
Hooson, D., ed. (1994) *Geography and national identity*, Blackwell, Oxford.
Hudson, B. (1977) The new geography and the new imperialism, 1870-1918, *Antipode* 9, no. 2, 12-19.
Johnston, R. J. (1991) *A question of place: Exploring the practice of human geography*, Blackwell, Oxford.
Jong, W. W. de (1974) Seventy Years H.J.K. and Bibliography H. J. Keuning, 1933-1973, *Tijdschrift voor Economische en Sociale Geografie* 65, 65-70.
Kan, C. M. (1876) *De Nederlandsche expeditie naar de Boven-Djambi en Korintji-Vallei*, J. L. Beijers, Utrecht.
Keuning, H. J. (1977) Aims and scope of modern human geography: A personal contribution to geographical theory, *Tijdschrift voor*

Economische en Sociale Geografie 68, 262-274.
Kleinpenning, J. M. G. (1975) Het Geografisch en Planologisch Instituut: Enkele hoofdlijnen uit de geschiedenis, in J. M. G. Kleinpenning et al. (eds.), *Na vijftien jaar Nijmeegse geografie: Opstellen over geografisch onderwijs en onderzoek*, GPI/KUN, Nijmegen, 17-25.
Livingstone, D. N. (1992) *The geographical tradition*, Blackwell, Oxford.
Livingstone, D. N. (2000) Putting geography in its place, *Australian Geographical Studies* 38, 1-9.
Mok, I. (1999) *In de ban van het ras: Aardrijkskunde tussen wetenschap en samenleving 1876-1992*, ASCA Press, Amsterdam.
Noyon, R. (1994) De laatste sociograaf, *Facta* 2, no. 2, 20-23.
Ormeling, F. J. (1998) KNAG stond vooraan bij openlegging van Tropisch Nederland, *Geografie* 7, no. 5, 22-28.
Pater, B. de (1999) *Een tempel der kaarten: Negentig jaar geografiebeoefening aan de Universiteit Utrecht*, Faculteit Ruimtelijke Wetenschappen, Utrecht.
Pellenbarg, P. H. (1999) Vijftig jaar geografie in Groningen, *Geografie* 8, no. 1, 43-44.
Schrader, R. (1974) Honderd jaar Koninklijk Nederlands Aardrijkskundig Genootschap 1873-1973, *Geografisch Tijdschrift*, new series, 8, special issue.
Stoddart, D. R. (1986) *On geography and its history*, Blackwell, Oxford.
Taylor, P. J. (1996) On the nation-state, the global, and social science, *Environment and Planning A* 28, 1917-1928.
Veen, H. N. ter (1925) *De Haarlemmermeer als kolonisatiegebied*, P. Noordhoff, Groningen.
Veen, H. N. ter (1950) *Van aardrijkskunde tot sociale wetenschap: Keuze uit de verspreide geschriften van prof. dr. H. N. ter Veen*, H. J. Paris, Amsterdam.
Velde, P. van der (1995) The Royal Dutch Geographical Society and the Dutch East Indies, 1873-1914: From colonial lobby to colonial hobby, in M. Bell et al. (eds.), *Geography and imperialism 1820-1940*, Manchester University Press, Manchester and New York, 80-92.
Veth, P. J. and C. M. Kan (1877) Adres aan den Gemeenteraad van Amsterdam over een leerstoel voor aardrijkskunde aan de Universiteit in die gemeente, *Tijdschrift van het Aardrijkskundig Genootschap*, part 2, no. 5, 368-369.
Vooys, A. C. de (1961) De sociografie in nieuwe gedaante (with an afterword by H. D. de Vries Reilingh), *Tijdschrift Koninklijk Nederlands Aardrijkskundig Genootschap* 78, 272-276.
Vries, J. de (1963) *Ontsloten poorten. Vijftig jaren volksuniversiteit in Nederland 1913-1963*, Van Gorcum, Assen.
Vries Reilingh, H. D. de (1961) De sociale aardrijkskunde als geesteswetenschap, *Tijdschrift Koninklijk Nederlands Aardrijkskundig Genootschap* 78, 112-122.
Vuuren, L. van (1935) De geschiktheid van den sociaal-geograaf voor beroepen in het practische leven, *Tijdschrift voor het Onderwijs in de Aardrijkskunde* 13, 177-202.
Wal, C. van der (1997) *In praise of common sense: Planning the ordinary: A physical planning history of the new towns in the IJsselmeerpolders*, 010 Publishers, Rotterdam.
Winsemius, J. (1949), *Vestigingstendenzen van de Nederlandse nijverheid*, 2 vols., Rijksdienst voor het Nationale Plan/Staatsdrukkerij, Den Haag.

Further reading. There are several English-language publications that survey recent developments in the practice of geography in the Netherlands. These overviews are not cited in the text, but they would be of interest to those seeking additional information. A brief selection of these works follows:

Bosman, J. and B. de Pater (1998) The international position of Dutch human geography: a quantitative exploration, *GeoJournal* 45, 85-96.

Dieleman, F. M. (1999) Research progress Dutch style, *Environment and Planning A* 31, 191-195.

Dieleman, F. M. and D. op't Veld (1981) Quantitative methods in Dutch geography and urban and regional planning in the seventies, in R. J. Bennett (ed.), *European progress in spatial analysis*, Pion, London, 146-168.

Dietvorst, A. G. J. and F. J. P. M. Kwaad, eds. (1988) *Geographical research in the Netherlands 1978-1987*, KNAG/IGU Netherlands, Amsterdam.

Dietz, T. and F. Kwaad (2000) *Dutch geography 1996-2000*, International Geographical Union, The Netherlands, Utrecht.

Ginkel, J. A. van (1996) Professional practice and the development of modern geography in the Netherlands, in V. Berdoulay and J. A. van Ginkel (eds.), *Geography and professional practice*, Utrecht, 31-45.

Ginkel, J. A. van and M. de Smidt, eds. (1983) 75 years of human geography at Utrecht, *Tijdschrift voor Economische en Sociale Geografie* 74, no. 5 (special issue).

Hoekveld, G. A. (2001) Christiaan van Paassen 1917-1996, in P. H. Armstrong and G. J. Martin (eds.), *Geographers: Biobibliographical Studies* 21, Continuum, London and New York, forthcoming.

Koster, E. A. and F.J. P. M. Kwaad (1993) Geomorphology in the Netherlands, in H. J. Walker and W. E. Grabau (eds.),*The evolution of geomorphology*, John Wiley, London, 299-309.

Paassen, C. van (1983) Highlights of the decades (1900-80) in nine countries (The Netherlands), in A. Buttimer (ed.),*The practice of geography*, Longman, London and New York, 261-274

Paassen, C. van (1984) Human geography in the Netherlands, in R. Johnston and P. Claval (eds.), *Geography since the Second World War: An international survey*, Croom Helm, London, 214-234.

Pater, B. de and M. de Smidt (1989) Dutch human geography, *Progress in Human Geography* 13, 348-373.

Smidt, M. de and E. Wever, eds. (1984) *A profile of Dutch economic geography*, Van Gorcum, Assen.

Weesep, J. van (1987) Academic urban geography in the Netherlands, *Urban Geography* 8, 65-76.

Weesep, J. van, J. H. J. Terwindt and P. G. E. F. Augustinus (1996) *The state of Dutch geography: A snapshot of work in progress*, IGU Nederland, Utrecht.

Zonneveld, J. I. S. (1979) Physical geography in the Netherlands, *Erdkunde* 33, 1-10.

Ben de Pater
Faculteit Ruimtelijke Wetenschappen
Universiteit Utrecht
Postbus 80115
3508 TC Utrecht
Nederland (The Netherlands)

7 Stories on the Making of Geography in Sweden

Anne Buttimer

Introduction

"Geography we learn on the march", King Karl XII remarked in the late 17th century, "history we make ourselves". The foundations of geography as a scholarly field in Sweden can clearly be traced to the expansionary ethos of the 16th and 17th centuries, particularly to the reign of Gustav II Adolph (1611-1632) and the daring deeds of his Chancellor, Axel Oxenstierna. Enlightenment themes have also left their mark on 20th-century practices and on the alternating tides of exploration abroad and inventory and planning at home through the 19th and 20th centuries. As a university discipline, geography in Sweden dates from the early 20th century, and its preferred styles of practice, substantive foci and scale of endeavour still shape and are shaped by the changing priorities of Swedish society.

Sweden's "Golden Age" (Vere Aureum Saeculum) (1520-1718)

Moments of national liberation or expansion have usually been creative ones for the practice of geography. In Sweden such a moment occurred in the early 16th century after Gustav Vasa had banished the Danes (1520), and it burgeoned through the reigns of Erik XIV (1560-1568) and Gustav II Adolf, who cherished magnificent dreams of empire. "Defender of the faith", Gustav II Adolf (Gustavus Adolphus) became Conqueror of the Baltic, opened windows on the North Sea and Atlantic, and also secured a firm hand on domestic affairs. The Swedification of conquered and colonized territory would be conducted "softly", e. g., by the founding (or takeover) of universities on the frontiers: Tartu (1631), Turku (Swedish Åbo) (1640), Greifswald (1648) and Lund (1668). Swedification of American Indians on the Delaware River (Clay 1914) and of Lapps and Finns on the northern fringe (Ruong 1975) was

G.A. Dunbar (ed.), Geography: Discipline, Profession and Subject since 1870, 191–223.

facilitated by Lutheran catechism, pastors, and bibles. The economic base of the empire was built on an astute coupling of Sweden's mineral resources with the burgeoning industrial markets of Europe. Walloon expertise in forge-and-blacksmith art was warmly welcomed (Arnstberg and Ehn 1976, Buttimer 1985). Traders and trading companies of Scottish or German origin were invited to live in Sweden, albeit in selected towns.

The largest single export item during the 16th century was tinned sheet metal (Mead 1981). In 1647 the Kungliga Bergskollegium (Royal Mines Authority) was set up to regulate details on mining and processing standards. Central and northern forests were cleared for agriculture, and Finns were allowed to practice the art of svedjebruk (slash-and-burn cultivation) they knew so well (Nilsson 1950). This demanded a fencing-in of Lapp territory, making sure, however, that their protein-rich products would remain available to sustain the army.

"An army marches on its stomach", the cliché ascribed to a later General, was well understood by this nation on the march. Via the Reduktionsverk, the legal body charged with reclaiming alienated Crown lands, moral and legal pressure was used to divest noblemen of wealth acquired in the form of land. In 1682 the farmers of Sweden received an offer-not-to-be-refused in the Indelningsverk (a system whereby each farmer could support one soldier) so that the burden of supporting a national army could be equitably distributed. Taxation of land itself was considered, and Rudbeckius (1643) surveyed all occupied properties with such an idea in view (Friberg 1954). From 1628 date the exhaustive Geometriska Jordeböcker, perhaps the earliest public registry of lands in all of Europe (Hannerberg 1971). As for demographic data-collection and account keeping on literacy, virtue or vice, household possessions or legacy throughout the realm, the State Church provided ideal service.

Liberation from foreign tyranny—ecclesiastical and imperial—was followed by a creative moment for science, its research horizons no doubt influenced by the late 17th-century imperial agenda. There was Christopher Polhem (1661-1735), father of mechanics, inventor of water pumps, automatic clocks and musical instruments, all of which won the favour of King Karl XII. With his student Emanuel Swedenborg (1688-1772), who regarded him as the "Archimedes of the North", Polhem founded the Collegium Curiosorum in 1710 and in the following year launched *Daedalus Hyperboreus*, Sweden's first scientific journal (Lindqvist 1984). Swedenborg made pioneering contributions to the measurement of degrees and distances on the globe, visited Greenwich, and shared his master's admiration for Isaac Newton (Swedenborg 1758).

The Golden Era also witnessed considerable self-confidence and even self-admiration. Olaus Rudbeck advanced the theory that classical art had originated in the north and that Plato's *Atlantis* was really a symbol of Sweden (Rudbeck 1679-1702). He drafted architectural plans for Uppsala in "hyperboreal" (a blending of Nordic, Greek and Babylonian) style, evidence of which may still be seen on the Castle terraces as well as in the Botanical Garden in Uppsala (See Eriksson 1969).

Geographical enquiry during this early period thus probed in two directions: toward the history, culture and topography of the home realm on the one hand, and toward astronomy and the challenge of global dynamics on the other (Richter 1959). Following the remarkable, albeit chauvinistically-inspired works of Olaus and Johannes Magnus (1539, 1555), Anders Bure wrote a patriotic historical geography of Sweden in 1626. Dahlberg's *Suecia antiqua et hodierna* (1661) praised that sensitivity to nature that he regarded as characteristic of Swedish people (Dahlberg 1898-1900). Already in 1630 Gustavus Adolphus inaugurated the Riksantikvarieämbetet (Royal Antiquarian Council) with explicit instructions for geographical reconnaissance of the land's resources, topography, and peoples. First fruits of this effort came in the *Epitome Descriptionis Suecia* by Mikael Wexonius, professor at Åbo and later entitled Lord Gyldenstolpe. Like the writings of Dahlberg and Rudbeck, the *Epitome* is permeated with a sense of cultural pride and patriotism. Sweden lies at the center, if not at the origin, of world civilization (Richter 1959, 94-117). Scholarly contacts abroad were cultivated too. Gabriel de la Gardie became Professor of Antiquity at Uppsala and invited Shefferus, a compatriot from Strasbourg, to study Uppsala (1666) and Lapp ethnography (1673). René Descartes came to instruct Queen Christina, and indeed many Enlightenment ideas were entertained in Stockholm (Lindroth 1978). Geopolitical concerns on all frontiers, too, stimulated field research especially on the eastern front, reports on Russia such as that of Petrejus (1615) continuing right up to the war of 1805.

As a university subject geography was a required course in the curriculum from 1655 on (Richter 1959). As in Harvard, Oxford, Leipzig and Paris at the time, geography was regarded as a science based on astronomy and mathematics, and the study of globes, Newtonian physics, map projections and navigation held great pedagogical apppeal (Warntz 1964). Unlike the North American and French cases, however, where intellectual energies were directed to speculations about analogies between natural law and the legal foundations for politics, Swedish science would bear its richest fruits in mechanics and engineering, ultimately applicable to industry, weaponry and geographical exploration (Söderberg 1960). The root metaphor of *mechanism* (Buttimer 1982, 1993) took hold of scholarly imaginations in all these settings and this opened wide horizons for research and experimentation in astronomy, medicine, hydraulics, chemistry and physics: fields within which Sweden would later achieve international fame.

Initial steps toward account-keeping, survey, and research on the people and landscapes of the homeland invited another kind of metaphorical imagination, viz., that of *mosaic*, which is best suited for meticulous data-gathering at local and regional levels. Nor was there any apparent conflict about espousing these two distinct world views. The success of Vasa's own military exploits was in large part due to the superiority of the Swedish account-keeping and supply inventories (ordning och reda—order and preparedness). From this period on in Swedish history, one could trace a deep-lying respect for "top-down" managerial decision-making, an image of the unimpeachable civil servant, and a consensus about public order. A Lutheran work ethic not only in the conduct of research but also in folkways generally may go a long way in accounting for the differences that survive between the practice of geography in Sweden and in other lands. Lutheran clergy down the centuries encouraged not only scholarship in the humanities and sciences but also supported the famous topographical surveys of the 18th century. They continued to serve the State with bookkeeping on vital statistics as well as military conscription throughout the imperial era. But they also discouraged humanism in its Renaissance spirit, viz., the kind of self-critical and reflective thinking that might question the status quo. Civil obedience and hard work were deemed virtuous for the higher cause of nationhood and humanity even if this required a suppression of pluralism and a discouragement of individual intiative and intellectual speculation.

Daedalus of the North (1718-1809)

After the defeat of Karl XII in 1718, Swedes had to tighten their belts, as it were, forego foreign luxuries, and develop domestic resources. Ninety percent of the Swedish population (1.8 million by mid-century) was still rural, and living conditions were poor (Guteland et al. 1974). War losses had devastated the economy;

every effort was needed to develop resources. The 18th century would be known as the Age of Freedom. The four estates—nobility, clergy, burghers and landed peasants—would assume power, and party politics flourished. Two leading parties, known as the Hats and the Caps ("Hawks" and "Doves", respectively), assumed radically different policies on economic and social life. The temporarily victorious Hats favoured mercantilism, the promotion of manufacturing and export-oriented industrial development, and were quite supportive of science and its potential applications. The founding of the Swedish Academy of Sciences in 1739 marked a major societal commitment to promoting research for national development. Among its founding members were Carl von Linné (Linnaeus), Anders Celsius and Christopher Polhem. Emphasis lay on developing the fatherland, patriotism being expressed through the study of natural elements, geography, and resources (Richter 1959, 139-140).

The predominant interests of this post-imperial period moved from geo-politics to geo-science, from potential challenges at the imperial boundary to a wholehearted commitment to applied science and surveys of resources for the economic development of the realm. At least three major lines of enquiry were fostered: local field surveys of resources and life, physical science, and natural history. The educational as well as pragmatic value of topographical suveys was recognized by both Academy and university: candidacy for fields such as law, medicine, economics and history all required the completion of such theses before continuing toward professional qualifications. Between 1700 and 1809 over two hundred such theses were completed (in Latin), all including an explicit concern for economic development. With support from local officials, clergy, writers and businessmen, this kind of survey became virtually a folk movement, precursor of the hembygdsforskning that would again recapture attention at the end of the 19th century. Natural science eventually freed itself from the exigencies of political arithmetic and the use of Latin, undertaking an experimental and theoretical course that yielded a rich harvest in the late 18th and throughout the 19th century (Söderberg 1960). For science this was indeed a glorious era, and departments of natural history were founded in several universities.

Central to these developments and perhaps ultimately most influential on Swedish scientific imaginations was Carl Linnaeus (1707-1778). With his disciple Per Kalm (1716-1799) he travelled all over Scandinavia, with some foreign travels as well, collecting innumerable samples from the earth's variety of botanical life. The

Linnaean spirit, besides evoking interest in nature, also shaped curiosities about classification, categorization and systematic (species-based) conceptions of natural elements. Its central root metaphor was that of the world as a *mosaic* of forms (Buttimer 1993), an approach that fostered a penchant for building systems based on elements shorn from their geographic surroundings, elements that could be neatly arranged in archives of museums and school-children's albums of dried flowers (Hägerstrand 1983).

The Academy itself laid down a durable infrastructure for continued monitoring of the land's resources and people. Detailed guidelines were set up for the lantmäteri (land survey) in 1741 and Statistika Tabellverket (statistical record of population) in 1749. Per Wilhelm Wargentin (1717-1783), secretary of the Academy for thirty years, astronomer and pioneer of actuarial calculations, could well vie with Linnaeus as the personification of the remarkable breadth of imagination and competence that characterized 18th-century scholarship in Sweden. Together with the Linnaean survey, Wargentin's systematic approach to national accounting had an enduring impact on the practice of geography. Agrarian reforms were also implemented during the 18th century, e. g., the Storskifte, proposed by Jacob Faggot (1699-1777) in 1757, were epitomized in Skåne with the Enskifte Reform (1803), whereby whole villages were dismembered and single farms spread out across a rationalized landscape. Prospects for a canal link between the North Sea and the Baltic, initially proposed in the 1400s, became much brighter after Polhem's solution to the locks at Trollhätte in 1793, and the Göta Canal was finally opened in 1830.

The 18th century was thus a foundational one for geography in Sweden, its research horizons rapidly expanding with the growth of natural science and the national drive for economic development. Courses were offered in all universities of the former Swedish empire within the framework of natural history, biology, or political economy. Research, sponsored mainly by the Academy, probed some unsolved puzzles from the Newtonian period, such as the shape of the earth (Maupertius 1738). Evidence of lowering levels of Baltic shorelines, coupled with the traces of marine life discovered by Linnaeus, Per Kalm, and even Celsius, suggested that Sweden's entire landscape had been formed under water. How this could be squared with biblical accounts was a conundrum, of course; even more seriously, how would it square with the Rudbeckian theory that classical civilization actually originated in the north?

Besides the Academy, geography found strong allies in the Bergskollegiet (Royal Council of Mines). Already in 1742 Daniel Tilas urged further geological mapping of Sweden, a challenge taken up by Wilhelm Hisinger (1766-1852) in 1790 and by Samuel Gustaf Hermelin (1744-1820) in 1804. In 1808 Hisinger proposed a regionalization of the country into "topographical provinces". In geology at the time, there were challenges not easily resolvable on logical or empirical grounds. Besides the pragmatic goals surrounding mineral resources, there were rival theories about the origins of the earth where science and theology were apparently in conflict: feuds abounded between "creationists" and "scientists" about the nature of the earth.

Geography was mainly a teaching subject during the 18th century. In 1738, Cosmografiska Sällskapet (Society for the Study of Cosmography) was founded in Uppsala, and high priority was given to the production of text materials. Torbern Bergman's (1735-1784) volume on physical geography was ready in 1766; Fredric Mallet's (1728-1797) text on mathematical geography and Stephan Insulin's (1726-1803) *Om folkslagens seder och lynne* ("On the Origins and Character of the Folk") followed in 1772. Anders Åkerman began constructing globes in 1759 and produced the famous *Atlas juvenilis* in 1769: a work that elicited remarkable efforts in cartography. Bergman's work was translated into several languages, serving not only as a basic text for Immanuel Kant's lectures at Königsberg, but also as a guide for Forster's journeys later (Bergman 1766). At Lund in the 1770s and '80s, several private colleges were set up for the study of geography, mathematics, and physics. The Fysiografiska Sällskapet (Society for the Study of Physiography), founded in 1772, adopted the agenda suggested earlier by Linnaeus himself in 1743: to study Sweden's coastal phenomena, and to examine the effects of erosional and depositional processes on the landscape. Natural history claimed keen interest, and William Derham's (1657-1735) *Physico-Theology* (Oxford 1713) was translated into Swedish in 1736 and read in physics lecture-halls at Lund.

The main thrust of human geography in the 18th century was toward issues of national economy (hushållning, lit. housekeeping). As a university subject, it covered domains later to be called economics, statistics, history or political science. Adequate text material, as distinct from statistical data, however, was lacking. Mikael Wexonius' *Epitome* (1650) was somewhat outdated. Insulin's book (1772) was basically a compendium of facts. The first major text on the human geography of Sweden, *Geographisches Beschreibung des Königsreichs Schweden*, written by the German Johannes Hübner in 1731, provoked a veritable storm among Swedish readers, who pointed out numerous inaccuracies in the text.

In 1741 Erik Tuneld (1709-1788, "Tunelius" after 1738) produced *Inledning til geographien öfver Sverige* ("Introduction to the Geography of Sweden"), with a preface by Linnaeus and Celsius (Richter 1959, 226-227). The German geographer Anton Friedrich Büsching later used Tunelius' text in his *Neue Beschreibung* (1754+), which was regarded as the model for geography texts throughout the remainder of the 18th century.

Sweden's mercantilist interests in the second half of the 18th century gave an important boost for human geography. The East India Company, founded in 1731, welcomed students—especially the disciples of Linnaeus—on board their many voyages. To each of fifteen disciples Linnaeus had assigned a special region of the world where those team research projects that had been so successful at home could be extended globally (Solander 1960, Gullander 1980). Several died while still trying to fulfill their master's commands.

The prospect of trade colonies (Delaware 1638, Cabo Corso 1649, Saint-Barthelemy 1754) had not vanished entirely from the Swedish imagination. Senegal, a free colony on Africa's west coast, was proposed, and several Swedes participated in the founding of Sierra Leone (1787), where measures were taken for the autonomy of native populations. As with other seafaring nations, Sweden was concerned about protection against piracy, especially in the Mediterranean. In many of these reports from foreign voyages, there was a decidedly moral tone, advocacy of civil justice, rationality and order, as well as an insistence on correct factual detail.

Toward the close of the 18th century, the mercantilist fervour became tempered somewhat. Under the regime of Gustav III (1771-1792), a wave of Francophilia swept through the literary, artistic and cultural echelons of society. Theatre, opera and ballet were in vogue; the salon replaced the laboratory (Frängsmyr 1989). The National Opera was opened in 1780 and the Swedish Academy in 1786. The glories of "pure reason" were sung, albeit in milder tones than those heard in other European lands. Johan Henrik Kellgren (1751-1795) satirized dogmatism and hypocrisy, and Nils Rosén von Rosenstein (1752-1824) ventured theories about civil law and freedom of thought; yet Sweden was dubious about the scientific humanism of the Encyclopédistes (Eriksson and Frängsmyr 1971). Tensions often arose between the guardians of cultural orthodoxy and the trailblazers in science and literature. Some support for the French Revolution was expressed among students at Uppsala, but the king planned an invasion of France to support his fellow royalty.

The assassination of Gustav III in 1792 closed the lid on frolic and Francophilia. Antipathies toward Napoleon and renewed threats from Denmark on the south raised Sweden's concern about its friends to the East. When Russian troops attacked Finland in 1808, however, they found a debilitated Swedish force. Finland was ceded to Rusia, and Sweden retreated to the geo-political limits that surround it today.

Pro Patria (1809-1860)

The loss of Finland in 1809 called forth a concerted effort to build a deeper sense of national self-confidence and patriotism in Sweden. In the same year a new Constitution, similar to the American one and due to last until 1975, guaranteed property rights and civil liberties. Political leadership was sought abroad, and Jean Baptiste Bernadotte became King Karl XIV Johan, first of the present royal line of Sweden. Wih Russian support, Karl Johan campaigned against Napoleon in 1813-1814 and thereafter conducted conservative policies at home.

Intellectuals rallied to the challenge of restoring national morale. In 1811 Götiska Förbundet (Gothic Union) set out to promote historical and geographical research along patriotic lines. Romanticism captured imaginations with an intense veneration of nature and pastoral scenes, e.g., in the lyrics of Carl Michael Bellman (1740-1795), and with laudatio for Sweden's great achievements in natural science. Literary and philosophical circles sprang up, and now their discussions were conducted in the vernacular.

As for the Academy of Sciences, the first fifty years (1786-1836) had indeed been crowned with success. By the end of the 18th century, however, many of its founders had died. Politically the "Caps" (liberals) could vie with the "Hats" (conservatives) as those grand promises about the potential economic benefits of Enlightenment science remained largely unfulfilled (Frängsmyr 1989). Strains of Romanticism and idealism were heard from Germany, and the image of Cosmos as organic whole expressed in the works of Goethe, Schelling, and Alexander von Humboldt impressed the botanists in Lund especially (Eriksson 1969, Eriksson and Frängsmyr 1971).

Carl Adolph Agardh (1785-1859), a contemporary of Humboldt, was struck by the holistic vision of Friedrich von Schelling (1775-1854), to whom he dedicated his textbook on botany. Nature and ecology, he believed, were sources of insight into connections between microcosm and macrocosm and into the organic bonds connecting reality as a whole. Agardh advocated a new philosophy of education based on the principles of Johann Heinrich Pestalozzi and lauded the virtues of scholarship for its own sake. The idea that scholarly research should be entirely justified in terms of its pragmatic usefulness was particularly loathsome for him. This romantic line was continued in Lund by the philosopher Christopher Jacob Boström (1799-1866), albeit more along stricter Hegelian lines, and eventually by Viktor Rydberg (1828-1895), whose poetry and radical theology clashed headlong with the rising tide of Darwinist orthodoxy among the natural scientists of his day (Liedman and Olaussen 1987).

Ambivalent attitudes toward nature and natural science became apparent during this "patriotic" period (Eriksson 1969, 81-111). Agardh's organicist approach contrasted starkly with the more mechanist one assumed by Jöns Jacob Berzelius (1779-1848), Sweden's renowned chemist, who advocated specialization, experimentation and the application of research results to practical ends. Secretary of the Swedish Academy of Sciences from 1818 to 1848, Berzelius kept abreast of scientific developments throughout Europe and defined "science" more rigorously, separating it from other branches of learning. Agardh envisioned broader horizons for knowledge and life, and warned about the ill-effects of separating humanistic and scientific enquiry. The legacy of this controversy, and the starkly opposed attitudes toward nature implied in them, endured right into the 20th century: on the one hand a reverence for natural (as opposed to artificial) beauty and a pragmatic view of resources, and on the other hand a virtually unquestioning commitment to mechanical modes of organizing science and social life.

It was in Uppsala that Romanticism made its greatest appeal. Close to geography's domain was the historian-poet-philosopher Erik Gustaf Geijer (1783-1857), for whom the love of one's own land and people did not necessarily imply self-adulating narcissism. Like Jules Michelet in France, he celebrated the landed folk, rooted in traditional soil, untainted by the quest for fame or empire (Landquist 1954). He set the drama of Sweden's history in its wider European context, each event construed as an interplay of tradition and creation. Geijer's historical writings offer a rare example of contextualism in the Swedish tradition, and his philosophy was echoed in that of Henri Bergson half a century later. His theory of personality formation and his ideas on co-responsibility and freedom in the building of community life were truly radical for that period (Henningsson 1961).

The stirrings of dissent and pleas for reform in religion and politics were already apparent in the 1820s and grew throughout the 1830s, despite some governmental opposition. Liberalism in social and economic life was one of the ardent longings of the day. Many idealistic reformers emigrated during the 1840s. That decade, in fact, witnessed far-reaching legislative measures at home: equal inheritance rights for men and women were acknowledged by 1845, and in 1848 the legal emancipation of women was won. Most significant for the practice of geography in this period was the introduction of compulsory elementary education in 1842.

There is little evidence that geographers of the day were very much involved in the ideological controversies then stirring within Academy and society. Their energies were directed toward the production of text materials for schools (Bergsten 1982). The 1842 declaration of universal mandatory primary education rallied efforts to procure a niche for geography in the curriculum together with the three Rs and the Lutheran catechism. The first basic school text, written by Daniel Djurberg of the Kosmografiska Sällskapet, was focussed on the physical and political map of the world (Djurberg 1806-1808). This endured through the 1870s, when eventually fresh ideas about curricular changes floated in from Denmark (Erslev 1871). At other levels, geographical writings tended to focus mainly on harvesting results from the topographical surveys, e.g., the 7-volume *Historiskt, Geografiskt och Statistiskt Lexicon öfver Sverige* ("Historical, Geographical and Statistical Lexicon on Sweden") (1859-1866) and Benjamin Höjer's (1767-1812) *Konungariket Sverige* ("Sweden's Imperial Realm") (1872-1883). Impressive catalogues of facts about the country had been assembled, but Swedish geographers seemed to be reluctant to write a comprehensive work on their own country (Kjellén 1900). Geijer's contribution to Olof Dalin's (1708-1763) *Svea rikets historia* (1853, originally published 1747-1762) demonstrates a conceptual sophistication comparable to that of Carl Ritter, and Agardh's ankologi—a proposed regionalization of the country into natural provinces—found a more receptive audience among botanists and historians than it did among geographers.

Geography in 19th-century Sweden was a subject marking time. The Linnaean tradition found continued expression in zoological, botanical, and topographical surveys in the field, and school-teachers sought to instruct pupils about the world map. As in other schools when the discipline set its focus on domestic housekeeping and on the instructional needs of society, its status was low. Then, as in the United States and France, it was from physical geography that impetus for wider horizons came in that impressive era of polar expeditions led by geologists and zoologists, the first major successs being achieved in 1861 under the leadership of Otto Torell. Not that societal conditions at home warranted much confidence for human geographers. Geijer may have celebrated the virtues of country living from his comfortable homes in Uppsala and Värmland, but agrarian life was anything but idyllic for many. By mid-century thousands would seek their fortunes across the Atlantic.

The predominant mood of geography during this patriotic period seems indeed to have been low key. Societal energies were directed toward social reform, e.g., the Constitution, the enclosure laws of the 1820s and the legal reforms of the 1840s. But it was a time when many opted to leave their country, not all of them for economic reasons. Research in the service of the fatherland belonged squarely within the spirit of Enlightenment. And it would be from among disciples of Linnaeus himself that those "natural scientists and engineers of rank" would also emerge.

Rerum Novarum (1860-1920)

The latter years of the 19th century might well be described as "the best of times and the worst of times" for Sweden. Large tracts of the country were ravaged by famine, poverty and ill health, and over a million emigrated between 1860 and 1920 (Guteland et al. 1974). By the end of this period, Sweden would again suffer much social and economic disruption related to World War I and the Bolshevik Revolution. Yet this was also a period when the Swedish economy awoke to the challenges of industrialization, readily capturing markets for its high quality steel and lumber products from the apparently infinite resources of Norrland (Sörlin 1988). A relative latecomer in the rise of capitalism, its remarkably quick successes were no doubt due to the creativity of its scientists, the 1842 law on universal primary education, and the work ethic fostered through generations of Lutheran discipline. Rapid industrialization, structural transformations of agriculture, and improvements in rail and canal transport all triggered profound changes in population distribution and mobility (Guteland et al. 1974). As thousands joined the queue for boat passage to America, life conditions at home were changing, patriarchal forms of employment weakening, and social classes drifting apart (Henriksson 1966). Political energies were vested in the promotion of commercial and technological development; the liberal policy of "every man for himself" prevailed.

Social movements were astir, some of them inspired by repatriated Americans, some by indigenous wrath. Swedes stirred to new challenges in athletics, folk education, the co-operative movement and, for the élite at least, the cult of the countryside. Though varying in ideology, these late 19th-century movements were to have profound effects on 20th-century policy and politics (Henrikson 1966). Some religious reform movements such as Svenska missionsförbundet (Swedish Mission Union) (1870) and others were directly imported from America, e. g., Örebromissionen (Örebro Mission), Filadelfiaförsamlingen (Philadelphia Community) and Godtemplarorden (IOGT) (1859). Many temperance movements, e.g., Nykterhetsorden Verdandi (NOV), were intimately associated with the growing labour movement in Sweden. The brutal suppression of a strike by 5000 sawmill workers at Sundsvall in 1879 added fuel to the fire. 1889 marked the foundation of two major labour organizations, Landorganisationen (LO) and Svenska Arbetareförbundet (SA—Swedish Worker's Union), and also the Swedish Social Democratic Party.

There were deep-seated ideological differences among the population generally, however, between those who favoured industrial development and liberal economic policy versus those who promoted rural agrarian and patriotic values (Sörlin 1988). On one side, there was evidence of science and engineering again recapturing their dynamism. John Erikson's invention of submarine monitors won applause during the American Civil War; Alfred Nobel (1833-1896), Lars Magnus Ericsson (1846-1926) and Gustaf de Laval (1845-1913) would personify the esprit that would place Swedish engineering in a leading world position. Whether in ball-bearings, mik-separators, the Aga lighthouse, pottery, barbed wire, dynamite, turbines or safety matches, few could compete with the Swedish patents from this period.

On the other side, there was a growing interest and concern about Swedish cultural identity (svenskhet) and a reverence of its rural roots, a trend that became enshrined in the Skansen museum folk festivals and the national parks. The wide popular appeal of this ideology later lent weight to the idea of Folkhem ("Home for the People"), first articulated in 1912 by Rudolf Kjellén, geographer and political scientist, and later to become a central icon of Social Democratic politics.

Tensions between these two apparently opposed movements found echoes in the literary world. In iconoclastic mood Strindberg, still venerated as the genius of modern times, penned merciless satires against hypocrisy, vaulting ambition and national pride. From

Mäster Olof to *Röda Rummet*, he voiced the poetry of a people on its way toward secular rationality, intolerant of class or rank distinctions. No ire was spared in dethroning potential heroes: Vasa, nobles, and patriots, including Sven Hedin, who had won a hearty welcome home from his Asian expeditions in 1893-1897 (Wennerholm 1978). Strindberg claimed that Hedin's alleged discoveries were actually faked. Hedin returned blow for blow and continued to stir popular opinion about the need for military defense before World War I. Hedin's apparent sympathy for German patriotism before World War I earned him some antipathy from young recruits to 20th-century social democracy.

It was in the era of Rerum Novarum that geography achieved its identity as a modern discipline. As in other European lands, it took wing through the exploratory urge, national commitment to universal education, the growing scientific interest in nature following Charles Darwin's *On the Origin of Species* (1859), and the same pragmatic interest in "natural resources" that characterized industrial nations of the period. Geographical societies and publications, professional and popular, began to appear. Sweden entered late into this general movement, but the spirit of exploration, no doubt dear to the hearts of a Viking people, did help to set the stage. Scandinavians held a leading rank in the world race to discover fresh insights on glaciers, peaks, drainage systems and climate, as well as to name and explain, albeit not necessarily to claim, all corners of the globe. The travels of Sven Hedin from 1893 to 1909 and from 1927 to 1933 captured the imagination and won applause at home and abroad. The conquest of the Northwest and Northeast passages in Arctic waters claimed high priority, the Vega voyages (1878-1880) stirring enormous popular enthusiasm. And it was in this era that Scandinavian earth scientists began to assume a leading position internationally in such fields as glacial morphology, mineralogy, hydrology and geology.

Axel Hamberg (1863-1933), professor at Uppsala from 1907 to 1928, was witness to an era filled with excitement about the scientific and national significance of these voyages, himself a consultant to Adolf Nordenskiöld's (1832-1901) expertly prepared expeditions, as well as that of Salomon August Andrée (1854-1897). Hamberg mapped the Sarekfjällen, assisted in summer terms by students from physics, geology, and meteorology from 1895 to 1931 (Hamberg 1901). Sven Lovén initiated the Marine Research Station at Kristineberg in 1887. Lapland was perceived as ideal ground for research on nature, and already by the turn of the century the Swedish Academy was promoting the establishment of research stations there and also promulgating an environmental conservation policy in the establishment of national parks.

Long before the formal establishment of Chairs for geography in universities, enthusiasts gathered to form Svenska Sällskapet för Antropologi och Geografi (SSAG) (Swedish Society for Anthropology and Geography) in 1887. In 1886 Svenska Turistföreningen began to publish its Yearbook, which was to exercise wide appeal to the entire population (Thaning 1986). *Ymer*, SSAG's own journal, and later the *Geografiska Annaler* (from 1919) continued to reach a broad international readership. The SSAG not only nurtured interest in exploration and travel accounts but also proposed a national atlas, convincing authorities and the general public that an education in geography was a sine qua non for every citizen. The contemporary multi-volume *Sveriges National Atlas*, which has won applause globally, draws its inspiration from this initiative (Aldskogius 1993, Castensson 1992, Helmfrid 1996).

In the early development of human geography as a university discipline, the same ambivalence between urban industrial and rural agrarian ideologies then current within Swedish society was apparent (Sörlin 1988). At the Handelshögskolor (Commercial Colleges) of Stockholm and Göteborg, the spirit of the Daedalus period was expressed in economic geography, and it won sponsorship and audience among the nation's trading, commercial, and diplomatic corps. In other universities, interests seemed to resonate more to those of the Pro Patria period, focus resting on historical subjects and on the regional geography of the world (Dahl 1973, Helmfrid 1977, Bergsten 1982). In Göteborg Rudolf Kjellén, a brilliant lecturer in geography and political science (later member of the Swedish government), evoked something of the Vere aureum saeculum in his ideas about power, territory, and national boundaries. Like Halford Mackinder in Britain, André Siegfried in France, and Isaiah Bowman in America, Kjellén stood out as a scholar of international vision at a time when most of his peers were pursuing specialized research on farms and houses, icecaps and lakes (Kjellén 1900, 1911, 1916). It was he who finally wrote an introductory text on Sweden itself, and in it he showed deep acknowledgement of Geijer's radical humanism. After his alleged influence on Karl Haushofer and the exploitation of his geopolitical ideas in Nazi Germany, Kjellén's name, and to a lesser extent that of Sven Hedin, fell under a curtain of silence among academic geographers (Söderquist 1986).

Historians and geologists were the first to champion geography's case in university curricula, and most of the first two generations had training in both. The historian Martin Weibull taught courses in geography at Lund already in 1877, as did Johan Fredrik Nyström (1855-1918) at Uppsala from 1892. Chairs were assigned to the Count von Schwerin, also an historian, at Lund in 1877 and to Karl Jacob Mauritz Ahlenius (1866-1906), disciple of Friedrich Ratzel, at Uppsala in 1902. Both wrote on the history of geographic thought: von Schwerin on Herodotus (1884) and Ahlenius on Olaus Magnus (1895). Indeed, a major focus of research attention for the first generation was the history of exploration, cartography, and scientific thought, following the lead of Adolf-Erik Nordenskiöld's *Periplus* (1887) and *Facsimile Atlas* (1889). Soon energies were directed toward the genesis of agrarian landscapes facilitated by the wealth of documentary evidence readily available in the archives. At the research frontiers of glaciation, geomorphological processes, and Arctic expanses, too, doctoral dissertations usually included "human" dimensions of regional life (Sjögren 1909, Nelson 1913, Nordlind 1911).

From the start, the making of geography in Sweden was influenced by German precedents. German ideas, practices, and research frontiers were transplanted onto Swedish soil. Debates were entertained over relationships beween systematic versus regional emphases, historically-oriented versus contemporary approaches to morphology, and the relative salience of "natural" and "cultural" influences on the shaping of landscape and regional identity. On the whole, however, such issues, so passionately pursued in other European countries, seemed to claim little attention in this post-imperial land. Boundaries of regions and home areas were, after all, settled; issues of territory and reach could scarcely evoke as much emotion as they did elsewhere. Culturally speaking, the Swedish landscape had far more homogeneity than other European lands. Immigrants of the vere aureum saeculum had long since become Swedified; landscape as mirror of civilisation could scarcely excite the kind of pays studies that attracted students on the Mediterranean, Celtic, or Baltic fringes of Europe.

Geography teaching in schools meanwhile appeared to resonate more to the interests of the Pro Patria period. There is little evidence of interest in folk movements or the political turmoil of the turn of the century. School texts encouraged map recognition exercises, now and then rejuvenated with fresh pedagogical ideas from Germany and Denmark (Bergsten 1982). They showed little awareness of the exploratory spirit about polar research or concurent events at home; teacher trainees were required to read Selma Lagerlöf's *Nils Holgerssons underbara resa genom Sverige* (Title of English translation, "The Wonderful Adventures of Nils") (Lagerlöf

1906-1907), Kjellén's *Inledning i svensk geografi* ("Introduction to Swedish Geography"), and Hedin's *Från pol till pol* ("From Pole to Pole") (1911) right through the mid-20th century. And every student was supplied with Carlsson's *Lärobok i Geografi* ("Learning Guide to Geography") and Roth's *Världsatlas* ("World Atlas") (Bergsten 1982).

Ever since 1871, there were patriotic elements also in the hembygdsforskning (home area studies) that were conducted in the Folkhögskolor (People's colleges, a uniquely Scandinavian institution, popular also in Baltic lands). Inspired by the Danish clergyman Grundtwig, the folk college movement responded to the educational needs of a rural population, stimulating not only handicrafts and skills but also keeping the radically humanist spirit of Geijer and even Waldenström alive (Nelson 1913). One of Lund University's first professors of geography, Helge Nelson, formerly rector of a folkhögskola, emphasised the importance of regional studies in his inauguration address (Hägerstrand 1982). Otherwise the courses of university geography and of the schools would diverge for a few generations.

Saecuritas (1920-2000)

Insecurities and civic unrest following the famine years of 1917-1918 eventually subsided and by the late 1920s there was some optimism about economic growth in the land. When the Social Democratic Party won a national majority in the early 1930s, political stability was realized. Per Albin Hansson, "father of the country", heralded Saecuritas for all. With only brief intermissions, this party held the reins of power for over half a century, their zeal eventually touching all aspects of Swedish life and thought. In science as well as in enterprise, social policy and external affairs, committees and councils sought to reconcile the values of egalitarianism and social justice with those of sustained economic growth. The practice of geography during the 20th century was intimately involved in this dream, shaping and being shaped by the political and social goals of the nation.

Sweden's formally neutral stance on world wars afforded time and energy to build an elaborate welfare system at home, while continuing to hold a leading position internationally with top-quality industrial exports. From the 1920s on, education at all levels sought to promote social responsibility and solidarity; teamwork toward common goals was deemed more valuable than individual merits. In the 1930s, Gunnar Myrdal and Bertil Ohlin expounded theories about welfare state economics—with due caution about the consequences for intellectual freedom—whereby the hitherto irreconciliable goals of economic growth and social security could be harmonized.

The postwar years witnessed growing tensions between these two goals of economic growth and egalitarianism. By the early 1960s Sweden's competitive edge in international markets was challenged by the relatively more robust industrial capacities of the reconstructed postwar economies of West Germany and Japan. Meanwhile, its virtually sacred (but expensive) policy of full employment and social security programmes at home curtailed possibilities for industrial re-investment. Labour unions, among other voices, called for government action. By the mid-'60s, enthusiasm grew about geographically-based approaches toward reconciling these apparently conflicting goals through regional policy and growth pole theory (Kristensson 1972). Here was a domain in which geographers would become especially involved, and in the process transform traditional definitions of region and discipline.

On the international front, especially since World War II, Sweden has earned an impressive profile, contributing substantially to various UNESCO and OECD projects on economic and social planning throughout the postwar period. Swedish scholars were among the founding members of the European Science Foundation and the Academia Europaea. Raoul Wallenberg's daring wartime deeds in Hungary and Dag Hammarskjöld's service within the United Nations illustrate the leading rôle that Swedes assumed in promoting peace and justice throughout the world. Together with the Netherlands, this country still ranks highest in per capita aid to developing societies. Various projects sponsored by Swedish International Development Aid (SIDA) and the Dag Hammersjöld Foundation promote dialogue between North and South, East and West.

University life in Sweden, as elsewhere in Europe, seems to have passed through its own cycle of post-1968 restructuring. Committees and boards comprising representatives from any and all conceivably interested parties parlayed over the substance and routine-operational procedures of academic life. The Ministry of Education proclaimed U68, a radically altered definition of university curricula, which, quite reminiscent of Auguste Comte's *Église de l'humanité* (1824), promoted training in specialised lines of expertise to replace the more conventional discipline-based ones, so that the production of knowledge could be more patently adjusted to societal goals. Functional specialization was thereby encouraged, and Anglo-American ideas would definitely supersede continental European ones.

Apart from athletics, ABBA, and anti-nuclearism, the 1970s seemed to be years of slumber and/or critical reflection. The energy demanded from students and scholars in adapting to many new structural arrangements probably impaired intellectual creativity. At the level of research policy, numerous attempts were made to transcend sectorization, again through a re-definition of boundaries. Projects on future studies rallied expertise from scholars of various backgrounds to study issues such as health, energy, and natural resources. Integration of research councils and grant-giving bodies was also planned, leading to umbrella councils such as Forskningsrådsnämnden (Swedish Council for the Planning and Coordination of Research). New university curricula organized around themes rather than disciplines were established, e.g. at Linköping, where students were trained in substantive areas such as water, technology and society, education, and other themes. Interdisciplinary research would become a favoured practice of the 1970s in Sweden.

As the fruits of these various projects were being harvested and evaluated in the 1980s, university life regained its dynamism and the values of scientific research were more openly acknowledged. Swedish scholars pondered their future rôle in a fully integrated Europe, and students were encouraged to avail themselves of research and study opportunities abroad. A major challenge of the 1990s was that of global environmental change. It was in Stockholm that the United Nations International Environmental Conference was held in 1972, and the Swedish Academy of Sciences launched the international journal of environmental research, *Ambio*. The Beijer Institute (International Institute for Energy and Human Ecology), founded in 1977, continued throughout the '80s to lead research on natural resources, climate change, energy and sustainability: themes that crystallized in the Brundtland Commission Report *Our Common Future* in 1987. Saecuritas at century's close implied more than efficient housekeeping and husbandry at a domestic scale; it now involved the entire globe. And indeed many geographers rallied energies to this cause.

Disciplining Geography in 20th-Century Sweden

By 1920 academic geography was well established in Sweden. Space does not permit a detailed account here (but see Dahl 1967, Helmfrid 1977, Hägerstrand 1982, Bergsten 1982, Godlund 1986). Suffice it to sketch, in very broad strokes, some enduring features of the cognitive styles adopted and the main vocational lines pursued by practitioners of the art during most of the 20th century (Buttimer 1983, Buttimer and Hägerstrand 1988).

In terms of cognitive style, throughout the early decades particularly, the preferred root metaphor was that of *mosaic.* This recalls the strong Linnaean roots of survey and mapping from the Daedalus Hyperboreus period on. Long-standing practices of field observations and excursions in Sweden no doubt yielded an array of methods for exhaustive inventory, classification, and categorization of phenomena, the horizons of conceptual imagination being co-terminous, as it were, with those of method and the national interest (Odén 1991, Eriksson 1978). The enduring appeal of *mosaic* reflects not only the historical legacy, but also the ideological opposition to its main rival metaphor, *organism,* in the early 20th century. The romantic flair of Humboldt's *Cosmos* or even the teleological elements of Ritter's *Erdkunde* were not exactly appealing for the pioneers of Sweden's exploratory era. Twentieth-century geographers inherited a well-rehearsed attunement to form, pattern, stratification, and ingenious codes and techniques for a careful mapping of nature and landscape. Instead of speculating over some "whole picture" or artistic synthesis, one aimed at comprehensive inventory. The model for synthesis would be that of Länderkunde (Hettner 1905, Hartshorne 1939) rather than pays-based géographie des lieux (Vidal de la Blache 1922), Germany supplying the paragons of disciplinary orthodoxy (Bergsten 1982, Odén 1991). Even the Danish geographer Conrad Malte-Brun's(1775-1826) pioneering *Précis de géographie universelle* (1810-1829) attracted little attention among Swedish geographers.

The conflation of organicist political geography and Geopolitik in the works of Rudolf Kjellén offers another major explanation for the early dismissal of *organism* as root metaphor in Swedish geography. What was applauded in the writings of Sven Hedin was the wealth and detail of geographical facts he uncovered rather than the anthropogeographical flavour of his cognitive style. In the early 20th century, Romantic idealism and patriotism were deemed to be conservative and reactionary attitudes: it would take another half-century before organicist ideas would again re-emerge in the context of environmental movements, whose echoes in the practice of geography were only dimly felt even by the 1980s.

Among the models and debates imported from Germany during the first decades of university geography in Sweden, the one that stirred most passion was concerned with alternative conceptual bases for integrating human and physical dimensions of regional enquiry. On one side was the Ratzelian (anthropogeographic) solution, where relationships between societies and milieux were to be studied in ecological ways; while on the other hand there was the Hettnerian (chorological) solution where the isomorphism of patterns—physiographic, architectural, and demographic—might suggest potential explanations or at least co-variations. Both approaches found expression in "regional" geography; both admitted an historical dimension. The former directed metaphorical imagination toward process, relationship, and organic connections among physical and human phenomena; the latter directed attention toward form, distribution, location, and inventory.

At Uppsala Ahlenius was a staunch defender of anthropogeography, and he illustrated this stance in his own study of population and settlement in northern Sweden (Ahlenius 1903). At Lund, however, it was the Hettnerian approach that was encouraged. Helge Nelson, the great advocate of regional geography, spoke approvingly of Ratzel and even of Vidal de la Blache, but he was very wary of environmentalism (Nelson 1913). Länderkunde and Siegfried Passarge's tradition of landscape study were what pleased him as adequate formulae for arriving at the regional Gestalt. J. G. Granö's *Reine Geographie* (1929), and even the Heimatkunde then flowering in Baltic lands, while admired by the Swedes, did jar some sensitivities (Granö 1929). Hettner's regional works would be required reading, but little attention was paid to his methodological works (Hägerstrand 1983).

Torsten Hägerstrand has described the competition held in 1913 between Helge Nelson and Sten De Geer for the Chair of Geography at Lund. Nelson argued for an integrated and historically-oriented field, whereas De Geer argued for a science of spatial distributions oriented to contemporary rather than historical patterns (Hägerstrand 1982). Nelson won this battle, but, in the long run, De Geer may have won the war. Chorological science would capture the imaginations of 20th-century Swedish geographers. It opened up a vast challenge for the refinement of analytical procedures and eventually pointed beyond formism itself, seeking insight into mechanisms and processes underlying those forms (De Geer 1908, 1923). Ecological or organic metaphors and speculations about the whole as being something more than the sum of the parts remained little more than casual footnotes (see, however, Enequist 1937, Dahl 1942).

Mechanism had exercised appeal much earlier within the physical aspects of geographic enquiry. Process geomorphology, encouraged through its links with hydrology and climatology, had been gaining ground on structural geomorphology ever since the days of William Morris Davis (Hjulström 1935, Rapp 1961, Lidmar-Bergström et al. 1993). Rhetorics of "system efficiency" also afforded a common language among those who advocated the divorce of physical and human geography. It was clearly the metaphor used to justify the merging of human geography with the newly-constituted Faculties of Social Science in Swedish universities in 1947. Thereafter, dissertation writers in human geography would face examination committees that might include economists rather than historians; those in physical geography would be encouraged to measure their contributions against natural science or geology. *Mechanism*, theme song of Social Democracy, would continue to thrive until the late 1970s, when a contextualist turn became apparent in virtually all departments of geography.

One important catalyst for paradigmatic change was certainly the emergence of "time geography", particularly at Lund (Hägerstrand 1970, 1974, 1983; Buttimer 1976, 1979; Wallin 1982; Carlstein 1984). In-depth studies of everyday life (Mårtensson 1979, Friberg 1990, Åquist 1992) evoked attention to the multi-facetted aspects of interaction in local settings. The metaphor of *arena* appeared with increasing frequency during the 1980s and 1990s (Löfgrén 1990, Ellegård 1993, Törnqvist 1993a, 1993b). Historical geography, too, opened up fresh perspectives on landscape (Berglund, ed., 1991, Lewan and Lewan 1992, Sporrong et al. 1995). As students from both kultur- and naturgeografi engaged with issues of environment and development—at home and abroad—the stronger this contextual orientation has become (Rundquist and Hatti 1969, Holmén 1991, Anderberg 1996, Jirström 1997).

These changing trends of metaphorical style can scarcely be understood without reference to changes in the context within which geography was practiced during the Saecuritas period. The vocational appeal of geography for the first generation lay in its obvious importance for the development of the country. Empirical explorations of landscapes and life, coupled with ever sharper analytical acuity on resource inventory, far outweighed theoretical arguments over distinctions between pure and applied research. Most of the early generation moved back and forth between school teaching, local field work, and keeping counts on weather or lake shorelines, while completing their doctoral dissertations. Societally speaking, geography's chief contributions, apart from fundamental education, were those of resource inventory and spatial organization: interests that geography continued to serve throughout the 20th century.

In terms of substantive focus, researchers had begun, already in the 1930s, to shift from regional to systematic (thematically-oriented) foci of interest. William William-Olsson explored the dynamics of urban functions—manufacturing, retail, services and transportation—activities that could explain the formal patterns of Stockholm city and region (William-Olsson 1937). Gunnar Alexandersson worked out the minimum requirements method for defining the urban economic base, and researchers at Stockholm's Handelshögskola studied the functions of ports and freight arrangements in the dynamics of international trade (Alexandersson 1956). From several sources, the prospect of functional analysis had begun to hold stronger appeal than the formal mapping of patterns. The overall tendency toward thematic specialization also brought geographers into closer contact with colleagues in other fields such as architecture and engineering, where the ideas of Lewis Mumford, the Bauhaus Bewegung, and eventually Le Corbusier enjoyed great popularity after the Stockholm Exhibition in 1930.

Already in the 1930s Swedish geographers had thus become involved in applied work. Traditional foci of concern such as questions of regional identity, sense of place, and relationships between economy and natural environment now lost their appeal. The demise of regional geography and the growing popularity of thematic lines of enquiry eventually dismembered classical approaches to environment. Instead of holistic regional descriptions of life and landscape within one area, e.g., in Anna Kristoffersson's 1924 study, one sought insight into the spatial forms and processes surrounding particular types of resource use, e.g., fishing, forestry, mining or manufacturing. Yet, up to the 1950s, dissertations on population and economy maintained broad historical perspectives and also sought to relate modes of economic life to the bio-physical environment. Early studies in economic geography, too, e.g., on forest products and their circulation, on the sawmill industry in Norrland, and on the herring industry in Bohuslän, all include explicit concern for the physical environment. After the 1950s, and perhaps not independently of the institutional split between kultur- and naturgeografi, the meaning of environment had changed. Issues of spatial order, the impact of policy and social structures on localization of industry and commercial activities, took over as central concerns. From the 1950s on, too, the primary employment opportunities for university graduates in geography would be in municipal and regional planning rather than in teaching.

Few topics illustrate these changing emphases better than transport and communication. Hoppe's 1945 study of circulation patterns in Norrbotten and Bylund's 1956 study on the colonization of Pitemark both emphasize the interaction of culture, technology, and environment in the evolution of landscape patterns. By the 1950s, transportation systems became foci of research as phenomena in their own right. By the early '60s, Gunnar Törnqvist's *Studier i industrilokalisering* made a persuasive case for industrial location strategies that virtually ignored details of history, culture, environmental conditions or geodesic distance between places.

Changing emphases in the practice of geography reflected radical shifts in the ideological and material context of postwar Sweden, by now rapidly transformed from a rural-agrarian to a post-industrial society. The intellectual and political climate encouraged research oriented toward the reshaping of landscapes and life; predictive models of social and spatial "processes" were quite as welcome as descriptions. The arrival of Edgar Kant in 1946 was to mark a decisive injection of fresh energy into the ranks of young scholars at Lund (Kant 1934, 1948, Hägerstrand 1978, 1983, Buttimer 1987, 1994). Two doctoral dissertations that he inspired and directed were to epitomize the cutting-edge of modernist geography internationally: those of Torsten Hägerstrand (1953) and Godlund (1954). For the practice of geography, a fundamental shift was underway from liberal education and formal scholarship anchored in inductive empirical research within the root metaphor of mosaic, to a deductive and theoretical approach to model-building and hypothesis-testing in the service of State and municipal planning. The net product of the halcyon days of regional science, particularly at Lund, was an ingenious convergence of *mosaic* and *mechanism*, keys to the metaphorical imagination of the "New Geography". During the 1950s and 1960s Swedish geography was held in high esteem internationally, particularly among enthusiasts of spatial science and applied scholarship. Hosting the International Geographical Congress in Stockholm in 1960 enhanced the visibility of Swedish research, which was, in many ways, on the cutting edge of disciplinary development. Enthusiasm for spatial modelling of macro-urban, economic and demographic phenomena flowered at Lund during the '60s (Améen 1985).

At home, too, geographers were highly regarded in government circles. They counted among that expert group of specialists (ERU) who were appointed in 1965 to gather evidence to support a policy of regional concentration in order to optimize conditions for economic growth, also recommending measures to "balance" the welfare effects among regions that would likely become depopulated as a result. The balancing act would be accomplished by the "rational choices" of individual firms, households, and the public sector. Platonic notions of the benign Artisan, so characteristic of regional planning internationally, found perhaps their best expression in Sweden's applied geography of the '60s and '70s: first through an optimizing strategy for industrial location balanced by a "rational" array of health, educational, retail and other functions; then through a decentralization of governmentally-sponsored activities; and finally through a restructuring of administrative boundaries and regions. The chess-board involved highly specialized sectors, however, each with distinct kinds of location logic and rationality; decentralization of functions not necessarily leading to a re-centering of lived realities within Swedish regions (Hägerstrand 1974).

Even in the 1960s there were tensions among different geographical perspectives on form and process. There was the primarily chorological view epitomized in the Christaller model, where spatial form was explainable in terms of centrifugal and centripetal forces, eventually crystallizing in hexagons of varying dimensions: a world view with appeal to the mineralogist, geometrician, or archaeologist desirous of capturing form, type, range and nature of spatially-perceptible phenomena. There was also the more ecological view epitomized in the work of Edgar Kant, which sought out processes analogous to the rhythms of biotic life: a view that appealed to enthusiasts for process, relationships, and interactions. The societal context in which these models were to be applied demanded solutions to the challenge of reconciling form and process: the interests in socio-spatial order for which municipal and regional authorities would be responsible versus the expanding scales of economic growth then promoted via market forces and entrepreneurial life.

In cross-cultural perspective one could indeed ask whether there may be some consistent associations between root metaphor, the scale of intellectual curiosity, and eventually the status of geography in the political economy of research sponsorship (Warntz 1964, Buttimer 1985, 1993). In many cases indeed, the root metaphors of *organism* and *mechanism* opened horizons to national and international questions, whereas those of *mosaic* and *arena* tended to deal with issues of predominantly local interest. In the early 1920s there were theses on geographical exploration and the history of ideas that were conducted within the metaphorical genre of organicism and these entertained international horizons. Then came the 1930s with a strong preference for local research concerned with economic resources and in these the *mosaic* became the favoured metaphor. The advent of *mechanism* already in the '30s marks a growing interest in issues at a *national* scale; here, in fact, human geographers appear to take the lead. The regional (Scandinavian) scale has been more favoured by physical geographers, and this reflects the solid foundations for field laboratories and shared research curiosities dating from the early days of the discipline in Sweden. The revival of interest in international (or global) scale questions has been apparent in both natur- and kultur- branches, reflecting not only the rapid introduction of computer-based techniques in GIS and remote sensing, but also the involvement of young geographers in projects dealing with the developing world.

Concluding Reflections: From Vasa Nation to Welfare State

The practice of geography at any moment may be interpreted in terms of meaning, metaphor, milieu and scale horizon (Buttimer 1993). Yet the making and breaking of succesive paradigmatic waves are not so easily explained. At times of national self-confidence and sometimes o'erleaping ambition, geographers strode forward toward higher levels of integration in the understanding of society and milieu; at times of national stagnation or decline, they often became culturally myopic and self-preoccupied. In Sweden, it seems, expansionary phases in politics were associated with or followed by broader horizons of science. At times when national preoccupations were more oriented to domestic affairs, and when Romanticism was in the air, geography entertained smaller-scale ambitions of topographical survey and school texts. Down through the centuries, Swedish geographers have shown a strong commitment to ordning och reda (order and preparedness), to careful empirical investigation, and to consensus rather than conflict. There has been little evidence of introspection, critical reflection, or worries over identity.

Ideas may spring up anywhere. Often it is on the periphery, in marginal or previously colonized lands, that innovative thought has flowered. Heralds of new theoretical ideas about humanity and environment were often generalists, not specialists; the double background in physical and human geography was no doubt a source of inspiration. Diversity of cultural background may also have been significant. Several luminaries in Swedish science and literature were descendants of immigrants. Norwegian, Finnish, Danish and Icelandic people and landscapes have played important rôles in physical geography; much talent crossed the Baltic also, particularly after World War II.

The making of geography in Sweden may be read as an array of impressive products and techniques for socio-spatial order at home and for the promotion of rationality and justice internationally. Twentieth-century policy-makers all over Europe and North America still consult the Swedish patents and implement them. The irony of the tale is that the making of geography in many cases heralded its breaking. After the successful accomplishment of various tasks or domains of research—whether in polar exploration, land surveys, municipal planning or geocoding—once the challenge was met and appropriate methods forged, these domains passed on to more specialized fields. Twentieth-century Social Democracy's encouragement of functional specialization was eventually to undermine geography's traditional rôle as an integrative science.

The institutional separation of physical and human branches of the discipline at mid-century is without question the most dramatic event in 20th-century Swedish geography. The "missing links" apparent after the divorce were many, and among the most significant ones were those with the biological sciences on the one hand, and those with history and the humanities on the other. Mainstream practitioners of kultur- or naturgeografi, construing their practice in the primarily positivist mode of 20th-century social or physical sciences, found less room for history, even less for aspects of the environment and its human significance that were not amenable to observation and measurement. When concern about milieu had become virtually a social movement in Sweden by the 1960s and '70s, geographers apparently paid little attention. Throughout the 1970s and '80s when milieu became a priority for sponsored research, it was not at all clear that monies should flow to either branch of geography. A growing awareness of the "missed opportunities" for environmental research at home eventually joined the long-standing concern about "developing societies", especially among physical geographers, during the 1970s. For some human geographers, these two streams converged in the conviction that the time had come to revive not only the bio-physical dimensions of disciplinary enquiry but also its traditions of historically-grounded research (Germundsson 1993, Bladh 1995, Holmén 1995).

The growing diversity and contextual sensitivity of Swedish geography during the last two decades of the 20th century is impressive indeed (Olsson 1979, Wallin 1980, Forsberg 1989, Löfgrén 1990, Ellegård 1993, Gren 1994, Clark 1995, Gren and Hallin, eds., 1998, Gren, Hallin and Moline 2000). The numbers of geographers have increased substantially and so have the number and variety of academic positions (Öhngren 1999). These trends are likely to continue now that Geography is back in the curricula of schools and teacher-training institutions (Wennberg 1990, Mårtensson 1992). There is Viking zeal behind recent accomplishments in remote sensing and joint (physical and human) research effort on the Sahel and Southeast Asia. At Stockholm and Uppsala, as indeed also at Lund, historical geography re-anchored itself once again within wider international networks of exploration (Berglund, ed. 1991). Research on smaller settlements (glesbygd) has regained appeal, the "insider's" perspective offering exciting challenge (Wahlström 1984, Mårtensson 1979). Revitalized collaboration with colleagues in other Scandinavian countries is apparent, especially among those keen on critical and humanist approaches (Öhman, ed. 1990). One of the most elegant products of recent years is the 17-volume *Sveriges National Atlas*, each volume of which is edited by an internationally renowned Swedish geographer. By century's close there were signs of a new kind of regional thinking in the making, not only as a framework for multi-disciplinary investigation, but also as a response to the serious problems of rural regions whose demographic and economic horizons were so profoundly altered after the rationalization plans of the 1970s (Törnquist 1998).

Each country has its own story to tell about occasions when geography has led, and when it has followed, trends among its audiences and sponsors. Tensions between intellectual integrity and social integration have not lessened over the centuries. As with many other academic fields today, geography faces ambivalent challenges. In the past, it seems, as long as the geographer could supply fresh ideas about avenues for exploration, the discipline's status and integrity remained virtually guaranteed. Success in the accomplishment of each task spelled dispensability. The challenge, in Sweden as elsewhere, is that of openness to the everchanging shapes of humanity's challenge to dwell more wisely on Planet Earth. Among those who hold this challenge dear, not only in theory but also in lived experience, Swedish geographers surely offer exemplary models.

References

N. B.--Swedish alphabetical order might be confusing to non-Swedes (e.g., Å and Ö follow Z at the end of the alphabet); thus, the names Åquist, Öhman, and Öhngren will be found at the end of this bibliography.

Ahlenius, K. (1903) *Ångermanälvens flodområde. En geomorfologisk-antropogeografisk undersökning*, Uppsala.

Aldskogius, H. (1993) Cultural life, recreation and tourism, *Sveriges National Atlas* 8, Stockholm.

Alexandersson, G. (1956) *The industrial structure of American cities: A geographic study of urban economy in the United States*, Almquist & Wiksell, Stockholm.

Améen, L. (1985) Mot en Linneansk stadsgeografi, *Svensk Geografisk Årsbok* 61, 230-239.

Anderberg, S. (1996) Flödesanalys i den hållbara utvecklingens tjänst. Reflektioner kring en "metabolism" - studie av Rhenområdets utveckling, *Meddelanden från Lund Universitets Geografiska Institutioner* 118.

Arnstberg, K.-O. and B. Ehn (1976) *Etniska minoriteter i Sverige förr och nu*, Liber Läromedel, Lund.

Berglund, B. E., ed. (1991) The cultural landscape during 6000 years in southern Sweden - the Ystad project, *Ecological Bulletin* 42, Munksgaard, Köpenhamn.

Bergman, T. (1766) *Physisk beskrifning öfver jordklotet*. I: *Verldsbeskrifning*, 2, Joh. Edman, Kongl. Acad. Boktr., Uppsala.

Bergsten, K. E. (1982) Geografen och geografer i Lund under 1900-talets första hälft, *Geografiska Föreningen* (mimeo), University of Lund.

Bladh, G. (1995) Finnskogens landskap och människor under fyra sekler - en studie av natur och samhälle o förändring, *Meddelanden från Göteborgs Universitets Geografiska Institutioner*, Serie B, 87, Göteborg.

Buttimer, A. (1976) Grasping the dynamism of lifeworld, *Annals of the Association of American Geographers* 66, 277-292.

Buttimer, A. (1979) Insiders, outsiders and the geography of regional life, in A. Kuklinski, O. Kultalahti, and B. Koskiaho (eds.), *Regional dynamics of socio-economic change*, Finnpublishers, Tampere, 155-178.

Buttimer, A. (1982) Musing on Helicon: Root metaphors and geography, *Geografiska Annaler* 64 B, 89-96.

Buttimer, A. (1983) *The practice of geography*, Longmans, London.

Buttimer, A. (1985) Farmers, fishermen, gypsies, guests: Who identifies?, *Pacific Viewpoint* 26, 280-315.

Buttimer, A. (1987) Edgar Kant 1902-1978, *Geographers: Biobibliographical Studies* 11, 71-82.

Buttimer, A. (1993) *Geography and the human spirit*, The Johns Hopkins University Press, Baltimore.

Buttimer, A. (1994) Edgar Kant and Balto-Skandia: Heimatkunde and regional identity, in D. Hooson (ed.), *Geography and national identity*, 161-183, Blackwells, Oxford.

Buttimer, A. and T. Hägerstrand, eds. (1988) *Geographers of Norden*, Lund University Press, Lund.

Carlstein, T. (1984) Time resources, society and ecology, *Meddelanden från Lunds Universitets Geografiska Institutionen* 88, Lund.

Castensson, R., ed. (1992) The infrastructure, *Sveriges National Atlas* 6, Stockholm.
Clark, E. (1995) The rent gap re-examined, *Urban Studies* 32, 1489-1503.
Clay, J. C. (1914) *Annals of the Swedes on the Delaware*, Swedish Historical Society of America, Chicago.
Dahl, S. (1942) *Torna och Bara. Studier över Skånes bebyggelseoch Näringsgeografi före 1860*, Ph.D. dissertation, Lund.
Dahl, S. (1973) Human geography, *Social science research in Sweden*, pp. 56-70, Science Research Council, Stockholm.
Dahlberg, E. (1898-1900) *Suecia antiqua et hodierna 1667-1716*, Wahlström & Widstrand, Stockholm.
Dalén, L. (1941) Den bohuslänska fiskelägesbygden, Ph.D. dissertation, Göteborg.
De Geer, S. (1908) Befolkningens fördelning på Gotland, *Ymer* 28, 240-253.
De Geer, S. (1923) On the definition, method, and classification of geography, *Geografiska Annaler* 5, 1-37.
Ellegård, K. (1993) Activities in their everyday context - using individual diary data to set forth the complex pattern of peoples' activities in their everyday life, in *Time-use methodology: toward consensus* (*Note e Relazione*, no. 2) Istituto Nazionale di Statistica, Rome, 29-42.
Enequist, G. (1937) *Nedre Luledalens byar. En kulturgeografisk studie*, Ph.D. dissertation, Uppsala.
Eriksson, G. (1969) *Romantikens världsbild speglad i 1800-talets svenska vetenskap*, Wahlström & Widstrand, Stockholm.
Eriksson, G. (1978) Kartläggarna, *Acta Universitatis Umensis* 15, Umeå.
Eriksson, G. and T. Frängsmyr (1971) *Idéhistoriens huvudlinjer*, Wahlström & Widstrand, Stockholm.
Forsberg, G. (1989) Industriombandling och könstruktur. Fallstuydier påfyra lokala arebetsmarknader, Ph.D. dissertation, Uppsala.
Friberg, N. (1954) Dalarnas befolkning på 1600-talet. Geografiska studier på grundval av kyrkböckerna med särskild hänsyn till folkmängdsförhållandena, *Meddelanden från geografiska institutet vid Stockholms högskola* 90, Stockholm.
Friberg, T. (1990) Kvinnors vardag. Om kvinnans arbete och liv. Anpassningsstrategier i tid och rum, *Meddelanden från Lunds Universitets Geografiska Institutionen*, Lund.
Frängsmyr, T. (1989) *Science in Sweden. The Royal Swedish Academy of Sciences 1739-1989*, Royal Swedish Academy of Sciences, Stockholm.
Germundsson, T. (1993) Landsbygdens Egnahem. Egnahemstörelsen, småbruklet och landskapet i sydsvenkst perspektiv, *Meddelanden från Lunds Universitets Geografiska Institutioner* 117, Lund.
Godlund, S. (1986) Swedish social geography, in J. Eyles (ed.), *Social geography in international perspective*, Barnes and Noble, Totowa, New Jersey, 96-150.
Granö, J. G. (1929) Reine geographie, *Acta Geographica* 2, no. 2, Helsinki.
Gren, M. (1994) Earth writing - exploring representation and social geography in-between meaning/matter, *Meddelanden från Göteborgs Universitets Geografiska Institutioner*, Serie B, Göteborg.
Gren, M. and P.-O. Hallin, eds. (1998) *Svensk kulturgeografi: en eskursion inför 2000-talet*, Studentlitteratur, Lund.
Gren, M., P.-O. Hallin, and I. Moline, eds. (2000) *Kulturens plats/maktens rum*, Symposium, Lund.
Gullander, B. (1980) *Linné i Dalarna: en antologi*, Forum, Stockholm.

Guteland, G. et al. (1974) *The biography of a people. Past and future population changes in Sweden. Conditions and consequences,* Allmänna Förlaget, Stockholm.
Hamberg, A. (1901) Geologiska och fysiskt geografiska undersökningar i Sarjekfjällen, Ph.D. dissertation, Uppsala.
Hannerberg, D. (1971) *Svenskt agrarsamhälle under 1200 år,* Läromedelsförlaget, Stockholm.
Hartshorne, R. (1939) The nature of geography, *Annals of the Association of American Geographers* 29, 171-658.
Helmfrid, S. (1976-1977) Hundra år av svensk geografi, *Ymer* 96/97, 360-372.
Helmfrid, S. (1996) The geography of Sweden, *Sveriges National Atlas* 17, Stockholm.
Henrikson, A. (1966) *Svensk historia,* 2 vols., Bonniers, Stockholm.
Hettner, A. (1905) Das Wesen und die Methoden der Geographie, *Geographische Zeitschrift* 11, 545-564, 615-629, 671-686.
Hjulström, F. (1935) Studies of the morphological activity of rivers as illustrated by the River Fyris, Ph.D. dissertation, Uppsala.
Holmén, H. (1991) Building organisations for rural development. State and cooperations in Egypt, *Meddelanden från Lunds Geografiska Institutioner* 113, Lund.
Holmén, H. (1995) What's new and whats regional in the 'new regional geography'?, *Geografiska Annaler* 77B, no. 1, 47-63.
Hägerstrand, T. (1970) What about people in regional science?, *Papers of the Regional Science Association* 24, 7-21.
Hägerstrand, T. (1974) The domain of human geography, in R. Chorley (ed.), *New Directions in Geography,* 67-87, Cambridge University Press, New York.
Hägerstrand, T. (1978) Edgar Kant 21.2.1902-16.10.78, *Svensk Geografisk Årsbok* 54, 96-101.
Hägerstrand, T. (1982) Proclamations about geography from the pioneering years in Sweden, *Geografiska Annaler* 64B, 119-125.
Hägerstrand, T. (1983) Searching for the sources of concepts, in A. Buttimer (ed.), *The Practice of Geography,* 238-256, Longmans, London.
Jirström, M. (1997), In the wake of the Green Revolution, *Meddelanden från Lunds Universitets Institutioner* 127, Lund.
Kant, E. (1934) Problems of population and environment in Estonia, *Publicationes Seminarii Universitatis Oeconomico-Geographici* 7, Tartu.
Kant, E. (1948) Den sociologiska regionen, den sociala tiden och det sociala rummet. Några kritiska tankar med anledning av en ny regional sociologi, *Svensk Geografisk Årsbok* 24, 109-132.
Kjellén, R. (1900) *Inledning i svensk geografi,* Wettergren & Kerber, Göteborg.
Kjellén, R. (1911) *Stormakterna,* Gebers Förlag, Stockholm.
Kjellén, R. (1916) Staten som livsform, *Politiska Handböcker* 3, no. 3, H. Geber, Stockholm.
Kristensson, F. (1972) *Människor, företag och regioner,* EFI, Stockholm.
Landquist, J. (1954) *Geijer, en levnadsteckning,* Norstedt, Stockholm.
Lewan, N. and L. Lewan (1992) Landskapets framtid - de stora jordegendomarnas betydelse. Synpunkter från markägare och myndigheter i Skåne, *Rapporter och Notiser* 105, Lunds Universitets Geografiska Institutioner, Lund.

Lidmar-Bergström, K., J. Kleman, and A. Rapp (1993) Geomorphology in Sweden, in H. J. Walker and W. E. Grabau (eds.), *The Evolution of Geomorphology*, 417-427, J. Wiley and Sons., Ltd., New York.
Liedman, S.-E. and L. Olausson (1987) General education, culture and specialisation, *Studies of Higher Education and Research*, 1987 vol., no. 6, Stockholm.
Lindqvist, S. (1984) Technology on trial. The introduction of steam power technology into Sweden 1715-1736, *Uppsala Studies in the History of Science* 1, Almqvist & Wiksell, Stockholm.
Löfgrén, A. (1990) Att flytta hemifrån. Boendets roll i ungdomars vuxenblivande ur ett situationsanalytiskt perspektiv, *Meddelanden från Lunds Universitets Geografiska Institutioner* 107, Lund.
Maupertuis, P.-L. (1738, 1977) *La figure de la terre* (trans. as *Jordens figur*), Faksimil, Stockholm, and Tornedalica, Luleå.
Mårtensson, S. (1979) On the formation of biographies in space-time environments, *Lund Studies in Geography*, Ser. B, no. 47.
Mårtensson, S. (1992) Huvudsakligt innehåll i kursplan för geografi Arbetsmaterial, juli 1992, *Geografiska Notiser* 3, 134-137.
Mead, W. R. (1972) Luminaries of the north, *Transactions of the Institute of British Geographers* 57, 1-13.
Nelson, H. (1913) Hembygdsundervisning i folkhögskolan, *Svenska Folkhögskolans Årsbok*, 25-35.
Nilsson, Y. (1950) Bygd och näringsliv i norra Värmland. En kulturgeografisk studie, Ph.D. dissertation, Lund.
Nordenskjöld, A. E. (1889) *Facsimile Atlas*, Wahlström & Widstrand, Stockholm.
Nordlind, L. (1911) En sydsvensk israndslinie och dess geografiska betydelse, Ph.D. dissertation, Lund.
Odén, B. (1991) *Forskarutbildningens förändringar 1890-1975*, Lund University Press, Lund.
Olsson, G. (1979) Social science and human action or on hitting your head against the ceiling of language, in S. Gale and G. Olsson (eds.), *Philosophy in geography*, 287-308, Reidel, Dordrecht.
Rapp, A. (1961) Studies of the postglacial development of mountain slopes, Ph.D. dissertation, Uppsala.
Richter, H. (1959) Geografins historia i Sverige intill år 1800, *Studier och källskrifter utgivna av Lärdomshistoriska Samfundet* 17, no. 1, Almquist & Wiksell, Stockholm and Uppsala; Lychnos-bibliotek, Uppsala.
Rudbeck, O. (1679-1702, new ed. 1937-1950) *Atlantica*, Lychnos-bibliotek, 2: 1-5, Uppsala.
Rundquist, F.-M. and N. Hatti (1969), *Cooperatives in rural development. Land ownership and cooperative membership in Sirsi Taluk, Karnataka State, India*, The Finnish Association for Development Geography, vol. 10, no. 3, Helsinki.
Ruong, I. (1975) *Samerna*, Aldus/Bonnier, Stockholm.
Sjögren, O. (1909) Geografiska och glacialgeologiska studier vid Torne träsk, Ph.D. dissertation, Uppsala.
Sporrong, U., U. Elstam, and K. Samuelsson (1995) *Swedish landscapes*, Swedish Environmental Protection Agency, Stockholm.
Swedenborg, E. (1758) *De coelo et ejus mirabilibus, et de inferno* (trans. as *Heaven and Hell*, 1958), The Swedenborg Society, London.

Söderberg, S. (1960) *Stora Svenskar i lärdomens och naturforskningens värld*, Almquist & Wiksell/Gebers förlag AB, Stockholm.
Söderquist, T. (1986) The ecologists. From merry naturalists to saviours of the Nation, Ph.D. dissertation, Roskilde, Denmark.
Sörlin, S. (1988) Debatten om Norrland och naturresurserna under det industriella genombrott, Ph.D. dissertation, Umeå (Carlssons, Stockholm).
Thaning, O. (1986) Hundra år med årsskriften, *Svenska Turistföreningens årskrift*, 100 års volume, 81-111.
Törnquist, G., ed. (1993a) Sweden in the world, *Sveriges National Atlas* 8, Stockholm.
Törnquist, G. (1993b) *Sverige i Nätverkens Europa. Gränsöverskridandets former och villkor*, Liber-Hermods, Malmö.
Törnquist, G. (1998) *Renässans för regioner*, SNS Förlag, Stockholm.
Vidal de la Blache, P. (1922) *Principes de géographie humaine*, ed. by Emmanuel de Martonne, Colin, Paris.
Wahlström, L. (1984) Geografiutveckling och geografisk utveckling some om platser betydde något, *Meddelanden från Göteborgs Universitets Geografiska Institutionen*, Serie B, no. 76, Lund.
Wallin, E. (1980) Vardagslivet generative grammatik - vid gränsen mellan natur och kultur, Ph.D. dissertation, Lund.
Warntz, W. (1964) *Geography now and then*, American Geographical Society, New York.
Wenngren, G. (1990) Geografi och skolgeografi, *Acta Universitatis Upsaliensis*, Almquist & Wiksell, Stockholm.
Wennerholm, E. (1978) *Sven Hedin: en biografi*, Bonniers, Stockholm.
Widmalm, S. (1990) *Mellan kartan och verkligheten. Geodesi och kartläggning 1695-1860*, Institutionen för idé och lärdomshistoria, Uppsala.
Wik, H. (1953) Norra Sveriges sågverksindustri från 1800-talets mitt fram till 1937, Ph.D. dissertation, Uppsala.
William-Olsson, W. (1937) *Huvuddrag av Stockholms geografiska utveckling 1850-1930*, Akademisk Avhandling, Stockholm.
Winberg, I. (1944) Flottningen i Sveriges allmänna flottleder fram till omkring år 1935, Ph.D. dissertation, Göteborg.
Åquist, A.-C. (1992) Tidsgeografi i samspel med samhällsteori, *Meddelanden från Lunds Universitets Geografiska Institutioner* 108, Lund.
Öhman, J., ed. (1990), *Nordisk Samhällsgeografisk Tidskrift* 11.
Öhngren, B., ed. (1999) *Swedish research in human geography*, HSFR, Stockholm.

Anne Buttimer
Department of Geography
University College Dublin
Belfield
Dublin 4
Ireland

8 Geography in Russia: Glories and Disappointments

David Hooson

Few countries have had a more tortured history over the past century or so than Russia (or the Soviet Union as it was dubbed for most of that time). However, its story is packed with intellectual excitement and has displayed an originality that has derived from its distinctive geographical and historical context. The evolution of Geography as an academic discipline has closely followed the discovery and exploration of this vast part of the world and the experience of living and working in its natural regions and its ambiguous position between Europe and Asia. Its medieval roots and its "frontier" continental spread mean that it has also been, in many ways, Europe and North America rolled into one over the last two centuries.

In the last half century of Tsarist rule Russia took its place as a fully-fledged participant in the literary, arts and scientific life of Europe, and was distinguished by a number of prominent and original geographers who stood up well in comparison with their contemporaries in other countries. This being so, it was surprising that until the present generation the Russian contribution to geography had not found its way into any of the general histories of the subject. This virtually complete silence about one of the most vigorous of the national schools of geography was, of course, partly a reflection of the "Linguistic Curtain," but by no means entirely, since some of the best work appeared in western languages. It took the sudden surge of curiosity about the activities of the Soviet scientists in the post-Sputnik era to evoke serious search for the pre-Soviet antecedents and for continuity with the present.

G.A. Dunbar (ed.), Geography: Discipline, Profession and Subject since 1870, 225–243.

Beginnings

While it can be said that, as with Germany and France, the institutionalization of geography as a coherent academic discipline took place in Russia in the last half century of Tsarist rule, roughly from 1870 to 1915, there were important precursors in the previous century and a half. The dawn of notable geographical endeavors came in the reign of Peter the Great in the early 18th century. He had an intense interest in geographical expeditions, appraisals of resources and map-making, and it seems that the achievements of geography outshone those of other sciences during his reign. Bernard Varenius' *Geographia Generalis* (1650) was translated into Russian, and this helped to stimulate many good regional studies of the parts of the expanding empire of the 18th century. I. K. Kirilov produced the first atlas of Russia (1734) and established good cartographic standards and coverage for the first time. This three-volume atlas, with 120 maps in each volume, covered even the most outlying parts of the Empire and included historical and economic plans and maps of various cities. V. N. Tatishchev was a historian as well as a geographer who was interested in the regional division of labor and began the first scientific questionnaire project in Russia, while at the same time organizing the new industries in the Urals (Vucinich 1963; Letiche, Dmytryshin, and Pierce 1964).

The Academy of Sciences was founded by Peter and originally staffed mainly by German and Danish scientist-explorers, such as Vitus Bering, Georg Wilhelm Steller, and Peter Pallas. They helped to put Russia and Siberia "on the map" as far as Europe was concerned. It would be a mistake, however, to think that this early history of geography in Russia was dominated by foreigners. In addition to Kirilov and Tatishchev, already mentioned, who were thoroughly Russian, two others from among a distinguished group should be singled out here. M. V. Lomonosov, growing up in Archangel on the White Sea, became a polymath scholar and scientist who, among other things, founded Moscow University in 1755, and he was also head of the Geographical Department of the Academy of Sciences. The poet Alexander Pushkin is said to have called Lomonosov "the first Russian university" for his mastery of so many fields. S. P. Krasheninnikov, a close associate of Lomonosov, produced a great work on Kamchatka and the Kuril Islands that was almost Humboldtian in its scope and was translated into western European languages at about the same time that James Cook was making his famous scientific journeys in the Pacific (Krasheninnikov 1764). It was the first Russian regional study to examine the connnections among the various natural phenomena themselves (e.g., volcanic activity and plant and animal life), as well as a concern for the improvement of the economic and social lot of the native inhabitants. In fact, this concern, together with a search for reforms, and the delimitation of a national system of regions, incorporating both the natural conditions and the search for optimum molds for human development, alongside basic exploration, had become implanted as the distinctive hallmark of Russian proto-geography even before the beginning of the 19th century.

The Geographical Society

The Imperial Russian Geographical Society was founded in 1845 in St. Petersburg in response to the build-up of ideas and activities and the urgent need for an institutional forum for the many converging interests. It quickly acquired a prominent and active rôle in Russian life of the mid-century and kept it until the end of the Tsarist era. It was generally considered to be the most successful and popular of the learned societies and clearly struck many chords in tune with prevailing national aspirations and ways of thought. As in comparable societies in Western Europe, a major preoccupation was with exploration, evaluation of natural resources and imperial grandeur, since this was a period of rapid expansion of the Russian empire with both colonialism, as in the Caucasus and Central Asia, and in the newly available grasslands of Ukraine and Siberia where an American-type colonization was underway. On the other hand liberal reformers formed an important faction within the Society and aimed to prepare the ground for social reforms, especially looking towards the emancipation of the serfs, which finally happened in 1861. In addition the Society assumed initial responsibility for geology, meteorology, anthropology and archaeology, and established regional branches in newly colonized areas like the Far East, Siberia and the Caucasus that issued their own periodicals. Some 400 volumes were published, including special reports and regional serials in the first fifty years of the Society's existence. In its vitality and range of interests, this record compares favorably with the contemporary equivalents in any other country.

The leading light of the Society during this period of exceptional growth and influence was P. P. Semyonov (1827-1914), who directed it for over forty years. He received the royal suffix to his name (perhaps an equivalent to a British knighthood) of Tian-Shansky, after the mountain range in Central Asia that he explored.

He was, however, a complex man of parts, ideally suited to the task of holding together the very diverse geographical interests. At the time when he was preparing for his Tian Shan expedition he was a leading member of the "Committee for the Emancipation of the Peasants from the Bonds of Serfdom". He had come under the influence of Carl Ritter in Berlin, where he also compared exploration plans with Friedrich von Richthofen. On returning to Russia at the time of great ferment for reforming ideas following the death of Tsar Nicholas I, he translated the part of Ritter's *Erdkunde* dealing with Asia and also started to compile his own monumental five-volume *Geograficheskо-Statisticheskii Slovar Rossiiskoi Imperii* ("Geographical-Statistical Dictionary of the Russian Empire") (1863-1883). The distribution of population, with all its ramifications and pointers to reformist policies, was a major focus of his thought, and he was actively involved in planning for the first all-Russian Census (1897). The pace of development in Russia was quickening rapidly in this period, with urbanization, railroad building, the beginnings of modern industry and international trade, which, together with the imperial expansion and settlement, obviously stimulated geographical ideas. Semyonov was the man who linked the traveller-explorers of the earlier period, many of whom were anyway recognizably geographers in the modern sense, with the institutional development of academic geography in Russia after 1880.

The flowering of Russian geography

In common with other emerging National Schools of Geography in the four decades or so before World War I, there were the fitful beginnings of the establishment of Geography, among other subjects, in the expanding universities. The age-old primary mission of the subject—basic exploration of the world—was drawing to a close, and the prime need thenceforth was for collating the mass of accumulated knowledge and assessing its significance. In turn this called for agreement and affirmation of new frameworks of thought for analysis and synthesis of the data and for the place and aims of geography among the other emerging disciplines (Hooson 1968).

Russian geography at this period can stand favorable comparison with its contemporaries in any other country. In spite of the obvious differences in political and cultural development, it should be emphasized that the Russian scholars of this period, in the face of considerable obstacles, were thoroughgoing participants in the intellectual and academic activities of the modern world, although they often did not receive their due recognition because of the language barrier. Academic freedom, especially after 1905, was comparable to that in Western Europe in many respects. The evolutionary and developmental concerns of Charles Darwin and Karl Marx, though widely known and seriously followed in Russia, were not at that time accepted and applied as uncritically as they were in countries like Germany and the United States. In particular this was evident in the case of the leading geographers, whose approval was generally functional and pragmatic—dynamic rather than evolutionary—and concerned as much with man's impact on nature as with environmentalism. They also had a vital interest in the practical needs of society.

The chief credit for the building up of geography in Moscow University and for popularizing the subject with a broad section of the educated public (which then comprised at most a quarter of the population) must go to D. N. Anuchin (1843-1923). He was head of the joint Department of Geography and Ethnography at the University from 1884. He was a liberal and humane scholar who was prominent in the general intellectual life of Russia, being, for instance, associated with one of the most influential newspapers, *Russkie Vedomosti.* His early research projects ranged from the bison to African flora and the Ainu people of Hokkaido, and he conducted several archaeological expeditions. Later he wrote a fine regional study of Japan and detailed analyses of the relief and drainage of European Russia. His main life-work, however, was in the dissemination of geographical and anthropological knowledge, not only in the university but in the production of semi-popular works, and, last but not least, the founding in 1894 of the most scholarly Russian geographical journal, *Zemlevedenie* ("Earth Science" or "Geography"), which he steered until his death in 1923. It was comparable with its contemporary, the French *Annales de géographie,* in scope, volume and quality. Anuchin himself wrote some two hundred articles for this journal on a wide range of subjects. Perceptive reviews, often of foreign works, reports of congresses and obituaries appeared above his initials in virtually every issue. All these activities contributed to the development of academic geography as an independent subject in Russia. At Anuchin's death in 1923, most of the occupants of chairs of geography (and anthropology) in Russia were his students, and the Geography Faculty at Moscow University has led the country thereafter. In this respect Anuchin bears comparison with Paul

Vidal de la Blache in France as the progenitor of a distinctive national School of Geography.

Of the many individual scholars active in this period I will select three for special mention. All attracted more attention in the outside world than did Anuchin. Alexander Voeikov (1842-1916) was the most widely known as early as the 1870s when his book *Meteorology in Russia* was published in Washington DC and he became a regular participant in the activities of the Smithsonian Institution. He had traveled widely in Asia and the Americas from an early age and had become known as an original climatologist focussing on heat and water balance studies, in particular in relation to agriculture. By the time his book *Klimaty Zemnogo Shara* ("Climates of the World") appeared in several languages (first Russian edition, 1884), his international authority had become recognized. But he was much more than a climatologist, and his later works were mainly concerned with the influence of man on nature, at a time when various forms of environmental determinism were flourishing in the West. He was also intent on introducing into the Russian Empire suitable crops that he had identified during his extensive travels. His regional monograph on Russian Turkestan is worthy of comparison with the best exemplars of the French School (Woeikof [sic] 1914). Although he taught at St. Petersburg University, his impact, unlike Anuchin's, was primarily as a researcher with a truly world view and a very broad geographical perspective. As an example of someone who started as a world traveler and grew into a world-famous international scholar and "compleat geographer", Voeikov can hardly be bettered. In terms of originality and range of expertise—theoretical and practical—he could be considered primus inter pares in the galaxy of memorable Russian geographers whose ideas and methods have not lost their resonance today.

V. V. Dokuchaev (1846-1903), unlike Voeikov, did not travel widely outside Russia, nor did he partake at all profoundly in the institutional development of geography as a discipline as did Anuchin. He was, however, very influential as the "father" of world soil classification, with climate recognized as the critical factor. He demonstrated the truly integrated approach to the study of the interacting processes of the natural environment that became the hallmark of Russian (and later Soviet) physical geography. He began with a classic work, *Russkii Chernozem* ("Russian Blackearth") (1883), which set out the basic principles and was based on meticulous fieldwork. His regions for analysis were the now familiar "natural zones" (steppe, forest, etc.) that have been so crucial in Russian history but had not been explained scientifically before. The impact of his theories on Russian physical geography produced a subject fundamentally different from the geomorphologically-focussed physical geographies of Western Europe and the United States of the time. His teaching in St. Petersburg was disseminated by a number of prominent students in Russia and by people like Curtis Marbut in America and A. J. Herbertson in Britain.

The final personality I have selected for mention here, Peter Kropotkin (1842-1921), was, paradoxically, absent from Russia from 1876 to 1918—i.e., during virtually all of the period in which university institutionalization was taking place in Russia and Western Europe. In the 1860s he did fundamental geomorphological research in eastern Siberia, following which he was offered the secretaryship of the Imperial Russian Geographical Society, headed by P. Semyonov. He refused the appointment, however, saying that he felt that he had no right to such "higher joys" while "all around him was misery and struggle for a moldy piece of bread" (Kropotkin 1899 [1962 ed.], 157). Thus was launched the career for which he became much more famous, that of political activist and anarchist. He was clapped into the Peter-Paul Fortress in St. Petersburg and while a prisoner there wrote his work on Siberian orography. He soon escaped from prison and started a new career in London. In addition to promoting his own brand of Anarchism, he also (like his fellow geographer and anarchist Élisée Reclus) kept up his interest in geography and its institutional development, from his base in the Royal Geographical Society where he was popular in spite of his politics. Apart from publishing his Russian research in English and French, he wrote many articles on Russia for the *Times* and other journals, as well as for the *Encyclopaedia Britannica.* He gave a speech at Oxford University, before Halford Mackinder went there, proclaiming that if Oxford had "had a Ritter" it would long since have taken the lead in university education. His influential paper, "What Geography Ought to Be" (Kropotkin 1885), published two years before Mackinder's more famous pronouncement, was more humane than the latter, though perhaps complementary in several ways. Kropotkin had renounced his previous predilection for "Physiography", asserting that "a study of nature without man is the last tribute paid by modern scientists to their previous scholastic education". This philosophy was in line with the Russian School of Voeikov, Dokuchaev, Anuchin, et al., and, had Kropotkin felt able to return to Tsarist Russia, he would without doubt have enriched that already distinguished academic community. As it was, he did not return until June 1917. Although he was initially welcomed by V. I. Lenin, Kropotkin died a disappointed man in 1921.

There were many other geographers in this Golden Age of Russian geography who would be well worth mentioning if space permitted. At this point, however, I should like to suggest some of the more powerful and enduring ways of thought that characterized that vital and productive period:

1. The way of looking at the natural environment was distinctly different from that of other contemporary schools of geography that were developing at the same time in Germany, Britain, France and the United States. The major emphasis was on bio-climatic rather than geomorphological features and on an integrated and functional, rather than a specialized, approach. Although this way of thinking can be traced back to the 18th century scientists mentioned above as well as to the works of Humboldt, the main scientific theories underpinning it derived from Voeikov's heat and water balance concepts and Dokuchaev's on natural zonation as a result of climate.
2. There was more concern with the impact of man on nature than vice-versa, again in contrast to the prevailing balance in America, and this was coupled with ameliorative schemes for modifying or restoring the natural balance.
3. There was a long-standing preoccupation with the delimitation of regional boundaries, both physical and human, usually with the explicit aim of application to the solution of practical problems, particularly towards the betterment of the life of the peasants—the overwhelming majority of the population.
4. The inventory, exploitation and mapping of natural resources, which resulted in a number of comprehensive atlases and hundreds of monographs and articles, mainly produced under the auspices of the Geographical Society. Underlying this was a persistent strain of philanthropy and patriotism.
5. Throughout this period, which was one that combined features of imperial expansion on the Western colonialist model with the frenetic settlement, railway building and resource exploration of the vast open spaces of the American "frontier" model, the geographers, along with most intellectuals, were preoccupied with a search for a new Russian "Eurasian" identity (Bassin 1999). This was partly in response to a long-standing love-hate affair with Europe and "the West", resulting in a split between "Slavophiles" and "Westernizers". "Visions"of newly encountered parts of this continental expansionism as extensions and modifiers of Russian identity were a recurrent feature of geographical interpretation at this period.

Those varied but interlocking aspects add up to a remarkably rich legacy bequeathed to the Soviet geographers. In spite of the massive disruptions of the first World War, a ruinous peace, revolution and civil war, these traditions carried over to the 1920s. Also, oddly enough, the achievements of the Russian school of geographers were recognized and praised, not only by Marx himself but also by Lenin and, more to the point, by Georgi Plekhanov, the prime philosopher of Russian Marxism, who also spoke warmly of Friedrich Ratzel's ideas (Plekhanov 1934, 34). In spite of all this, however, the rich Russian tradition in geography was submerged, like so much else, in the tidal wave precipitated by the totalitarian clamp-down after 1929.

The Stalinist period

The intellectual climate, in geography as in all other aspects of Soviet life, changed abruptly with the imposition of Joseph Stalin's first Five Year Plan. Geographers, like others, were fully mobilized and cadres of specialists were quickly trained and sent off on immediate practical tasks. This had the effect of seriously fragmenting geography and sidestepping the traditional broad, integrated philosophy. Cultural, historical, political and regional studies were discontinued, and even a narrowly defined economic geography fell afoul of the political directives, which involved serious risks for honest academic analysis. Some prominent geographers were arrested and even killed for using western location theories to advise against projects that turned out to be dear to Stalin's heart. Therefore, it is hardly surprising that many geographers opted for physical topics, partly for a quiet life. Something similar occurred in Germany in Adolf Hitler's time with the rise of *Geopolitik*, in the way of mind-control and tendentious, sycophantic reactions.

In any case, the upshot was that by the time of the death of Stalin in 1953, the character of Russian geography had been transformed. Publications had become overwhelmingly physical in orientation, and even these had become highly specialized and, surprisingly, given the expressed needs of the regime, often seemed divorced not only from relevance to economic planning but also to be showing signs of having appreciably moved away from the natural zonation or heat and water balance principles of Dokuchaev and Voeikov.

The elevation of the power of the Academy of Sciences vis-à-vis the universities accentuated the imbalance in the structure of geography, since the directors of the various institutes, including the Institute of Geography, were invested with almost dictatorial power over people in their profession. Two examples: director A. A. Grigoriev was able to publish an authoritative article in the *Geographical Journal* (London) on "The State of Soviet Geography", portraying the subject as almost entirely physical and oddly apolitical and parochial (Grigoriev 1955). Shortly afterwards, he was able, as "geographical editor", to depict geography in the latest edition of the *Great Soviet Encyclopedia* as, again, exclusively a collection of physical sciences, while economic geography merited just a sentence submerged in the Economics entry and other brands of human or regional geography were omitted altogether.

In fact, a respectable corpus of high quality work in various branches of human and regional geography had been done, largely contained in a vigorous journal called *Voprosy Geografii* ("Questions of Geography"), launched by N. N. Baransky and other Moscow University geographers in 1946. Throughout this fearful period there was a tacit agreement between the Regional School headed by Baransky and the Landscape School, headed by L. S. Berg, to keep alive some of the pre-revolutionary traditions (Weiner 1999).

Ferment and renewal around 1960

In retrospect, the period between the mid-1950s and mid-1960s seems like a Golden Age for the Soviet Union, after the death of Stalin and before the onset of the Leonid Brezhnev "era of stagnation", as it was subsequently dubbed. Essentially, these were the Nikita Khrushchev years, characterized by a new optimism, vigor, openness and improvement in living standards. The emergence of the Soviet Union as a "Super-power" was signalled dramatically by the launching of Sputnik in 1957, which electrified not only the United States but the many "new nations" that were breaking out of colonialism and looking for a new rôle model. The new "Heartland" power, apparently holding sway over Eurasia from Central Europe to the South China Sea, brought back Mackinder's theories to the Pentagon (as I can personally testify from teaching a short course on the Soviet Union there in 1958). Although these years were far from tranquil internationally (with the Hungarian suppression, the building of the Berlin Wall and the Cuban missile crisis) the national economy was stirred up by unprecedented reforms, such as the replacement of coal by oil and gas—stimulated by new discoveries—and by new possibilities for foreign contacts and the publication of previously (and subsequently) banned books.

In this new and exciting atmosphere, the stage was set for grand disputation within the discipline of geography, directed towards the humanizing of the subject and especially restoring the integration of its physical and human aspects, which had, as we have seen, been a fundamental tenet of the pre-revolutionary geographers.

The battle royal

One should have an understanding of the difficulty of overturning established power structures in the Soviet Union, even in the relatively propitious atmosphere of the Khrushchev period. The fact that the heads of the two most powerful institutions in the field, the Academy of Sciences' Institute of Geography and the All-Union Geographical Society (I. P. Gerasimov and S. V. Kalesnik, respectively) had come out unequivocally against the heresy of a "unified" geography and for the "legal" separation of physical from economic geography ("separate laws") meant that compromise seemed to be ruled out (Hooson 1959).

But the argument came to a head in 1960 with the publication of a book called *Teoreticheskiye Problemy geografii* ("Theoretical Problems of Geography") by V. A. Anuchin, a professor at Moscow University (Anuchin 1960). It was not only a scholarly historical and philosophical analysis of the evolution of geography, especially though not exclusively in Russia, but also hard-hitting, aimed directly at demolishing the prevailing doctrine of duality in Soviet geographical theory and practice. Anuchin was lucky enough to have the enthusiastic support of N. N. Baransky, then 80 years old and revered by the profession at large, who had built up economic and regional geography at Moscow University, and his influential successor, Yu. G. Saushkin. But even so it was an uphill struggle, involving a dramatic doctoral dissertation defense at Moscow University attended by hundreds of people. By this time the case had become a cause célèbre, but even so the final vote denied Anuchin the required two-thirds majority votes for his doctorate, reflecting the importance of essentially external votes.

Eventually the conflict became so intense that the Communist party itself stepped in and essentially adjudicated in favor of the integrated view, with a denunciation of the Stalinist definition of the environment as "a purely natural category". It was further alleged that this concept had become the pretext for the construction of "an insurmountable wall" between nature and society, damaging the search for solutions to pressing environmental and economic problems and distorting the planning process. Amazingly, Geography thereafter was featured in various national magazines, the most highly visible of which was a series of six articles in the influential *Literaturnaya Gazeta*, initiated by Anuchin with "A sad tale about geography" and rounded off by Saushkin with "The today and tomorrow of geography", a call for thorough restructuring, with the explicit support of the editors (*Soviet Geography* 1965). The ideological battle was henceforth over, and a new set of guidelines was set up by the establishment, supplanting the traditional topical specialties with a set of integrated topics focussed on specific problems and regions.

Thus the official line in Soviet geography in the late 1960s had become radically different from that of the late 1950s, so that it seems in retrospect to qualify as a genuine "revolution", consonant not only with pre-soviet, but also, remarkably, Marxist traditions as well.

Therefore, we can look back upon the years about 1960—when I was lucky enough to visit the Soviet Union for the first time and got caught up and involved in their intellectual disputations—as a significant watershed, or perhaps revolution, in the history of Soviet geography. Much the same period would be said by some people to constitute a watershed or revolution in American geography, but there were fundamental differences in what was going on in the two newly emerged rival super-powers at the time. In the mid-1950s, when Soviet geography, as we have seen, had become overwhelmingly physical and "legally" split into two (unequal) parts, American geography had the opposite problem, having by then become dominantly human in emphasis while still holding to the traditional philosophical axiom of the essential unity of the subject. Both national schools of geography were poised for a radical reorientation of their assumptions, theories and practice in the late 1950s, accompanied by bitter arguments and a propensity for polarization and even demonization. However, whereas the American "revolution" focussed on techniques, quantitative methods and model-building, and was definitely not harking back to earlier national traditions, the Soviet contemporary equivalent was more philosophical and much less technical and contained an important element of intellectual nostalgia for a perceived broken national heritage in Russian Geography.

"The era of stagnation"

The two decades between the geographical "revolution" and the dismissal of Khrushchev in the mid-1960s and the accession of Mikhail Gorbachev in 1985 were characterized by a duller, more repressive and authoritarian period, discouraging originality and controversy, with a corresponding increase in cynicism, opportunism and corruption in Soviet society. In retrospect, this "era of stagnation", coupled with the concentration of resources on armaments, led to the crumbling of the state, which Gorbachev tried, too little and too late, to save. The mania for secrecy led, amongst other things, to widespread deliberate distortion of maps, which obviously had a disastrous effect on geography, to say the least. Some good work was done, of course, and the shift of emphasis towards a more humane economic geography and away from physical geography was continued in some measure. But the power structure remained basically unreformed. Twenty years after the director of the Institute of Geography was publicly castigated and, by implication, threatened with dismissal, he was still in full charge and died on the job at the age of 80.

The Gorbachev period

Following the relative torpor of the rule of Leonid Brezhnev, Yuri Andropov and Konstantin Chernenko, a second time of renewal was ushered in with the accession to power in 1985 of Mikhail Gorbachev, who, sensing the sclerotic condition of the system, launched the twin national projects of *Glasnost* (Openness) and *Perestroika* (Restructuring), both of which slogans have now faded away. Alongside this was a stated intention to begin to re-write Soviet and Russian history with a regard for the truth, to expose mistakes and start a clean slate. Soviet ideology, while also originally proclaiming a clean slate, with its intention to build "the new society", had in fact consistently downplayed or distorted history, in spite of Marx's preoccupation with it.

The glasnost injunction got a boost with unexpected speed and in an unexpected direction in April 1986, with the atomic explosion at Chernobyl in the Ukrainian SSR. As it turned out, this detonated a series of explosions of what may be called "environmental nationalism" among the various "republics" of the Soviet Union, directed at, and blamed on, the Moscow overlords. The independence movement in Ukraine and thus eventually the collapse of the Soviet Union itself were hastened by the Chernobyl incident. Other festering environmental grievances surfaced all over the country, such as the oil pollution of sea and air around Baku, the pulp-mill at Lake Baykal and—worst of all—the shrinking of the Aral Sea to half the size it was in 1960. In the last case, the Russians were blamed by the Central Asian peoples for distorting their formerly diversified economies by forcing cotton monoculture on them, together with ill-considered and wasteful irrigation schemes. The final insult for these people was the cancellation in 1986 of long-laid plans to divert water from the Ob-Irtysh system in Siberia "to save the Aral Sea". This was seen as the final proof that the Russians, having raped their natural environment and economy, had callously left the Central Asian peoples to die. "Environmental racism" was charged.

The other half of this upsurge of environmental rage—that of nationalism—was rearing its head in various forms. There was resentment about Stalin's gerrymandering of the various ethnically designated republics in a classic exercise of "divide and rule" techniques. The assignation of the Armenians of Karabakh to Azerbaijan is one of the most intractable legacies of these policies to this day. Then, from the economic angle, resentment arose over the exploitation of natural resources, such as gold and diamonds in Yakutia, or oil in Tatarstan, with the local regions not seeming to benefit as they should.

The emergence of political geography

The most striking feature of the new geography that was precipitated by these convulsions was the emergence of political geography from the shadows (Hooson 1996). For more than half a century from the early 1930s to the late 1980s, this branch of the subject, which had been well developed before, was virtually banned. This may seem to be a paradox, in that most political of states, the Soviet Union, but in fact the dominance of the centralized command-administrative structure left no room for independent analysis of regional characteristics and interests. The old catchall economic geography, with its deadening emphasis on heavy industry and centralized Five Year Plans, as well as a physical geography still curiously unfocussed on environmental issues and human concerns, proved singularly unsuited to cope with the erupting problems of the Soviet Union in what proved to be its dying years between 1986 and 1991. Conversely, the new political geography suddenly proved to be a heaven-sent framework within which to come to grips with the real, urgent, problems facing the people and regions of this vast country, which had hitherto been swept under the rug.

I well remember the palpable atmosphere of excitement and anticipation at the second conference-seminar on political geography in June 1990 at the Geographical Society in Leningrad, which I had the privilege of attending. Led by Leonid Smirnyagin and Vladimir Kolosov from Moscow, with the active encouragement of Sergei Lavrov, it sent an unmistakable signal that political geography had at last been "legalized" and had struck a chord in tune with the aspirations of some of the brightest of the young geographers, and also with the urgent needs and problems of the country. The level of intellectual excitement was comparable to that of the late 1950s and early 1960s and contrasted with the regressive orthodoxy of the intervening quarter-century. However, as in the earlier period, the deadening persistence of the centralized administrative structure was a formidable impediment to effective application of the "new" ideas to the solution of urgent regional and national problems.

After the Soviet Union

One-sixth of the land area of the earth officially lost its identity at the end of 1991 and, at the time of writing, eight years on, has yet to develop a stable or recognizable identity. Within the newly "independent" Russian federation the first president, Boris Yeltsin, expansively recommended that the often quite dysfunctional administrative sub-regions "take as much autonomy as they could swallow". The new Russian president, Vladimir Putin, although anointed by Yeltsin as his successor, has publicly repudiated this invitation and has called for a restitution of a strong centralized state authority. The brutality of the war against the secessionist "republic" of Chechnya underlines the seriousness of this problem; however, in the anarchic conditions of the last few years many regional "governors" have in fact established semi-independent fiefdoms, often in the context of criminal activity. This confusing and menacing reality, coupled with the abysmal decline of the economy, public health and the financing of research and educational institutions, has made it difficult for the new geographers to work effectively, in the context of a shrinking economy and drying up of government subsidies. The end of the binding ideology, the collapse of key institutions and the abandonment of the framework of the (rarely fulfilled) Five Year Plans have created an organizational vacuum and therefore in theory great opportunities for a clean slate and application of new ideas. There is also an aching vacuum of the spirit, born of disillusionment not only with communism and the betrayal of its ideals but also with the way in which democracy and the market economy have been unfolding in the last few years, under Western tutelage, leading to a deepening poverty for the many and great riches for the very few. What would Kropotkin say today? The cynicism, malaise, anxiety and anger, together with the need for a complete re-writing of hitherto-approved Soviet history, make a new beginning far from easy.

Opportunities and setbacks for geography

Eventually the most fundamental requirement, following the political overturning, will have to be the return of Geography, with the spotlight trained on the real regions, which have a deep meaning for the peoples who have inhabited them, and which are bound up with an identity that includes a familiar and beloved natural environment that they will defend as much against despoliation as against invasion (Pryde 1995). The agenda facing young geographers fits well with the dominant themes of identity, political ecology and socio-economic regional and cultural studies (Hooson 1999). Equally urgent is the broadening of historical and international perspectives, which had been stunted and distorted for the lifetime of all but a very few living today.

But the pitfalls that still bedevil the revival of geography are indicated in an impassioned article published in a major Russian newspaper by a Moscow geographer (Shuper 2000). It began by announcing and lamenting "the recent elimination of the subject of geography from the standard list of required university courses", and went on to document the continuation even today of the old Soviet habit of deliberately altering published maps (supposedly to confuse "the enemy") or at least by omission, such as the deliberate absence of scale in a recent (1998) atlas of Moscow. A particularly gruesome practical example cited was that the officers who stormed the Chechen capital of Grozny in the recent costly battles were equipped with xeroxed Soviet-era maps that had intentionally excluded nearly half of the existing roads! More fundamentally, the author bemoaned the fact that the managers and entrepreneurs who are charged with planning have inadequate educational foundations for distinguishing between different levels of development and specific characteristics of the regions of Russia and for devising appropriate strategies for social and economic reform.

This dismal verdict recalls the "sad tale about geography" in the 1965 national press that began what turned out to be a rather short-lived reform. Incidentally, it may evoke recognizable echoes in the United States, now the sole "indispensable" super-power, which still does not have geography in most high schools or in several of the most prestigious universities, and is constantly being called upon to lead the world with inadequate knowledge about that world and its problems, which increasingly have key geographical dimensions.

Legacy and promise

Having opted to focus my scholarly attention, nearly half a century ago, on a country called the Soviet Union, then considered very important, I am painfully aware that, except for short periods, it has seemed like a study in morbidity. I have been buoyed up by my natural instinct to go historical and philosophical. As a result, when I was swept up in the ferment and disputation about the discipline in the fifties and sixties, and as a by-product was impelled to delve into the rich legacy of pre-revolutionary Russian geography, I was just able to bear the boredom of keeping up with the unnatural economic geography, with its illusory Plans, and the inhuman physical geography of most of the Soviet period.

Now that freedom of thought and expression seems genuinely to have arrived in Russia, at least the bright aspiring young geographers, who are still being nurtured by some fine leading scholars, especially in Moscow and St. Petersburg universities, have a chance to show their mettle, provided that the absurd and annoying hangovers from Soviet "secrecy" can gradually be rectified, and that the economy starts to pick up and allow for necessary spending for education.

Of the richness and rightness of the legacy of what may be called the Russian School of Geography, with its emphasis on meaningful integration of human, natural and regional themes, together with its persistent concerns for applications towards the betterment of life, there can be no doubt. Although it is hard to be optimistic about Russia at the outset of the twenty-first century, I do feel that these traditions, which over the perspective of the past two centuries stand up well in comparison with those of any other countries, will be enormously powerful and supportive to their heirs.

There is no shortage of talent or original ideas in Russia nor of urgent problems and interesting themes pressing for solution from a geographical perspective for the good of the whole. Russian geography has been revitalized and energized before, and can, with our sympathy and encouragement and with luck, do so again. Cautious optimism is still in order.

References

Anuchin, V. A. (1960) *Teoreticheskiye Problemy geografii*, Moscow. English version, *Theoretical Problems of Geography* (1977), edited by Roland Fuchs and George Demko, with an introduction by David Hooson, published in Columbus, Ohio.

Bassin, Mark (1999) *Imperial Visions: Nationalist Imagination and Geographical Expansion in the Russian Far East 1840-1865.* Cambridge, England.

Grigoriev, A. A. (1955) The State of Soviet Geography, *Geographical Journal* 121, 429-439.

Hooson, David (1959) Some Recent Developments in the Content and Theory of Soviet Geography, *Annals of the Association of American Geographers* 49, 73-82.

Hooson, David (1968) The Development of Geography in Pre-Soviet Russia, *Annals of the Association of American Geographers* 58, 250-272.

Hooson, David (1996) The New Political Geography in Russia: Antecedents and Applications, pp. 103-109 in Vincent Berdoulay and J. A. van Ginkel (eds.), *Geography and Professional Practice* (Nederlandse Geografische Studies 206), Utrecht.

Hooson, David (1999) National Identity and Environmental Outrage, with Special Reference to the Post-Soviet Lands, pp. 463-467 in J. R. Pitte and A. L. Sanguin (eds.), *Géographie et liberté: mélanges en hommage à Paul Claval*, Paris.

Krasheninnikov, S. P. (1755) *Opisanie zemli Kamchatki*, St. Petersburg. English edition, *The History of Kamtschatka and the Kurilski Islands, with the Countries Adjacent* (1764), Gloucester and London.

Kropotkin, Peter (1885) What Geography Ought to Be, *Nineteenth Century* 18, 940-956.

Kropotkin, Peter (1962, first published 1899) *Memoirs of a Revolutionist*, New York.

Letiche, J. M., B. Dmytryshin, and R. A. Pierce, eds. (1964) *A History of Russian Economic Thought—9th through 18th Centuries*, Berkeley and Los Angeles.

Plekhanov, Georgi (1934) *Fundamental Problems of Marxism*, London.

Pryde, Philip (1995) *Environmental Resources and Constraints in the Former Soviet Republics*, Boulder, Colorado.

Semyonov, P. P. (1863-1883) *Geografichesko-Statisticheskii Slovar Rossiiskoi Imperii*, 5 volumes, St. Petersburg.

Shuper, V. (2000) Nezametny Grabezh [Unnoticed Theft], p. 7 in *Moskovskiye Novosti* (8-24 April 2000).

Soviet Geography: Review and Translation, vol. 6 (September 1965),pp. 27-56. English translations of the six articles published in *Literaturnaya*

Gazeta (1965).
Vucinich, A. (1963-1971) *Science and Russian Culture*, 2 vols., Stanford, California.
Weiner, D. R. (1999) *A Little Corner of Freedom: Russian Nature Protection from Stalin to Gorbachev*, Berkeley and Los Angeles.
Woeikof, A. (1914) *Le Turkestan russe*, Paris.

David J.M. Hooson
P.O. Box 1286
Point Reyes Station, California 94956
USA

9 Academic Geography, American Style: An Institutional Perspective

W. A. Koelsch

The late 19th century (in the United States, the post-Civil War period) saw the emergence of permanent institutions created to sustain interests that had earlier been undifferentiated, erratic, or personal: symphony orchestras, comprehensive arts museums, research universities, academic disciplines. These new ventures owed much to rising levels of personal and national wealth, increasing cultural and civic consciousness in burgeoning American cities, the march of academic specialization and, not least, the career strategies and seized opportunities of individuals in an era of faith in science, progress and broadened access to education and culture (Dunbar, in Blouet 1981, 71-88; Koelsch, in Blouet 1981, 89-104). These developments form the context for the rise of geography as an American university discipline after 1870 and the emergence of the geographer as a credentialed academic amidst the mix of disciplines and scholars competing for newly available resources and status in a season of new beginnings and high expectations.

Pre-professional American geography

By the mid-19th century, geography in America had become an omnium-gatherum of ill-digested information concerning the earth and the characteristics of the places upon it. It had often been a part of the curriculum of the liberal arts colleges and private academies of the early American republic. But geography had largely been dropped from college programs by 1830 and, like arithmetic, had become a mere admissions requirement (Broome 1903, 42). To the extent that instruction in geography survived in the American college, it was as an auxiliary field, the "handmaid" of natural science, history or classics, taught by practitioners of those disciplines. J. D. Gross at Columbia University (1784), apparently the first American academic to have geography as part of his title, also taught moral philosophy, chronology and German (Dryer 1924, 143; Warntz 1964,

G.A. Dunbar (ed.), Geography: Discipline, Profession and Subject since 1870, 245–279.

90). In 1825 the United States Military Academy employed the Rev. Charles P. McIlvaine, an Episcopal priest (later Bishop of Ohio), as "chaplain and Professor of Geography, History, and Ethics." He was let go two years later, however, when the authorities feared his evangelical preaching would discourage the martial spirit of the cadets (Weld 1938, 108).

As these disparate examples illustrate, geography took a back seat to other concerns in the post-secondary institutions of the antebellum period. Widespread acceptance of the assumption that geography was (in Kantian terms) a propaedeutic or foundational discipline, however, meant that it was readily conceded a place in the pre-college curriculum. In the 1830s the Northeastern states, followed by those in the Midwest, began to mandate a standardized academic program, including geography, in the common or elementary schools. What was taught under that rubric, however, varied widely. In whatever form, geography lessons were often rote memory exercises: "Maine, Augusta, on the Kennebec River" and the like. Teachers of the common school subjects, especially in rural areas, were not often academically trained for the task. It was a sufficient qualification to know a little grammar and arithmetic, and to be of sound moral character. However imperfectly mastered by the teacher, geography became the first representative of either the natural or the social sciences to be required of all students.

Some eleven different types of geography were taught in American secondary schools prior to 1861, including ancient geography, modern geography, sacred geography, mathematical geography, and political geography (Roorbach 1937, 6). Geography as part of a more advanced, although at the time scarcely college-level, teacher preparation program was formalized in the so-called "normal schools," appearing first in 1839 at Lexington, Massachusetts. School geography, map drawing, lessons on the globes, and ancient geography were all parts of the normal school program. Methods of teaching geography were also transmitted through short-term "teachers' institutes" that sprang up in Connecticut, Massachusetts, New York and elsewhere in the following decades. Neither the teachers in such programs nor the teachers they prepared could be called, at this point, professional geographers.

To the degree that the numerous tribe of textbook writers in the late 18th and early to mid-19th centuries were professionals, they were learned in other fields. They might be clergy, such as Jedidiah Morse, called "the Father of American Geography" for his plethora of textbooks and gazetteers published between 1784 and 1826. Some were professional educators, such as Emma Willard, Principal of the

Troy (New York) Female Seminary, or her sometime collaborator William Woodbridge, both of whom tried to incorporate the methods of the Swiss educational reformer J. H. Pestalozzi in their textbooks (Walters 1987). Even the lexicographer Joseph Worcester had begun his career by writing geography and history texts and gazetteers before preparing his edition of *Johnson's English Dictionary.*

As a national educational market developed through population migration and growth and the spread of the common school idea, textbook entrepreneurs responded by creating their own stables of writers for series of general or "modern" geography texts. They also published texts for more segmented parts of the geography market, such as those in "ancient geography" or "classical geography," which supplemented instruction in classical languages (Koelsch 1992). Their content has been examined by several 20th-century scholars. But there is much primary work to be done on the biographical, historical, business and consumption aspects of geography textbooks before we can speak with the assurance that, for example, Spanish historians of geography can now provide as a result of their examination of geography's role in that nation's educational system.

Into this matrix, in the 1850s, came the Swiss geographer Arnold Henri Guyot, student of Carl Ritter and Alexander von Humboldt and arguably the first professionally trained geographer in the United States. Guyot's first American lectures, published in 1849 as *Earth and Man*, not only established his reputation among scientists and the intellectual élite of the Northeast, but immediately became a resource for teachers of geography wishing to go beyond the materials made available to them by learned amateurs. Guyot was soon employed in lecturing at normal schools and teachers' institutes, and his book made contemporary European notions of geography widely available to teachers and opinion leaders (Walters 1987).

Guyot's geography was global in its reach, explanatory in its method, scientific in a contemporary sense in its assertion of causal relationships, and meaningful in its teleology in that time and place. Guyot made a convincing case for geography as a coherent and unified academic discipline linking nature and human activity. *Earth and Man* was used as a text in several leading colleges, not least Harvard. Guyot's developmentally-based series of school texts and wall maps, largely produced after his removal to the College of New Jersey (now Princeton University) at mid-decade, set a new standard for instructional aids.

What became known as "physical geography" in the comprehensive mid-century sense began to appear in pre-college instruction in the early 1850s. It was legitimated as a subject for college admission in 1870 when both Harvard, the leading private university, and the University of Michigan, the leading public institution, recognized it as part of their admissions requirement (Broome 1903, 42, 49). Beginning in the Spring of 1871 Harvard also offered physical geography as a natural science course, a pattern soon followed by other colleges and universities. There is need for careful study of the relationships of these activities to the expansion of state requirements for universal schooling and teacher preparation.

The Beginnings of Professionalization, c.1870-c.1915

In addition to scholarly research and its support, professionalization implies the credentialing of future practitioners, mainly through the authority to supervise student research and to offer degrees in the discipline. That process was encouraged by the expansion of the curriculum through the elective system, by the competitive appeal of foreign study and particularly of the German Ph.D. degree (Martin 1988b), and by the impact of the new Johns Hopkins University and its successors, with their strong emphasis on graduate study and research, on the older universities. As in other disciplines, the late 19th century saw the slow ratcheting up of academic programs in geography. In the 1880s and 1890s numerous opportunities for the production of professional geographers emerged, particularly in the private universities of the Northeast (Dunbar 1986).

One such opportunity resulted from the new curricular emphasis on natural history, subsequently differentiated into biological and earth sciences. A second came about through internal changes within the field of political economy (itself becoming differentiated from history and moral philosophy), which first provided for and later released the arena occupied first by commercial and later by economic geography (Fellmann 1986a). A third, less well-marked but clearly evident by the 1890s, was an amorphous, often historically grounded territory that appeared initially as "anthropogeography" and incorporated concerns that would eventually emerge as cultural geography (Livingstone 1992). The rôle of geography as handmaid continued for a short while in classics, until philology came to dominate the field. In American history, the importance of geographical knowledge was reinforced by the emergence of frontier theory in the 1890s. While each of these streams was represented in academic programs, their locus, emphasis and prevalence varied. Only the first two were major elements in the production of geographers before 1915.

The most significant academic program in physical geography was that directed by William Morris Davis, the major organizer of the discipline in late 19th- and early 20th-century America (Beckinsale, in Blouet 1981, 107-122). Davis, whose own highest degree was master of mining engineering, apprenticed as a meteorologist under Benjamin A. Gould and as a geologist under Harvard's Nathaniel Southgate Shaler (Livingstone 1992), who himself had been exploring what an expanded natural history program might offer geography. Shaler hired Davis as an assistant in the Kentucky Geological Survey, then as Instructor in Geology at Harvard. In 1885 Davis was promoted and given his own course, Physical Geography and Meteorology, from which slender base he built a program of research, publication and student supervision in various subfields of physical geography before being promoted to the largely research post of Sturgis-Hooper Professor of Geology in 1898. His students in the Harvard programs (including summer courses) became some of the leading figures in early 20th-century American geography, although Harvard appears to have awarded only two doctorates in physical geography proper, those of F. P. Gulliver (1896) and J. W. Goldthwait (1906), until after Davis' retirement in 1912.

Other principal centers of research and graduate study in physical geography in the 1890s were Cornell University, whose program was directed by Ralph S. Tarr, a student of Shaler and Davis; Yale University, which expanded into geography significantly under Herbert E. Gregory (Martin 1988a); the new Johns Hopkins University, whose founding President was the geographer (and Guyot disciple) Daniel Coit Gilman; and, after 1892, The University of Chicago under Thomas C. Chamberlin and Rollin D. Salisbury (Pattison, in Blouet 1981, 151-163). Chicago's founding president, William Rainey Harper, had unsuccessfully attempted to recruit Oxford University's Halford Mackinder for his new institution in 1892.

At Hopkins, the geologist William Clark took over a required undergraduate geography course from the historians. Clark stripped the course of its historical content and, like Davis, built a physical geography program out from that base and from his relationships with surrounding institutions, eventually directing Maryland's Weather Service, Geological Survey, and Soils Survey. Clark and his geology colleague George Williams produced a small number of graduate physiographers and one climatologist before a younger and more specialized generation of Hopkins geologists cut back the scope of what Clark had termed the "New Geology" (Koelsch 1981).

These academic entrepreneurs, like those in other disciplines, faced a task of boundary-setting, particularly in distancing geography from geology. For Francis W. Parker, a Guyot disciple addressing the International Geographic Congress of the World's Columbian Exposition in Chicago in July 1893, the division was a simple one: "Geography is the science of the present appearance of the earth's surface. Geology is the history of the present appearance of the earth's surface . . ." (Parker 1893, 125). Yet the academy demanded not only definition but scientific legitimation. Davis' solution was to derive a timeless analytical model, the so-called "geographical cycle" of youth, maturity, old age, and rejuvenation, which he applied as a heuristic for later landform study around the globe (Beckinsale, in Blouet 1981). The Chicagoans, on the other hand, pressed for a method of analysis that included what Chamberlin called "multiple working hypotheses," emphasizing the particularities of earth history, a mode of analysis that was to have a longer life, albeit in other areas, than the Davisian cycle (Pattison, in Blouet 1981; Entrikin 1984).

A second locus of future professional strength was built up from geography's role in newly established schools of commerce and business. Economic geography framed the historicist approach younger economists had recently imported from their studies in Germany (Fellmann 1986a). The first separate department of geography in the United States was established within the University of California's recently opened College of Commerce in 1898. Its commercial geographer, Lincoln Hutchinson, often taught his courses under the aegis of economics or history (Dunbar 1981). A number of economists in the late 1890s and early 1900s taught the first economic geography in their home institutions or wrote high school or college commercial geography texts (Fellmann 1986a). In Clark University's newly established "Collegiate Department," for example, European-trained Frederick A. Bushee incorporated American and European commercial geography (as well as economic and industrial history) into his year-long introductory "Principles of Economics" course from 1903 until his departure in 1908. But his wholly American-trained successor omitted such topics, mirroring the general discard of the field by economists as historicism lost its brief hegemony.

More significant for the growth of that option within academic geography, however, were developments in the Wharton School of Finance and Commerce at the University of Pennsylvania. Although not without dissent, the first generally recognized Ph.D. in economic geography is credited to partly German-trained Emory W. Johnson, who was awarded a Ph.D. in Pennsylvania's economics program in 1893 and taught commercial geography courses in the Wharton School one-third time for several years. In 1901, J. Paul Goode, a new Pennsylvania Ph.D. in economics who had previously studied with Salisbury, began teaching courses in physical and economic geography there. Two years later he was succeeded by J. Russell Smith, who, after study with Friedrich Ratzel at Leipzig, had also taken his doctorate through the economics program at Pennsylvania and eventually headed a new Department of Geography and Industry in the Wharton School before joining the faculty of Columbia University's School of Business in 1919 (Rowley 1964; DeBres 1989). The University of Pennsylvania produced seven geography doctorates by 1918; of them, Goode, Smith, and Walter S. Tower were to be significant figures in early 20th-century geography.

Teacher education in geography was often conducted in normal schools or university departments in education, not necessarily by professional geographers. The first American doctorate in geographic education was awarded in 1906 to David Gibbs, a student of G. Stanley Hall in education at Clark University and later a professional school administrator. Some of the first geography professionals, however, did find positions in teacher education programs. Most notable were two students of Davis and Shaler, Mark Jefferson at the Michigan State Normal College, Ypsilanti (now Eastern Michigan University) and Richard E. Dodge at Teachers College, Columbia (DeBres 1989). Dodge established the *Journal of School Geography* (later *Journal of Geography*) in 1897. But the rôle of geography in the preparation of American teachers is still inadequately studied historically.

In the first decade and a half of the 20th century, however, American geographers were to build a set of institutions in which professional practitioners took ownership and control. The first independent geography department capable of producing doctorates on a sustained basis was established in 1903 at the University of Chicago under Salisbury's chairmanship. Only one doctorate was awarded in the department's first decade, but beginning in 1914 a small but steady stream of new Ph.D.'s in geography began to emerge, including many who would become leaders in the discipline in the 1920s and 1930s. In 1913 Salisbury instituted a justly famous graduate seminar in which all faculty and graduate students participated, and in the summer of 1915 Chicago offered what was probably the most comprehensive summer field course in geography to that time. Goode had built on Salisbury's "earth-referring social science" to organize the program intellectually around the concept of human ecology, defined as the study of human-environmental relations. This intellectual rationale was carried into the past by Harlan H. Barrows, who gave expression to the concept in an oft-cited Presidential Address to the Association of American Geographers in 1922, and by Ellen Churchill Semple, an independent scholar who had studied with Friedrich Ratzel and selectively brought his anthropogeographic method and insights into American human geography via her writing and her lecturing at Chicago (Pattison, in Blouet 1981).

Davis had shifted his own intellectual rationale for geography during the 1890s from an almost exclusive emphasis on physical geography to the causal study of the relationship of the earth to its inhabitants (Hartshorne, in Blouet 1981, 139-149; Martin, in Blouet 1981, 279-289; Martin 1985). In this respect he was probably influenced by his association with prominent schoolmen such as Charles McMurry and Parker, themselves heavily committed to the Ritter-Guyot tradition. By 1902 Davis saw geography as constituted by a physical half and, potentially, a human or "ontographic" half. The first was an independent entity based in the interrelationships of surface geology, meteorology, and oceanography. The second was seen as geographic only if the phenomenon being studied could be causally related to the first. Furthermore, Davis insisted that each half be studied thoroughly before these relationships could be identified, a method that enabled him to deny professional legitimacy to much current commercial geography and, indeed, to its practitioners.

Davis' second major contribution in this period was to call for and found a new professional entity, the Association of American Geographers (1904), whose members would meet a standard of original research productivity, at least according to his concept of geography (James and Martin 1979). The new society thus differentiated the discipline institutionally from the older Section E of the American Association for the Advancement of Science, an important forum for earlier research in geographical geology. Interestingly, only one of the charter members, Martha Krug Genthe, held the doctorate in geography, earned at the University of Heidelberg in 1901. In 1911 the Association began publishing its *Annals*, carrying articles and discussions representing professionally acceptable disciplinary concerns.

Twenty years after the founding, Charles R. Dryer opined that the combination of meetings and journal had "effectively maintained the dignity of geography, as a subject worthy of serious attention in college and university education" (Dryer 1924, 148). Davis' boundary-setting activity was a strong early influence on the group, however. Initially Davis' strongly physically based criteria for membership in effect defined what a professional geographer was (and, by implication, was not). For a much longer period Davis exercised a kind of intellectual veto over papers presented by younger geographers, often summoning their authors to breakfast or luncheon the next day to feed on his personal critique.

The period also saw the initiation or transformation of other geographical societies and their publications. The oldest of these, the American Geographical Society (1851), established initially as a merchants' information bureau in New York City, was professionalized after 1915 under a new Director, Isaiah Bowman, formerly of Yale (Wright 1952; Martin 1980). Its *Bulletin* was closed out in favor of a new professional quarterly, the *Geographical Review* (1916). The Society had given early support to the new Association and its *Annals*, and in 1912 had sponsored a "Transcontinental Excursion," led by Davis, which brought European geographers from fourteen countries to meet and travel with American professional geographers. Teachers of geography and those who prepared them organized as the National Council of Geography Teachers (1915), which in 1920 became the sponsor of the *Journal of Geography* and various state and local councils—the latter also an unresearched area.

Some local laymen's geographical societies, such as those in Philadelphia and New York, were founded or partly professionalized with the aid of nearby academics, who then used the society's publications as an outlet for their own research and that of their students (Wright 1951; Martin 1985, 223). Another local group was organized in Washington as the National Geographic Society (1888), with a core membership of government scientists. Its limping scientific journal, the *National Geographic Magazine*, was successfully reorganized after 1900 by a new editor, Gilbert Grosvenor, to focus on articles of popular appeal. The magazine pioneered in the use of photography, including, in 1903, a now infamous shot of a pair of bare-breasted Filipinas, one of many photos of naked or lightly clad "natives" that once prompted comedian Richard Pryor to characterize the magazine as "the black man's *Playboy*." The transformation of the National Geographic Society from a "scientific" to a "popular" organization was unacceptable to Davis and other new professionals, who responded by forming the Association. The National Geographic Society did, however, issue the initial invitation for the first Congress held in the United States (1904) of what was to become the International Geographical Union.

The growing sense of professional identity among geographers was well expressed by Dryer at a meeting of the Association held at the American Geographical Society's headquarters in New York in April 1914: "Here our science is not the vassal of geology, history, or commerce, but an independent, sovereign state" (quoted in Wright 1952, 167). By the time of American entry into World War I, geography had emerged both as an independent academic discipline and as a profession. It had built a series of institutional bases, produced a small set of disciplinary leaders, by and large developed an intellectual rationale (although the exact formulation of the relations between man and environment was to be vigorously contested over the next forty years), and structured recognized graduate curricula. A group of new professional journals had appeared, as well as new or renovated geographical organizations through which further professional identities could be forged. Some professional geographers had also produced school and college texts embodying current disciplinary perspectives.

There were, however, losses as well. Although most of the new graduate programs and allied institutions, and the overwhelming majority of professionals (as measured by members of the Association), were in the Northeast, by 1915 geography at Yale was nearly extinct, it was diminished at Cornell, and it disappeared entirely at Johns Hopkins until 1943. The older humanistic branches of geography, once the "handmaid" of classics and history, also lost ground as classical studies declined. Davis, always ready to relegate economic geography to the economists, was equally ready to leave historical geography to the historians (Davis 1924, 197, 199). But the historians had begun to draw on the newer social sciences and increasingly deplored geography's sometimes strident environmental explanations of cultural phenomena (Herbst 1961).

In the schools, geography remained a part of the elementary curriculum. But the institutionalization of Davis' recommendation that the high schools require a field- and laboratory-based course in physical geography as the first-year science course, embodied in the report of the National Education Association's Committee on Secondary School Subjects (the so-called "Committee of Ten") in 1892, had become a flat failure by 1905. Untrained teachers gave up the attempt to interest reluctant students, and natural scientists saw little utility in physical geography as a propaedeutic for later study in biology, chemistry and physics. The peak year for high school physical geography was 1896, when a little under a quarter of all students took it, although the absolute number of students enrolled continued to rise until about 1915. Physical geography was replaced by "general science" as the foundation high school science course. By 1928, physical geography drew slightly over two and a half per cent of high school enrollments (Fellmann 1986b). High school physical geography continued to decline until after World War II.

Commercial geography seemed to offer more possibilities for expansion as high schools incorporated business education programs to compete with commercial training schools and to attract a larger and more diverse student body. The teachers, however, were either underutilized physiographers or teachers of other commercial subjects, self-taught in this area (or who had a period free). Although this curricular thrust had initially come from parents and professional educators rather than geographers, some geographers realized that high school physical geography was a non-starter and that commercial geography might become a growth area (and textbook market). The number of students taking commercial geography rose from a little over 13,000 in 1900 to nearly 180,000 in 1934, or four per cent of all high school enrollments. But by 1949 the numbers had dropped to 90,000, or 1.7 per cent (Fellmann 1986b).

Dodge's attempts to move into the preparation of secondary school geography teachers had brought conflict with Teachers College authorities and ultimately forced his resignation in 1916 (DeBres 1989). In the same year the influential Council for the Social Studies was formed, but geographers largely declined participation in the design of social studies curricula on the grounds that geography was a natural and not a social science (James 1969, 478-479). This stance effectively insured that the geography of professional geographers would be absent from the programs of most college-bound high school students. After 1921 there were some attempts to get human or world geography into the high schools (often serving as fare for weaker students), but after World War II these courses were largely absorbed into or replaced by social studies, world history, or world cultures. Largely limited to place-name and map study auxiliary to other concerns, what passed for high school geography would continue to be taught by professional educators with little or no formal geographic training.

The search for holism

The 1920s and 1930s saw substantial growth in the discipline both in numbers and professional development. American participation in the World War had furnished new opportunities for physical geographers, especially in terrain analysis and mapping and in meteorology (Koelsch 1996). There was also a substantial demand on the part of American government agencies for persons knowledgeable about commodities (i.e., commercial geography). Some schools taught the geography of the European conflict in the context of current affairs. Also during the war the American Geographical Society under Bowman's direction had assembled a team of geographical and other experts, "The Inquiry," selected for their knowledge of particular areas, to provide intellectual guidance on territorial matters for the American delegation to the Paris peace conference in 1919 (Gelfand 1963; Martin 1980, Ch. 5). Although the actual postwar territorial settlements owed far more to secret commitments made by European politicians during the war itself and to demands for reparations than they did to geographical research, the impact of wartime service on American geography was substantial. It gave great impetus to the shift, already *in utero* by 1915, from a primarily physically-oriented discipline to one in which human geography took center stage.

Between the wars Latin America became a growth area, and the American Geographical Society began to focus its efforts on that region, especially on a major cartographic project, the 107-sheet "Millionth Map" of Hispanic America, completed between 1922 and 1945 (Wright 1952, 300-316). The commercial geography of the 1890s grew in the 1910s into the more general economic geography, whose coming of age was to be marked, in 1925, by the first specialized American geographical research journal, *Economic Geography*. Enrollments in economic geography boomed during the Great Depression of the 1930s, as many college students sought practical, business-oriented programs. Semester- or year-long service courses in economic geography abounded, often accounting for the largest enrollments in geography departments. In the same period Chicago's Charles Colby began to train a long line of able graduate students in urban geography (Spencer 1974, 28-29).

Geographers both before and after the war pointed to rising enrollments as a sign of geography's increased acceptance by the academy. But there is some evidence that, in a period of great interdepartmental competition for students, the academic standards of these service courses were not always particularly high, an appealing prospect to many undergraduates. At Chicago, the Dean advised entering freshman Burton Rascoe, who like many students was working his way through and looking for an easy science course, that "Commercial geography has a reputation for being a snap. Most of the football players go for it." But to Rascoe's chagrin, a new instructor had been assigned to the course with instructions to "tighten it up," and he and about fifty others failed utterly. No one received a higher grade than C-minus except two Asian students, the only ones able to answer such questions as "How much timber was exported from Vladivostok in 1910?" and "What is the annual export of hemp from India?" (Rascoe 1937, 172-174).

Political geography also emerged as an empirical, often policy-oriented subdiscipline, thanks in great part to Bowman's widely used text, *The New World* (1921, 4th ed. 1928). Prewar ties between American and German geography were succeeded by the increasing intellectual influence of the "French School," represented by Jean Brunhes and other important figures in French human and regional geography, several of whose works received English translations. And the holistic region, a mélange of human and physical phenomena related to particular places on the earth, emerged as a unifying concept for American geographers.

The 1920s saw an expansion of graduate programs in geography in the state universities of the Middle West (Rugg, in Blouet 1981, 245-263), as well as shifts in the topics and methods of research. In the East, however, most of the universities that had pioneered advanced work in geography during the 1890s had cut back or eliminated the professional preparation of geographers by 1920, although in 1928 Harvard began an advanced program of French-informed human geography under Raoul Blanchard and Derwent Whittlesey. Except for Chicago, which went from strength to strength during this period, most of the doctoral programs in the interwar period were developed in institutions that had not been major centers for advanced instruction and research in geography before the war (Dunbar 1986).

In 1921 Clark University opened its Graduate School of Geography, immediately accepting candidates for the doctorate and rapidly becoming the leading producer of geography Ph.D.s (Koelsch 1980, 1988). The earliest American doctorate in geography awarded to a woman appears to be Ruth Harvey's Ph.D. at Yale in 1908 (Martin 1998, 9). But Clark's first Ph.D. in geography (1922) went to a woman geographical educator, Mary C. Clune, and the Clark program quickly assumed leadership in advanced degrees for women

(Monk 1998). In 1923 the Society of Woman Geographers was formed, although women were seldom employed in the major graduate departments. Women in this period were often advised to stop at the M. A. level and teach, either in the public schools, in the teachers' colleges, or in colleges exclusively for women.

In 1923 a second important new doctoral program was established at the University of California under the direction of Carl Sauer, a Chicago Ph.D. who had previously led the geography segment of Michigan's geology department (Speth, in Blouet 1981, 221-244; Speth 1999). At Wisconsin, another pioneer in the field, geography demonstrated an institutional pattern more typical of Midwestern state universities: beginning in a geology department, moving to autonomous status for a distinctively geographic program within the geology department (perhaps by now renamed "geology and geography"), and eventually achieving full administrative independence as a department of geography, as Wisconsin's did in 1928 (Trewartha 1978). Wisconsin, Ohio State (1922), Michigan (1923) and, to a lesser extent, Minnesota (1925) and Nebraska (1928) emerged as new powerhouses in geography. In 1924 members of Midwestern geography faculties took control of the Association of American Geographers and its *Annals* from Davis and his students (James and Martin 1979, 69).

Davis himself had become an adherent of regional geography by 1911, although most of his own late research centered on the geology of coral reefs. Davis' expansive vision for an ideal graduate geography department specified as many as nine full-time geographers exclusively devoted to research, each a specialist both in a topical subdivision of geography and a major region of the world, along with supporting undergraduate professors, librarians and technical staff (Davis 1922). Davis' ideas were never to be implemented in any real-world department, although Yale before 1915 had developed somewhat along these lines. But the rapid development of regional courses after 1919 suggests the increased interest in international affairs that undergirded, in part, the expanding scope of geography programs. The regional concept, articulated in a series of Presidential Addresses to the Association for twenty years from 1918 onward, also led to greater breadth of course coverage in related topical areas, with human (usually economic) geography prominent among the new offerings. Substantial work begun before the war in the delineation of "physiographic provinces" and climatic regions continued, and these often became the containers into which human phenomena were distributed.

As physical geography contracted into geomorphology (a field uneasily shared with geologists), and meteorology's geographic residue became descriptive regional climatology (Koelsch 1996), the environmental side of geography was diminished, although in the environmentally conscious 1930s courses in the conservation of natural resources became a drawing-card for enrollments. For most budding geographers, however, physical geography became something to be gotten through on the way to the economic and regional courses. Indeed, in his 1928 AAG Presidential Address, Columbia's geomorphologist, Douglas Johnson, advocated excising geomorphology altogether from geography. The major sections of Wallace Atwood's 1940 text, *The Physiographic Provinces of North America*, all ended with a brief segment entitled "An Approach to the Human Drama." But fitting a few general, often dubious statements about historical and economic phenomena into the glass slipper of supposed "natural" or "geographic" regions proved less than appealing, intellectually or pedagogically, to many students and faculty.

While certain continuities with pre-war programs were evident after 1920, such as the use of plane-table surveys and campus weather stations, the new graduate departments of the Midwest began to develop field techniques and approaches more suited to the new discipline than those of traditional field geology. Cooperative relationships among the Midwestern departments, especially the Spring field meetings among faculty from 1923 to 1938, led to quantitative methods for sampling and mapping both forms and functions in small unit areas (James and Mather 1977). The new methods were in turn taught to graduate students in that long-lived masculinist rite of passage, the summer or early fall "field camp," and "field work" became a required part of graduate theses. Land uses and land quality studies, particularly in rural and small-settlement areas, became a popular genre among mature scholars as well, reflecting in part their Midwestern rural and small-town origins, The British adage that "muddy boots beget wisdom" found many adherents on the western side of the Atlantic.

American geography in the interwar period was an increasingly pragmatic and technique-oriented discipline whose major concern was how to study and represent actual land use (both in form and function) in the here and now. This orientation was to reap substantial benefits in the 1930s as the nation began to struggle with major environmental problems, and many geographers found employment or consulting niches with government agencies during that period. Although continued by the work of isolated individuals such as Ellsworth Huntington, now a Research Associate and the only geographer left at Yale in the interwar period (Martin 1988a), the great historically based, environmentally determined master narratives of the prewar era were rejected by the younger generation of geographers in favor of small-scale, contemporary, field-based studies of real-world landscapes. All in all, it was a period of significant disciplinary growth on the academic and applied sides.

Sauer's new program at Berkeley was a significant exception to the presentist and pragmatic approach of most geography departments. The Berkeley program was isolated both intellectually and spatially from the Midwestern heartland of geographic research. Field observation (rather than field survey) and a strong historical and morphological orientation, backed up by substantial library research and often directed toward Latin American problems, characterized the early work of the so-called "Berkeley School" (Speth, in Blouet 1981; Speth 1999). The program's emphases owed much to Sauer's involvement in a group of historically-minded Latin Americanists from several disciplines and with the anthropologists Alfred Kroeber and Robert Lowie. Sauer's culture-historical approach reflected his incorporation of the methods and orientations of several 19th- and early 20th-century German geographers. It was reinforced in the 1930s by his "discovery" of George Perkins Marsh's *Man and Nature* (1864), which had emphasized the human transformation of landscape. Sauer's methods and, indeed, his statements about them were of less importance than his lively curiosity and his (too much under-read) substantive research. The small Berkeley program attracted more than its share of able, independent-minded doctoral candidates.

The idea of the comprehensive region as the Holy Grail of the discipline, although providing professional geographers with a distinctive raison d'être in this period, proved an elusive project. The nature of the region was argued by precept, small-scale example, and methodological discussion at professional meetings. Clearly the concept had a variety of uses. In its systematic or topical mode, the region could differentiate the geographical geomorphologist from the geologist or the climatologist from the meteorologist. In its more comprehensive or holistic version, the region could serve as a rationale for joining human and physical phenomena in the same discipline. The concept provided new Ph.D.s in geography with a professional identity, while retaining close ties with the "folk model" of geography, the popular understanding that geographers are knowledgeable about places. It was also a concept around which both the remaining environmental determinists and the newer generation of "possibilists" could unite, since its character could be explained to the satisfaction of either simply by adjusting the mix of human and natural factors.

By 1939, however, the real-world holistic region had become something of a chimera. In that year a key methodological document by Richard Hartshorne, entitled "The Nature of Geography," was published in the *Annals*. Based on substantial, if somewhat selective, reading of the German methodological literature, the lengthy essay argued that the object of geography's interest was "areal differentiation." Subsequently published separately, and rethought to some degree twenty years later in a successor volume, Hartshorne's work was to exercise substantial influence on American professional geography and on the training of the immediate postwar generation of graduate students, although many (like Sauer himself) felt "read out of the profession" by Hartshorne's treatise (Spencer 1974, 30 and note 22).

The pedagogical expression of the regional concept also often proved less than satisfactory. Undergraduate and graduate programs lifted up "regional synthesis" as the ultimate goal and highest work of the geographer. But the continent-sized regional courses and the textbooks written in support of them were hardly models of synthesis. Most, indeed, with the principal exception of Russell Smith's *North America* text of 1925 (rev. 1940 and 1942), were encyclopaedic in scope and virtually unreadable in expression. Looking back at them, one is reminded of Germaine Greer's comment, "The world's a wonderful place; how can geography teachers make it so boring?" Nor were the topical or "systematic" texts any better. Introductory courses and texts in the "Principles of Geography" incorporated few if any principles, except perhaps those recycled from other disciplines. The most popular interwar text (which lived on in various revisions until the 1960s), Vernor Finch and Glenn Trewartha's *Elements of Geography: Physical and Cultural* (1936), was basically a checklist of geographical legitimacy, Midwest style, and would discourage curiosity in the most intrepid. Trewartha's ponderous climatology text (1937), based on the work of Wladimir Köppen, was a more effective soporific than anything sold by the local pharmacist. Nearly all advanced regional and systematic texts were oriented to contemporary phenomena, although Semple's somewhat anachronistic *Geography of the Mediterranean Region: Its Relation to Ancient History* (1931) and Whittlesey's *Earth and the State* (1939) were singular (and scholarly) exceptions.

The old normal schools, after World War I moving toward three- and four-year programs and a new status as teachers' colleges, continued to provide most teachers for the elementary grades. But in the interwar period professionally trained geographers (often with master's degrees only) began to be regularly appointed to their faculties. Some universities in this period also added a specialist in the teaching of geography (often the only woman in the department), either as a departmental appointment or in allied schools of education. The 1920s and 1930s saw substantial research in the teaching of geography by geographers, of whom the most prominent was Edith Putnam Parker of Chicago. This research was made available to the larger educational community in 1933 by the publication of *The Teaching of Geography* by the National Society for the Study of Education.

Several professional geographers in major graduate departments launched textbook series to meet the need for up-to-date teaching materials, most conspicuously at Chicago, where Barrows and Parker produced a widely used elementary series, Colby and Alice Foster a high school text, and Wellington Jones and Whittlesey a text for the college level. In 1923 Goode published the first edition of his long-lived *Goode's School Atlas.* Soon after moving to Columbia, J. Russell Smith began a series of elementary texts organized around human use, as well as an industrial geography for junior high school students (Rowley 1964). At Clark, in addition to his college physiography book and a popular work on the Rocky Mountains, Atwood produced several widely adopted series of school texts. At Berkeley, Sauer produced a single school text, *Man in Nature* (1939), noteworthy for its unusual focus on Native American cultures and habitats.

Coursing toward the Millennium

The experience of World War II, like that of World War I, powerfully affected the future course of American academic geography. Augmenting the cluster of geographers already employed in the Federal government, academic geographers found themselves hard at work on practical problems in various intelligence-gathering and mapping agencies. Although some had demonstrated even during the 1930s that they could readily cross the boundary between the academic and the applied, in other cases the unspecialized and sometimes quite dated knowledge of regional generalists proved less than useful in wartime circumstances.

As a result of their wartime experience and observation, a number of younger geographers became convinced that a more quantitative, interdisciplinary, problem-focused, and above all more specialized training was necessary in the contemporary world. Once back on campus, having brought with them their new skills, knowledge, and enthusiasm, they began to argue for the development of more "systematic" (as opposed to regional) research in geography, particularly in human geography. This generation, now members of new or renovated graduate departments during the period of postwar academic expansion, largely led developments in industrial, transportation, and urban geography, the cutting edges of the field in the 1940s and 1950s. Like their predecessors, the early postwar generation used the construction and analysis of maps as their primary quantitative method. Especially noteworthy was Edward Ullman's mapping of American commodity flows, through which he was able to develop several new concepts concerning the relations between places in what came to be called the study of spatial interaction. These initial empirical studies were the first tentative steps toward the development of new spatial theory, influenced by the work of German locational theorists Walter Christaller, August Lösch, Johann Heinrich von Thünen, and Alfred Weber.

A new professional body, the American Society for Professional Geographers (ASPG), had been organized in 1943. Its members were largely younger geographers, many working in government agencies, who had not yet published much or whose research did not permit academic-style or independent publication. Five years later, after much discussion and mutual reservations, the ASPG merged with the Association of American Geographers (AAG), thus opening the older organization to many new members. The ASPG brought with it a new journal, *The Professional Geographer*, and the structural innovation of regional divisions (James and Martin 1979). Under the reorganized AAG two of these were constituted by the affiliation of two older independent regional organizations, the New England Geographical Conference (1922) and the Association of Pacific Coast Geographers (1935). Others grew out of ASPG divisions or were created after the merger. Such regional gatherings became useful venues for discussion of research and teaching among professionals and students who might not be able to attend national meetings.

The postwar expansion of American colleges and universities, fueled by the educational provisions of the "G. I. Bill of Rights" (legislation providing benefits to veterans of military service) and later by rising national prosperity, coupled with the assumption that most middle-class students would go to college, benefited enrollments in geography courses and therefore departmental growth. Many academic institutions responded to a renewed interest in international affairs by establishing or expanding area studies programs, initially adding Russian and East Asian affairs to prewar Latin American foundations, then expanding into African and Middle Eastern studies. Major universities such as Yale, which brought back geography in 1945 (Martin 1988a), and a number of élite private colleges added geographers, often primarily to service such programs, with mixed results. In the immediate postwar period the demand for qualified geographers who could function equally usefully in cross-disciplinary situations exceeded the supply, and more than once (e.g., at Amherst College) the experiment ended with dropping geography from the institution's program.

The old normal schools, now teachers' colleges, developed into comprehensive state colleges, often of great size by the latter years of the century. Some moved into graduate instruction, usually at the master's level only. Several states, notably California (which had begun the process in 1919), expanded their flagship state universities by establishing branches or independent campuses. They also developed or expanded public community college systems, in which some geographers found employment. By the late 1970s, three-fourths of all academic geographers were working in public institutions, often in quite large departments, mostly in the Middle and Far West. A growing number of Southern state universities expanded both in size and in capacity to award graduate degrees as well. During the early 1960s the four largest Ph.D. programs had together produced one-third of all American doctorates in geography. But the rise of new and expanded departments in the 1960s and 1970s meant that no longer could the traditional providers monopolize professional training.

During the 1920s and 1930s many American graduate departments of geography had hired their own best Ph.D.s. This practice (known as "inbreeding") gave a certain cohesiveness, even distinctiveness, to their programs. A number of graduate departments that had achieved full independence during the prewar period now saw their faculty cohort grow old together, often with some stagnation of programs and points of view: e. g., Berkeley, Chicago, Clark, Illinois, Indiana, Michigan, Nebraska, and Ohio State. As older faculty retired or died during the postwar period, however, they were replaced by younger scholars with more varied institutional and professional backgrounds, viewpoints, and aptitudes. Retirements and departmental expansion during the

1960s, coupled with substantial renovation of graduate curricula, put most of these older departments back into the front ranks in the 1970s and 1980s.

New graduate departments, or the expansion into nationally significant doctoral programs of older ones, occurred in the University of California, Los Angeles, Northwestern University, Syracuse University, and a number of state universities outside the traditional Middle West heartland. These newly established or newly expanded graduate departments were often committed to advancing newer techniques and points of view, especially during the so-called "Quantitative Revolution" of the late 1950s and early 1960s. This "revolution" was led by geographers trained at Iowa, Northwestern, and especially the University of Washington, where new approaches to human geography had been stimulated by a lengthy visit from the innovative Swedish geographer Torsten Hägerstrand in 1959.

Graduates of these programs went out to colonize others (Michigan, Ohio State, Chicago), which then sent out their own graduates trained in the spatial viewpoint and in its new methods. The "space cadets," as they were jocularly known, were bright, eager, productive, aggressive, and committed to changing the discipline radically, away from what they regarded as "descriptive," "unscientific," "idiographic," and otherwise unsatisfactory modes represented by the older generation of geographers. They instituted spatially and quantitatively, often theoretically, oriented courses in human geography, but most had little time for physical geography. Brandishing the mantra of "scientific geography" (implying, of course, that no other way of doing geography could be scientific), they hoped to remake the discipline into a generalizing, law-giving social science on the model of 1950s economics and sociology. If geography as a discipline were to disappear into a unified social science—well, no great loss. They were self-consciously participating in a "revolution," although precisely *which* revolution—quantitative, theoretical, models, conceptual, behavioral, or whatever—was not always clear. As Peter Taylor wryly observed after tracing similar developments in Britain, "Seven revolutions in one generation, makes geography 'the Latin America' of the scientific community!" (Taylor 1976, 141, note 4).

This latest "new geography" was spread through summer institutes in quantitative methods, by cognate institutions such as the Regional Science Association, by the circulation of discussion paper series, and by special sessions and caucuses at professional meetings. By the late 1960s the new "spatial-analytic" orientation had spawned a new journal, *Geographical Analysis* (1969). About the same time, *Economic Geography* and other professional journals began to reflect the new spatial turn.

The "new geography" did not, however, sweep all before it. No sooner had the "space cadets" reached positions of power and influence in the profession than they were challenged, first by events, then by alternative alignments within the profession. Deep divisions in the United States over the Vietnam War and the accompanying radical protests of the mid-1960s called into question the wisdom and social purpose of governmental decision-making, centralized planning, and Western "ethnoscience." Within geography, the critique was focused on urban redevelopment projects, superhighways run through poor neighborhoods for the benefit of more affluent suburbanites, industrial parks and regional shopping malls built on the outskirts of cities, "top-down" international development schemes, and in general the legitimacy of those activities and modes of thought with which the "space cadets" were identified and funding for which had fueled their academic programs.

Soon there were new caucuses: the Social and Ecologically Responsible Geographers (SERGE) and the Union of Socialist Geographers (1974). In 1969 a new journal, *Antipode: A Radical Journal of Geography*, was begun by graduate students and faculty at Clark, and by 1972 an explicit turn to Marxist analysis had appeared among a handful of academic geographers. Many of these were of British origin and had brought with them lenses shaped by their early experience of a rigid class system. Although critical of the spatial-analytic school, Marxist geographers shared with them certain common characteristics, such as the identification of geography's purposes with a future unified social science (albeit "revolutionary" in a different sense), great and often controversial productivity, and a tendency to appropriate self-serving mantras. But capitalism did not collapse as predicted, the revolution did not appear on schedule, Marxist geographers got promoted and tenured on the same old academic basis as their predecessors, and "socialist geography" became yet another AAG specialty group, on the same plane as "Bible geography." By the 1980s, as the legitimacy of the Marxian "master narratives" came under attack from post-modernists, various neo-Marxists adopted the comforting term "critical" to characterize their current modes of analysis. The modes of non-disciples of the Gospel According to St. Marx and its numerous preachers and exhorters were thus, a priori, *un*critical.

The environmental movement of the late 1960s and early 1970s had found many departments of geography quite unprepared for an upsurge of popular concern about the natural environment and human interactions with and stewardship of the earth's resources. The environmental and physical dimensions of geography had been largely dismissed by both spatial analysts and Marxists as fetishes of a determinist and discredited past. Indeed, as early as 1949 Edward Ullman had argued before a Harvard faculty committee studying the future of geography there that the study of the relations between man and environment was "no longer acceptable" as a task for geography. Two years later, in a farewell exchange with President James B. Conant (who reminisced about a teacher who had taught him where rivers and mountains were), Ullman informed him that "we don't do rivers and mountains anymore" (Glick 1988, 51-52). By 1970, just as it looked as though the man-environment theme of previous generations of geographers had been safely buried, there was a public demand for something that many American geographers were unprepared to supply, and other disciplines moved rapidly to surf the environmentalist wave. Not for the first (or last) time in the history of American geography, it was raining duck soup, and American geographers were out there with forks.

The environmental movement was a powerful impetus to the return of physical geography, the stepchild of the discipline for so many years. It also gave opportunity for several older departments—Chicago, Clark, and California (Berkeley) among them—to rediscover or reinvent the meanings of "environment" as a geographic tradition. Barrows' student Gilbert White and his colleagues at Chicago in the late 1950s and early 1960s had begun to turn out a number of Ph.D.s interested in problems of natural hazards perception and management, considered in the light of recent developments in economics and psychology, especially cognitive psychology. The focus on real-world environmental problems, to some a backward turn in the discipline, nevertheless appealed to the activist strain of the times.

Historical and cultural geographers (especially those influenced by the ideas of Sauer, Ralph Brown, and John K. Wright, now reexamined in the light of recent developments in psychology and philosophy) began to look at past cognition of environment as a focus for research into and understanding of past geographies. An exciting intellectual symposium sponsored by the Wenner-Gren Foundation, on "Man's Role in Changing the Face of the Earth," held in Princeton, New Jersey in June 1955 (with subsequent publication by the

University of Chicago Press), was to have a large, if delayed, effect on the revival of the human-environmental theme in American academic geography. In the 1970s and 1980s, another offshoot of this tradition emerged as "cultural ecology," with close links to anthropology and archaeology (Turner 1989). Children's perceptions of the environment became for a time a special focus at Clark, where cross-disciplinary work with psychologists and others was stressed during the 1960s and 1970s. Cartographers found the psychology of perception valuable as they reevaluated map symbolism and traced the mental maps of students and other captive audiences. While each of these groups had different agendas, and although no common theoretical or methodological ground emerged from this "Cognitive Reformation" (so termed by Robert Kates), these various movements once more legitimated the nature-society tradition within American geography. They also legitimated exploring the ideas and values of their subjects, rather than simply tracing their spatial behavior.

No one persuasion held sway by 1970 in most American graduate departments, and certainly not in the still largely male-dominated profession. Indeed, in the 1970s additional "revolutions" were identified, often resulting in either proclamations of victory or, if not a victor, in claims of epistemological or ontological superiority. Adherents of phenomenology, existentialism, idealism, poststructuralism, structurationism, and realism proffered various claims, and sometimes real-world examples. By the 1980s most of these points of view had more or less become artificially grouped, along with historical geography, into the general category "humanistic geography," seen as part of a broader "post-positivist" critique of the spatial-analytic school. Feminist approaches became prominent in the geographic literature in the 1980s as women began to occupy positions of authority in the profession and in the major graduate schools. Gender-related themes of "sexuality and space" also incorporated lesbian and gay male subjects and scholars. Geographers were less successful in recruiting from among America's many ethnic minorities, however, in spite of a significant inter-university project in the early 1970s to expand graduate opportunities for African-American students.

Before World War II cartography had largely been a service skill (or perhaps a career-opener for geography undergraduates and master's students). The first doctorate in geographic cartography per se was awarded only in the 1940s. After the war the field emerged in several major departments (Kansas, Washington, Wisconsin) as a distinct specialization and analytic discipline. A major scholarly project based in Wisconsin's geography department began publishing a multi-volume, multi-cultural history of cartography in 1987, and other innovative works in the mapping sciences have also appeared in recent years. In the wake of increasing use of computer cartography, part of the general field morphed into Geographic Information Systems, or GIS, a significant intellectual and technological advance quickly adopted by other academic disciplines. By the 1990s GIS had become a desideratum for entry-level positions in geography departments, much as cartography had been in the 1950s, or statistics in the 1960s and 1970s. So strong was its attractiveness that some enthusiasts even proposed to ditch academic geography entirely and replace it with departments of GIS.

American academic geography had seemed, in the 1960s and early 1970s, to be in flourishing condition as the "Baby Boomers" (born in 1946 and later) flooded American colleges and universities and, as one byproduct, expanded geography enrollments and positions. But there were also significant losses in the last half of the century. The emphasis on foreign field research that had blossomed in the early postwar period, and the area studies programs and regional geography courses that had been the curricular expression of geography's global reach, began to fade out sharply, although some departments and individuals (many of them culturally oriented geographers) continued to support such programs. This decline was not unrelated to the "positivist" search for laws derived from an American base but presumed to be universally valid, as well as to the decline in foreign language competence (and degree requirements) at both undergraduate and graduate levels.

American geography's increasing pull-back from the world at large was also reflected in the declining proportion of articles dealing with foreign areas in the major professional journals. These developments made for an even wider separation of professional geographers from public perceptions of and support for geography as a discipline concerned with a world mosaic of places. By the late 1980s, however, a "new regional geography" had begun to be advocated, one which was to be more process-oriented and more embedded in its global context than the regional geography courses of the 1950s. It is yet too early to tell how this will play out in terms both of liberal education and of public and administrative support.

A second collective loss occurred as a number of graduate departments that had traditionally been prominent in research and graduate training were defenestrated, particularly in leading private universities, and especially in the cradle of American academic geography, the Northeast. Some of the older surviving programs had either faded away or were terminated in the early postwar period, notably those at Harvard, Stanford University, and Washington University in St. Louis. The University of Pennsylvania closed out geography in 1963, although it retained some geographers in its Regional Science Department, and Yale dropped its geography department yet again in 1967.

One scholar found that, between 1970 and 1976, the number of geography departments in private institutions dropped by one-sixth and, despite expansion in the South and West, there had been an overall net loss of 32 departments in that limited period (Fink 1979). During the 1980s the procession of lost graduate departments or programs included Michigan (1982), Pittsburgh (1983), the University of Chicago and Columbia University (both 1986), and Northwestern (1987), although in the same decade there was some expansion of doctoral programs in other institutions, notably so at the University of California, Santa Barbara. The geography department of the leading African-American institution, Howard University, was closed out in 1991, and there were other closures of master's or bachelor's programs during this period. But even the loss of small departments (e.g., Bradley University, in 1984), or of the one or two geographers in combined programs, was serious enough, since it was opportunity foregone.

In the Northeast, a small symbolic victory had come in 1973 when Princeton University, gifted by history with two chairs in geography (normally occupied by a geomorphologist in the geology department and by a political scientist in the Woodrow Wilson School of International Affairs), actually awarded one of them to a geographer, Julian Wolpert. At the century's close, however, Dartmouth College had the only remaining geography department in the so-called "Ivy League" of élite Northeastern institutions, although Brown University had a geographer attached to sociology. Although geography departments in some of the state universities in the region continued to flourish (e.g., Pennsylvania State University), the closing of geography in the major private universities sent a powerful signal that geography is no longer valued by academic administrators in the institutions that traditionally have turned out the country's economic decision-makers and its cultural and political élite. That message is not lost on numerous excellent private colleges, many of which are "feeder" institutions for the élite universities, and most of which presently offer no courses in geography. The negative consequences are large, both for the professional recruitment of future geographers and for the background that other students, as influential alumni, will lack as they make decisions about public policy and private initiative.

Another historically important institution, the American Geographical Society, racked by financial problems, was radically cut back and decentralized in the 1970s. The Society's renowned cartography department and its research programs were eliminated, and its library was transferred to the University of Wisconsin-Milwaukee. Although it continued through the 1990s to develop innovative programs, its major publications were outsourced to cooperating geography departments distant from its New York headquarters. The smaller geographical societies founded near the close of the 19th century and of some importance in the early stages of the discipline's growth had largely disappeared from view by the latter part of the 20th.

The discipline's premier professional organization, the Association of American Geographers, also saw setbacks toward the end of the century. In 1974 it instituted an "open convention" policy, under which all proposed papers were accepted for its annual meetings, a device that brought higher and higher attendance and many more concurrent paper sessions. The results were predictable: a decline in quality, programs resembling a metaphorical gopher field, and further disciplinary fragmentation. It is thus no surprise that as early as 1978 nearly a quarter of the faculty members teaching in American graduate programs in geography were not members of the AAG (Bodman 1982, 43). Many were from programs ranked in the top ten. In 1980 the most productive American geographer since Davis, Brian Berry, complained in his Presidential Address of a "new tribalism" that had made the annual meeting "a locus without a focus" (Berry 1980, 451). Still a major social event, the annual meeting has become an extravaganza whose success is measured in figures for attendance, sponsor income, and simultaneous sessions, and is thus perhaps more akin to a medieval fair than to a scholarly organization. In the words of the old African-American Christmas carol, "De people keep a'comin' and de train done gone."

In the late 1990s, widespread dissatisfaction with the review and selection process of the Association's flagship journal, the *Annals*, resulted in the astonishing decision to divide it into four equal sections under different editors. It was a remarkable testimony to the fragmented state of the discipline, confirmed by subsequent bickering over the titles and scope of the favored foci. Tensions over allegations of theoretical or political biases on the part of editors and boards, the overwhelming size of annual meetings, and issues around the quality of programs were not unique to geography in the 1990s, of course. Conflicts over editorial policies or practices in professional journals had also emerged in sociology and modern languages, and professional disputes in other disciplines (such as history and psychology) had resulted in the secession of more purely "academic" groups. For a small discipline like geography, however, internecine warfare is not an affordable luxury, and at the end of the century these problems and those of membership loss and fiscal crisis remained. It may be that the best way of marking the Association's centennial in 2004 would be to scupper the whole thing and start over.

Professional concern about the gap in geographical instruction between grammar or junior high schools and the freshman year of college was valiantly but unsuccessfully addressed in the High School Geography Project (1961-1970). That program developed curricular materials designed to teach students how to think like geographers, a variant on the inquiry-based pedagogical strategies then coming into play in history and other disciplines. Despite the expenditure of over 2.5 million dollars in development costs, however, the year-long "Geography in an Urban Age" curriculum attracted little attention, and in 1977 it was estimated that less than five per cent of all school districts were using the materials.

During the 1980s, under the leadership of Gilbert Grosvenor, the well-financed and politically wired National Geographic Society, troubled by reports of surveys revealing widespread ignorance of world place knowledge by Americans, moved into the improvement of school geography in a substantial way. For almost the first time since 1904 (except for joint sponsorship of the 1952 International Geographical Congress in Washington), the Society sought links with academic geographers and with other professional organizations. A new periodical, *National Geographic Research* (1985), was a worthy, albeit ultimately unsuccessful, attempt to provide an academic outlet, although few professional geographers took advantage of it. In 1986 the Society launched an 80-million-dollar Geographic Education Program, through which it financed a grass-roots network of state "geographic alliances" (in all 50 states by 1995) of university and community college faculty, public school teachers, and educational administrators. The Society also sponsored a series of summer training institutes and produced attractive visual teaching materials based in its substantial traditional strengths in photography and mapping (Grosvenor 1995).

Grosvenor also began working through the National Governors' Association to reach key political decision-makers, and in 1989 President George Bush and the nation's state governors called for the adoption of national standards in a number of core disciplines, specifically including geography. A coalition drawn from the four major geography organizations and largely funded by the National Geographic Society responded by developing "Geography for Life" standards for the fourth, eighth and twelfth grades of school. Despite the fact that Governor Bill Clinton of Arkansas had chaired the committee identifying the core disciplines, there was little or no national leadership in this area once Clinton himself became President in 1993. Indeed, in May 2000 the White House web site was running a crude map showing the cities President Clinton would visit in a "School Reform Tour"; Owensboro, Kentucky was located by these would-be reformers in the middle of western Tennessee. Many if not most states appear to be developing curricula and examinations based on the social studies standards rather than on those for geography.

The Society's work of reform in school geography was accompanied by more than a little headline-hunting, which in turn raised eyebrows among some professional geographers. One embarrassing moment occurred after the Society commissioned the Gallup polling organization to conduct an international survey to determine how many people could locate places on a blank map. As expected, Americans came in at or near the bottom of various lists. At the press conference Grosvenor called to denounce the geographic ignorance revealed in the findings, an inquisitive reporter asked him to identify the states bounding Texas, which he was unable to do. Commenting on the event in the light of more general educational trends, educational historian Lawrence Cremin suggested that "we should not be surprised when the young are occasionally skeptical of the standards laid upon them" (Cremin 1990, 11-12).

In 1987 Senator Bill Bradley of New Jersey and others sponsored legislation calling on Americans to observe a "Geography Awareness Week" each November. In 1989 the National Geographic Society began a game-show-style competition, the National Geography Bee, which by 1995 was attracting six million student competitors a year. (It once prompted an unsuccessful call to the Graduate School of Geography at Clark University by an anxious father wanting to hire a professor to coach his son for the event.) A popular educational television show, "Where in the World Is Carmen Sandiego?," sought to get younger children excited about geography by tracking a team of more than usually dodgy thieves around the world. Perhaps the most inventive response to the widely deplored American student deficit in place knowledge appeared in 1990 at Davidson College (North Carolina), when a volunteer group styling itself the "Bathroom Brigade for Geographical Literacy" taped world maps to 150 stalls in campus bathrooms. Although these and other such efforts publicized place awareness in education (and provided fodder for numerous cartoonists), they may have done academic geography more harm than good by reinforcing the popular impression that geography is little more than place names or map games for eight-year-olds. Such efforts seem a curious throwback to geography's pre-professional, pre-Civil War rôle, under the now discredited aegis of "faculty psychology," as the school subject most able to train the "faculty" of memory.

It is far too early to assess the results of all this flurry, of course. But so far there seems to be little articulation of the newly available materials, standards and practices between school and university. And short of celebrity pop singer Ricky Martin abandoning "La Vida Loca" for geography, American teenagers have little in the way of role models in the discipline, unlike their British counterparts after the telegenic Prince William chose geography as one of his three A-level examinations.

Whither American Academic Geography?

American geography at the beginning of the 21st century remained a small, often beleaguered academic discipline. Except at tiny Clark, where geographers comprise ten per cent of the entire faculty and where the doctoral program is one of the institution's flagship programs, geography is not central to the academic program or scholarly visibility of any American research university. Furthermore, it would be difficult to identify any other university study that in the course of the 20th century had achieved full departmental independence and programmatic recognition at the doctoral and research level in so many leading American universities, only to lose both by the end of that same century. At the pre-college level, in spite of substantial efforts in the 1960s and again in the 1990s, geography remains largely a minor component of social studies or of global education, taught most often by teachers with little or no background in the discipline. That American academic geography is institutionally in a parlous state is frequently lamented. That American geographers have only themselves to blame is less frequently asserted (but see Turner 1989, 89-90 and note 17; Gade 1999).

Not the least remarkable aspect of geography's state as an American academic discipline has been the fracturing of any consensual view of what the discipline and its practitioners should be about—a consideration explicit in the external reviews preceding the closure of several major graduate departments. Factionalism, excessive and almost idiosyncratic specialization, and a persistent inability to articulate the case for academic geography successfully to non-geographic audiences (particularly university administrators) are certainly parts of the problem. The "delayed bandwagon" effect of jumping for the latest theories in another discipline just about the time the theorist has either reconsidered his position (e.g., E. P. Thompson) or has been thoroughly dissected and set aside on his own turf (e.g., Michel Foucault) has also been less than helpful. One scholar has tartly described the period as one of "transatlantic spasms of sheeplike behavior" (Gade 1999, 124). The internal focus of American geography in the 1920s and 1930s had been criticized by recent theoretically inclined geographers for its alleged parochialism. But the game of "my imported theorist is more intellectually rigorous than your imported theorist" in the internal debates of the last forty years hardly seems an improvement.

Early attempts on the part of both the spatial behaviorists and the Marxists (albeit from different political perspectives) to make geography over into the spatial component of a generalized social science were not successful. Current proposals that geography departments should be made over into GIS programs seem equally ill-advised. The logic of each of these purported solutions would once more confine American academic geography to its 19th-century rôle as the handmaid of more important academic disciplines. Even some self-proclaimed postmodernists are not free of triumphalist rhetoric, predicting a hegemonic role for space (presumably the geographer's métier) as the central mode of analysis in all of social science. Such claims for the primacy of geography are not new; Davis asserted something like it in 1912, the year that the new White Star liner RMS *Titanic* was proclaimed "unsinkable."

In the immortal words of the common-sense philosopher (and baseball player) Yogi Berra, "The future [of American academic geography] ain't what it used to be." Fortunately, there are currently voices within the profession (e.g., Hanson 1999) asserting that space, place and environment—all three—are complementary rather than competitive notions within geography. They are new names and meanings for concepts that, although differently cognized, continually recur in the history of geographic discourse. To throw out any one of them is to undercut the intellectual and pedagogical rationale for disciplinary independence. On the other hand, a case can often be made for the "handmaid" strategy in particular institutions, especially where geography, once strong, has been dropped as a major field; sheer pragmatism suggests that any representation is better than none.

The world is not kind to our theories of it, and there is nothing more evanescent than the latest social science theory. The celebrated "postmodern turn" of the *fin de siècle* suggests skepticism regarding the false comfort of fast-disappearing methodological absolutes and so-called "master narratives." It may also take us nearer the worlds (and wisdom?) of ordinary folk and what they might ask of geography and geographers beyond games and place names. As Barbara Kennedy has cogently put it, "The need is to convey a lively sense of the world's variety, as well as an appreciation of how that has come to be as it is and how it may change" (Kennedy 1987, 246). An intelligent, articulate, professional response to that need, informed by theory but driven by real-world experience, might give American academic geography a second chance in the 21st century.

Note

To limit the number of references, I have omitted many standard sources, including the indispensable *Geographers: Biobibliographical Studies* series and recent surveys of geographic thought since 1945. I have normally cited only one or two germane sources, often choosing unfamiliar ones, when I have in fact consulted many more. This essay is grounded in my own published and unpublished research, bibliography developed during two decades of teaching undergraduate and graduate courses in geographic thought, and half a century as a bemused participant-observer of and in American academic geography. I am indebted to the editor and to Professors Geoffrey Martin and Paul Frederic for their suggestions on an earlier draft.

References

Berry, B. J. L. (1980) Creating Future Geographies, *Annals of the Association of American Geographers* 70, 449-458.
Blouet, B. W. (ed.) (1981) *The Origins of Academic Geography in the United States*, Archon Books, Hamden, CT.
Bodman, A. R. (1982) Academic Origins, *Geographical Perspectives* 50, 43-49.

Broome, E. C. (1903) *A Historical and Critical Discussion of College Admission Requirements*, Macmillan, New York.
Cremin, L. A. (1990) *Popular Education and Its Discontents*, Harper and Row, New York.
Davis, W. M. (1922) A Graduate School of Geography, *Publications of the Clark University Library* 6, no. 6, 1-26.
Davis, W. M. (1924) The Progress of Geography in the United States, *Annals of the Association of American Geographers* 14, 159-215.
DeBres, K. J. (1989) An Early Frost: Geography in Teachers College, Columbia and Columbia University, 1896-1942, *Geographical Journal* 155, 392-402.
Dryer, C. R. (1924) A Century of Geographic Education in the United States, *Annals of the Association of American Geographers* 14, 117-149.
Dunbar, G. S. (1981) *Geography in the University of California (Berkeley and Los Angeles), 1868-1941*, DeVorss and Company, Marina del Rey, CA.
Dunbar, G. S. (1986) Geography in the Bellwether Universities of the United States, *Area* 18, 25-33.
Entrikin, J. N. (1984) Carl Sauer, Philosopher in Spite of Himself, *Geographical Review* 74, 387-408.
Fellmann, J. D. (1986a) Myth and Reality in the Origin of American Economic Geography, *Annals of the Association of American Geographers* 76, 313-330.
Fellmann, J. D. (1986b) Rise and Fall of High-School Economic Geography, *Geographical Review* 76, 424-437.
Fink, L. D. (1979) The Changing Location of Academic Geographers in the United States, *The Professional Geographer* 31, 217-226.
Gade, D. W. (1999) Gallic Trajectories of Geography and the Self, *Geographical Review* 89, 122-128.
Gelfand, L. E. (1963) *The Inquiry*, Yale University Press, New Haven, CT.
Glick, T. (1988) Before the Revolution: Edward Ullman and the Crisis of Geography at Harvard, 1949-1950, in J. E. Harmon and T. J. Rickard (eds.), *Geography in New England*, New England/St. Lawrence Valley Geographical Society, [New Britain, CT], pp. 29-62.
Grosvenor, G. M. (1995) In Sight of the Tunnel: The Renaissance of Geographic Education, *Annals of the Association of American Geographers* 85, 409-420.
Hanson, S. (1999) Isms and Schisms: Healing the Rift between the Nature-Society and Space-Society Traditions in Human Geography, *Annals of the Association of American Geographers* 89, 133-143.
Herbst, J. (1961) Social Darwinism and the History of American Geography, *Proceedings of the American Philosophical Society* 105, 538-544.
James, P. E. (1969) The Significance of Geography in American Education, *Journal of Geography* 68, 473-483.
James, P. E. and E. C. Mather (1977) The Role of Periodic Field Conferences in the Development of Geographical Ideas in the United States, *Geographical Review* 67, 446-461.
James, P. E. and G. J. Martin (1979) *The Association of American Geographers: The First Seventy-Five Years*, Association of

American Geographers, Washington, DC.
Kennedy, B. A. (1987) [Book Reviews], *Institute of British Geographers, Transactions and Papers*, N. S. 12, 244-248.
Koelsch, W. A. (1980) Wallace Atwood's 'Great Geographical Institute,' *Annals of the Association of American Geographers* 70, 567-582.
Koelsch, W. A (1981) Pioneer: The First American Meteorological Doctorate, *Bulletin of the American Meteorological Society* 62, 362-367.
Koelsch, W. A. (1988) Geography at Clark: The First Fifty Years, 1921-1971, in J. E. Harmon and T. J. Rickard (eds.), *Geography in New England*, New England/St. Lawrence Valley Geographical Society, [New Britain,CT], pp. 40-48.
Koelsch, W. A. (1992) Ancient Landscapes Revisited: Classical Geography in the American University, 1870-1930, in Fritz-Peter Hager et al. (eds.), *Aspects of Antiquity in the History of Education*, Verlag August Lax, Hildesheim, Germany, pp. 177-182.
Koelsch, W. A. (1996) From Geo- to Physical Science: Meteorology and the American University, 1919-1945, in J. R. Fleming (ed.), *Historical Essays on Meteorology*, American Meteorological Society, Boston, pp. 511-540.
Livingstone, D. N. (1992) "A Geologist by Profession, A Geographer by Inclination": Nathaniel Southgate Shaler and Geography at Harvard, in C. A. Elliott and M. W. Rossiter (eds.), *Science at Harvard University: Historical Perspectives*, Lehigh University Press, Bethlehem, PA, pp. 146-166.
Martin, G. J. (1980) *The Life and Thought of Isaiah Bowman*, Archon Books, Hamden, CT.
Martin, G. J. (1985) Paradigm Change: A History of Geography in the United States, 1892-1925, *National Geographic Research* 1, 217-235.
Martin, G. J. (1988a) Geography, Geographers and Yale University, c. 1770-1970, in J. E. Harmon and T. J. Rickard (eds.), *Geography in New England*, New England/St. Lawrence Valley Geographical Society, [New Britain, CT], pp. 2-9.
Martin, G. J. (1988b) On American and German Geography, circa. 1850-1940, *Proceedings of the New England/St. Lawrence Valley Geographical Society* 18, 15-25.
Martin, G. J. (1998) The Emergence and Development of Geographic Thought in New England, *Economic Geography* 78, Extra Issue, 1-13.
Monk, J. (1998) The Women Were Always Welcome at Clark, *Economic Geography* 78, Extra Issue, 14-30.
Parker, F. W. (1893) The Relation of Geography to History, *National Geographic Magazine* 5, 125-131.
Rascoe, B. J. (1937) *Before I Forget*, Literary Guild of America, New York.
Roorbach, A. O. (1937) *The Development of the Social Studies in American Secondary Education before 1861*, n. p., Philadelphia.
Rowley, V. M. (1964) *J. Russell Smith: Geographer, Educator, and Conservationist*, University of Pennsylvania Press, Philadelphia.
Spencer, J. E. (1974) The Evolution of the Discipline of Geography in the Twentieth Century, *Geographical Perspectives* 33, 20-36.
Speth, W. W. (1999) *How It Came to Be: Carl O. Sauer, Franz Boas and the Meanings of Anthropogeography*, Ephemera Press, [Ellensburg, WA].

Taylor, P. J. (1976) An Interpretation of the Quantification Debate in British Geography, *Transactions, Institute of British Geographers*, N. S. 1, 129-142.
Trewartha, G. T. (1978) *Geography at the University of Wisconsin-Madison*, [University of Wisconsin Department of Geography, Madison, WI].
Turner, B. L. II (1989) The Specialist-Synthesis Approach to the Revival of Geography: The Case of Cultural Ecology, *Annals of the Association of American Geographers* 79, 88-100.
Walters, W. D. Jr. (1987) Voices for Reform in Early American Geographical Education, *Journal of Geography* 86, 156-160.
Warntz, W. (1964) *Geography Now and Then: Some Notes on the History of Academic Geography in the United States*, American Geographical Society, New York.
Weld, R. F. (1938) *Brooklyn Village, 1816-1834*, Columbia University Press, New York.
Wright, J. K. (1951) The Field of the Geographical Society, in T. G. Taylor (ed.), *Geography in the Twentieth Century*, Philosophical Library, New York, 362-367.
Wright, J. K. (1952) *Geography in the Making: The American Geographical Society, 1851-1951*, American Geographical Society, New York.

William A. Koelsch
The Versailles, #3C
3310 First Avenue
San Diego, California 92103
USA

10 The Development of Geographical Study in Canada, 1870-2000

John Warkentin and Paul Simpson-Housley

Geography in Canada before 1870

Across northern North America aboriginal people had an astute knowledge of their land, and European explorers from the early 16th century to the end of the 19th depended on aboriginals to be guided over land and on water from the Atlantic seaboard, progressing in stages to the shores of the Pacific and Arctic oceans. Aboriginals even drew maps for Europeans. Fur traders especially built up an extensive knowledge of the land. In the 1840s, for example, David Thompson, distinguished surveyor and trader, wrote a description of the country from Lake Superior to the Pacific Ocean, first published in 1916, that is now recognized as one of the finest regional geographies ever written on any part of North America (Thompson 1962). In the North reconnaissance explorations were undertaken by British naval men seeking the Northwest Passage, and they prepared valuable reports of natural features. Canada's premier scientific organization of the 19th century, the Geological Survey of Canada, was founded in 1841. After Canadian Confederation in 1867 its field of operations covered all of northern North America except, of course, Alaska and Greenland (Zaslow 1975). Slowly, and in many different ways, a rudimentary but sound factual information base on Canada accumulated. An unconscious grace and intelligence infuses much of this early geographical writing; the geography emerges from the unforced flow of narratives describing new areas experienced through hard travelling and seen through fresh eyes.

For Québec in the early 19th century, France was a source of educational material for the schools. In English-speaking Canada schools were highly dependent upon British and United States textbooks in many subjects. Concern about use of U. S. textbooks in Canada was and is always present, particularly in geography and history. In the 19th century, U. S. geography texts were criticized not only for their brief treatment of British North America, but for anti-

G.A. Dunbar (ed.), Geography: Discipline, Profession and Subject since 1870, 281–315.

British sentiments. The first substantial Canadian geography texts to block the use of foreign textbooks appeared in the mid-19th century, but the problem of American educational material persisted.

Despite these promising beginnings the task of geography remained immense in the 1870s. Knowledge of most parts of the country was thin and sketchy. Not only that, the geography of the country was changing very rapidly as settlers from many lands arrived, pushing aboriginals aside as their life of movement gave way to a sedentary life based on individual pieces of property.

Geographical Exploration and Interpretative Studies, 1870-1930.

1 Discovery

In late 19th- and early 20th-century Canada an ebullient spirit was pervasive as huge new territories were opened, expanding imaginations. Resources alone did not bring fresh visions: entirely new spaces entered the Canadian mind, especially the West and the North, with new horizons, new opportunities, and occasional disappointments.

Canada in this period began to know itself. The country had to be created in people's minds. The challenge was variously met. Visitors could not resist the temptation to write general books; scientists and adventurers went to the frontiers, recording what they found. New mapping was carried out, intellectuals interpreted regional tensions, and in their works, novelists, poets, and painters created images of the land. Geography was taught in the schools, but there were no university geography departments to forward this massive educational mission.

After 1870, adventurers could still hope to perform great feats of travel and achieve fame. Scandinavians led the way. In four Arctic winters, 1898-1902, the Norwegian Otto Sverdrup mapped the unknown western coast of Ellesmere Island and discovered large islands farther west, Axel Heiberg, and Amund and Ellef Ringnes (Sverdrup 1904). In the period 1903-1906, Roald Amundsen successfully completed the first voyage through the Northwest Passage. In 1913-1918 Vilhjalmur Stefansson, born in Manitoba of Icelandic origin, and probably the last explorer to achieve enduring fame in the Canadian Arctic, finally ended the classic era of exploration based on long sledge journeys (Stefansson 1921). However, the last unknown land in Canada, Air Force Island in Foxe

Basin west of Baffin Island, was not discovered until 1948 (by aerial survey). Another great goal of adventurers was to be the first to the North Pole. The American naval officer Robert Peary was given this credit in 1909, although it was debated whether he or Frederick Cook could claim primacy. Current evidence would suggest that neither actually achieved their goal. Ralph Plaisted is now credited as being the first at the North Pole by ice as late as 1968.

South of Arctic Canada large areas also remained little known for many decades after 1870. Almost half of Canada is difficult to interpret geologically because it consists of Precambrian rocks that lack index fossils. In 1883 the famous Austrian geologist Eduard Suess gave that ancient nucleus of the North American continent a name that became widely accepted—the Canadian Shield—but it would be another hundred years before its true geological complexity was revealed.

The rôle of ice in shaping land is of great significance in understanding the geography of Canada. It is ironic that in the late 19th century, American glacial geologists pushed their field surveys into Canada at approximately the same time that Canadian geologists were exploring far to the north. In 1893 and '94, J. B. Tyrell of the Geological Survey of Canada deduced from the orientation of glacial striae that the region west of Hudson Bay was the western, or Keewatin, centre of North American continental glaciation (Zaslow 1975, 162-165). In the period 1892-1899, A. P. Low, also of the Canadian geological survey, on his exploratory traverses in Ungava, identified the Labradorean centre of the continental ice sheet. Early American and Canadian geological investigations in southern and northern Canada laid the classic groundwork for many subsequent studies of glacial landforms and the history of glaciation in Canada. These interpretations have not been superseded, only refined.

2 Maps, Atlases, Journals, and Documents

For much of the 19th century Canadians were dependent upon English map-makers for maps of British North America. In southern Canada surveys under the Surveyors-General of the various colonies and later provinces laid out crown land for settlement. Government survey information was accessible to private mapping firms to prepare maps and county atlases for public use, prepared for many counties in eastern Canada in the 1870s and '80s. To comprehend and picture their own country, people need atlases of the entire nation. It was not until 1875 that Canada got its first atlas of the country as a whole. Surprisingly, two rival substantial commercial atlases were published that year (Tackabury 1875, Walker and Miles 1875). Through their presentation of essays and maps, both publications can be considered to be introductions to the geography of Canada. They probably for the first time gave many Canadians a feeling for the spatial complexity of their great land, and helped build the imaginative geography of the country.

Soon after Confederation various federal government departments gradually became important map-producing agencies (Nicholson and Sebert 1981). Beginning in 1892-1893 Canada's first extensive topographic map series, a survey of the prairies at 1" - 3 miles (1:190,080), was developed. By far the most important series of topographic maps was the contoured 1" - 1 mile (1:63,360) series, with the first sheet published in 1906. In the 1920s, another great series, the contoured 1" - 4 miles (1:253,440) series, was initiated, with the first sheet published in 1926. These map series have been converted to 1:50,000 and 1:250,000 scales, respectively, and the two series remain the basic maps for studying the natural and human features of Canada. As the country was occupied, the challenge of producing small-scale reference maps of Canada, and the need to record systematically place names, names of natural features, and elevations, led to the appointment in 1880 of a geographer in the Department of the Interior. James White, a surveyor who began in government service with the Geological Survey of Canada, was appointed geographer in 1899, and as the staff grew his title was elevated to Chief Geographer. Under his direction, the first *National Atlas of Canada* was published in 1906, considered to be the second official national atlas in the world, preceded only by that of Finland (Canada 1906). A new expanded edition of the National Atlas followed in 1915 (Canada 1915).

Non-governmental organizations also took up the cause of geography. La Société de géographie de Québec was founded in 1877 in Québec City, the third geographical society in North America. A mainly professional group in Québec City founded the society to popularize geography, and study Québec and make it known. The Canadian Geographical Society was founded in Ottawa in 1929 by a few federal civil servants, some of whom had spent much of their careers surveying distant parts of the country. We had thus come full circle: the very men who in their youth were classic scientific explorers of Canada, now in the prime of their careers, created an organization to bring geographical information to the general public. In 1957 it became the Royal Canadian Geographical Society. The *Canadian Geographical Journal* first appeared in 1930, and is prominent on newsstands today as the *Canadian Geographic*. The Champlain Society was founded in 1905 in Toronto to publish important geographical and historical documents relating to Canada. Many of the documents are explorers' accounts of their journeys.

3 Geography Books for the General Public: Imperialism, Description, Interpretation and Creative Works

In Great Britain, in the nineteenth-century hey-day of imperial power, geographical compendia were published on distant parts of Queen Victoria's empire. Edward Stanford, a well-established mapmaker and publisher, based in London, England, produced a famous multi-volume series of geography books on the Empire and other parts of the world. Stanford's *Compendium of Geography and Travel, North America*, first printed in 1883, was published in new editions with new authors in 1897 and 1915, ending in the last version as a vast reference tome on Canada and Newfoundland over 1,000 pages in length (Hayden and Selwyn 1883, Dawson 1897, Ami 1915). Geography books need not be lengthy to be good. G. M. Dawson, the brightest light in the Geological Survey, in 1892 contributed to a much livelier book for Macmillan covering the entire Empire, except India and Ceylon, which he co-authored with an Australian, Alexander Sutherland (Dawson and Sutherland 1898). The section on Canada and Newfoundland, half of the book, is a brilliant summation of both the physical and the human geography.

Between 1870 and 1930 Canadian geography benefited greatly from many writers, ranging from adventurers to public intellectuals, who prepared accounts of Canada. A prime example is Arthur Buies, long-time secretary of La Société de géographie de Québec, who wrote well illustrated geographical descriptions of pioneer sub-regions of the province in the late 19th century (Buies 1880, 1889, 1890, 1895). During the long European imperial period many parts of the world were the subjects of brief geographical sketches by visitors from the European metropolises. Canada received its share from a steady stream of European observers, mainly British. Well-known literary figures visited Canada and wrote their impressions. Rupert Brooke described a quick tour in 1913-1914 from Montréal to the Rockies and Selkirks, with many epigrammatic observations on the regions and cities visited (Brooke 1916). To this day, Canadian journalists use them as a source of provocative quotations to set up their own essays on the country.

Many books discussed Canada's frontier areas. As settlement progressed, and the agricultural ecumene was occupied, European and Canadian interest steadily shifted westward, toward what were considered to be romantically remote areas to visit. At the turn of the century, these southern journeys paled as settlers moved in and ended the excitement. Attention shifted to the Arctic, particularly to the Barren Grounds west of Hudson Bay. These were accessible to individual adventurers through the transportation system and trading posts run by the Hudson's Bay Company (Ingstad 1933). The early 20th century was also a time of looking benignly, almost with a promoter's eye, at the North's settlement potential. This is reflected in titles such as *The Arctic Prairies* and *The Friendly Arctic: Five Years of Polar Exploration* (Seton 1911, Stefansson 1921).

Among the abundant travel and adventure books appeared a few written by well-informed acute observers who incisively interpreted the geographical development of the country. Some authors gave a fine insight into the country in this period. Themes discussed include Canada's relationships with the United States (Smith 1891), the regional character of Canada (Parkin 1895), the European occupation of Canada (Rogers 1911), and English Canadian-French Canadian relationships (Siegfried 1906). These books remain well worth reading. A few novels written of this period gave a profound insight into the geography of particular parts of Canada (see Simpson-Housley and Norcliffe 1992), including Ontario (Duncan 1904) and Québec (Hémon 1916), and helped create feelings of local regional identity (Dunham 1924). Regional landscapes made themselves felt in poetry, which caught the spirit of different regions. For instance, realistic or not, the raw frontier atmosphere of the Klondike Gold Rush of the late 19th century was fixed for all time in the popular verse of Robert Service (Service 1940). Painters, however, have had a wider, more profound and lasting influence on the perception of the Canadian land than the writers. In the late 19th century, rural scenes of settled Canada were popular subjects. Painters then ventured into the Canadian Shield and the Rocky Mountains, some such as Lucius O'Brien sponsored by the CPR, to depict spectacular vistas, fiords or mountain ranges (Reid 1990). But the startling, lasting, impact was that of Tom Thomson and the Group of Seven painters, of the 1910s-1930s, whose paintings of the Canadian Shield enabled Canadians to see that rugged landscape through new eyes (Mellen 1970). Indeed, these had such an impact that the Shield, and the Canadian North, became symbols of Canada, and much has been written on that theme. Paintings by Emily Carr of British Columbian forest and aboriginal landscapes have had a particularly profound impact on how West Coast landscapes and aboriginal peoples are perceived (Shadbolt 1979).

Beginnings of Academic Geography, 1870-1950: Schools, Universities, Ideas

1 Geography in Schools

Education has always been a provincial responsibility in Canada and the educational system in each province has a tightly integrated hierarchical structure. Strong central control of curricula in each province, and of textbooks authorized for each subject in every grade, stems in English-speaking Canada from the powerful influence of the famous Ontario educator Egerton Ryerson, and in French-speaking Canada of the Roman Catholic Church.

In the late 19th century, Canadian publishers actively promoted geography textbooks written in Canada. Ontario and Québec took the lead, but textbooks were also published in Nova Scotia. From them we can learn about the contemporary state of geography in Canada and prevailing geographical ideas. More fully illustrated U. S. textbooks with improved diagrams and photographs were occasionally adapted for the Canadian market. A Canadian author would be hired to insert Canadian examples and illustrations where possible throughout the text. Of course, as well as posing a serious problem of content in adapted texts, living next to the influential English-speaking U. S. could also be an advantage, a stimulus and source of educational ideas for English-speaking Canada.

In Québec various authors and a teaching order, les Frères des Écoles Chrétiennes, produced substantial French-language geography textbooks. Some books went into many revised editions, and contained considerable up-to-date detail on Québec. In the twentieth century, a series by another teaching order, Les Frères Maristes, became dominant; after 1920 their publications included some effective volumes combining maps and texts in atlas-geographies.

Occasionally an exceptionally good book appeared—for example, the outstanding regional textbook, *The Canadian West: A Geography of Manitoba and the North-West Territories* (McIntyre 1904). However, educators not only recognized that the quality of geography textbooks was generally poor, but the stimulus that geography be recognized as a university subject was badly needed, not only to improve texts but to train competent teachers. There were simply no academic geographers who could write textbooks or serve as teachers' consultants. Hence the occasional turn to American textbooks, published in steadily improving formats and contents. Although wariness prevailed. In the early 1920s a new *Ontario Public School Geography* was prepared by the Department of Education, with the intention "to utilize Canadian talent, to encourage Canadian enterprise, and to secure a thoroughly Canadian outlook, however formidable and hazardous such an undertaking might prove" (Anonymous 1922, 3). It was authorized as a text from 1921-1922 to 1950-1951, with only slight changes. It stretches credibility that no substantial revisions were made, especially in a geography book that was to affect minds for more than a generation.

Nevertheless education and geography are living things. Teaching changes continued to be made, as various educational currents stirred geography in Canada. At the turn of the 19th century the American educator John Dewey, in his efforts towards a more inquiry-oriented approach to education, sometimes used geography to illustrate new methods. In the 1920s and '30s, many of his ideas on teaching, particularly the inquiry method in 'school projects', were incorporated into what was called 'progressive education' in the United States. These American ideas came to Canada in two ways: directly from across the border, but also indirectly from Great Britain, where American ideas of progressive education were well known. In Britain the 'school project' became known as the 'enterprise method', and when the idea recrossed the Atlantic it was more acceptable in Canada under that name, simply because now it was not imported directly from the United States. Canada never fought the fierce battles over progressive education that raged in the United States, but changes there were, mostly associated with the introduction of what was called 'social studies', a term first used in Canada in the 1920s. The provinces most affected by American concepts were Alberta and Saskatchewan. In the elementary grades in some provinces geography and history were fused into 'social studies'. In most Canadian provinces limited geography was offered in high school. Physical geography was covered in some schools, but courses in the natural sciences or commercial geography gradually tended to absorb physical geography in the curriculum, and in the 1930s it faded away as an independent course in some provinces. Geography was probably strongest in high schools in Ontario, where commercial geography was taught after 1910 as a required subject.

Canada was very fortunate when on occasion a university-trained geographer from Europe would immigrate and teach in a Canadian school. A few Canadian teachers were impatient with the lack of university training in Canada, and in the early 1930s at least one ambitious teacher took courses in geography in a United States university. That would not do. What was lacking in Canadian school geography was strong guidance from universities, and when it did come it was warmly welcomed by teachers.

2 Early Geography Courses in Universities.

Some disciplines have been late in coming to Canadian universities. Geography is one. This does not mean that institutions were unaware of geography. Tantalizingly, geography began to surface in isolated university courses in English-speaking Canada. In the late 19th century, geology students at Queen's University and the University of Toronto were introduced to courses in physical geography. Considering the close relationship of geology and geomorphology these innovations are not surprising. Of more long-term importance, however, was the enthusiasm of economists for geography at the University of Toronto, Queen's, and McMaster, where courses titled commercial and economic geography were mounted in the early 20th century. In 1906 James Mavor of the University of Toronto was the first to introduce economic geography. He later contributed several chapters on the economic geography of Canada to *The Oxford Survey of the Empire* (Herbertson and Howarth 1914, 96-198). Later, internationally known economists such as H. A. Innis (Toronto) and W. A. Mackintosh (Queen's) supported geography at their institutions. There were similar stirrings at the University of British Columbia, where in 1922-1923 the name of the Department of Geology and Mineralogy was changed to Geology and Geography, the first recognition in Canada of 'geography' in the name of a university department. At UBC, economists as well as geologists supported the subject, and by 1938 it became possible to major in geography by selecting appropriate courses.

In Québec in this period there was more directly geographical dialogue and writing than elsewhere in Canada. About 1910, Abbé Adélard Desrosiers, the principal of l'École Normale Jacques Cartier in Montréal after World War I, studied under Paul Vidal de la Blache and Marcel Dubois. He received a diploma in history and geography, and was a strong supporter of geography. Geography was considered an important part of the curriculum at l'École des Hautes Études Commerciales (HEC) in Montréal. Henry Laureys (1882-1958), a native Belgian, trained at the Université de Louvain, came to Canada in 1911, and joined the HEC. He was very much concerned with applied economic geography, and wrote essays and books on the topic (Laureys 1913, 1914, 1926, 1929). A contemporary, Émile Miller (1885-1922), geography teacher in l'École Normale Jacques Cartier and in the Faculty of Letters, Université de Montréal, had a much more humanistic approach to the subject. Here was a Canadian scholar who through wide reading had a firm intellectual grasp of the scope of the discipline and saw geography as a way of viewing the world (Miller 1912, 1921). Active in improving geography in the schools of Québec, and in popularizing it, and probably the first academic to hold the title Professeur de Géographie in a Canadian university, Miller's contribution to Canadian geography deserves further study.

The development of geography in Québec was stimulated in a distinctive way by a direct and strong two-way connection with France. Geographers trained in France taught geography courses in Québec universities in the 1920s and '30s, and Québec-born geography students such as Benoît Brouillette (1904-1979) and Pierre Dagenais, who later became important geography pioneers in Québec, went to France for advanced degrees. Brouillette was, to our knowledge, the first Canadian to obtain a doctorate in geography, from the Université de Paris in 1931. In the grand tradition of European geographers doing research in other parts of the world, leading French geographers visited and wrote about Québec. Raoul Blanchard (1877-1965), Professor of Geography for many years at Grenoble, was particularly influential. Warmly welcomed and highly respected, Blanchard became a vital intellectual stimulus and rôle model for geography in Québec. Geographers from both France and Québec bore the load of developing geography at the university level in Québec. Brouillette, who taught at l'École des Hautes Études Commerciales, Université de Montréal from 1931 to 1969, and also instructed at Université Laval in Québec City in 1941-1945, was an indefatigable supporter of geography in Canada and in international organizations. Pierre Deffontaines (1894-1978), a French geographer who taught in Barcelona for much of his career, continued the professional work of Blanchard in Québec. From 1948 to 1952 he taught geography courses at Laval, where he was the first professor of geography.

3 Geographical Writing

In Québec, Raoul Blanchard wrote classic regional syntheses of different parts of the province based upon field observation, many interviews, and thorough library research (Blanchard 1930, 1935,

1938, 1947, 1949, 1954, 1964). Before academic geography departments got well underway in Canada, important geographical ideas were being applied to the study of the country by geographers from other countries, as well as by Canadian academics in other disciplines. Movement over long distances will always be vital in Canada. Marion Newbigin, a British geographer, examined the shaping influence of the St. Lawrence River on the growth of Canada (Newbigin 1926), and in the hands of economic historian Harold Innis and historian Donald Creighton the rôle of the St. Lawrence in North America evolved into a grand overarching interpretation of Canadian development, with the St. Lawrence River valley and Lower Great Lakes viewed as the organizing axis of Canada (Innis 1930, Creighton 1937). A few Canadian historians tested Frederick Jackson Turner's frontier thesis on Canada. They concluded that it failed to throw light on Canadian development.

The idea of the ecumene is a seminal concept in studying the geography of Canada, as first demonstrated by an American geographer, Mark Jefferson (Jefferson 1934). Closely related as a topic of great interest to Canadians is the geographical limit to agricultural settlement. In the 1930s a nine-volume series was published on Canadian pioneer problems, edited by the Canadian economic historian W. A. Mackintosh and the American geographer W. L. G. Joerg (Mackintosh and Joerg 1934-1940). Leading Canadian economists, historians, and sociologists did the archival and field investigations. The project is of interest here, not only because of its geographical nature, but, paradoxically, for the fact that *not a single* Canadian geographer was involved! Their absence once again highlights the lag in developing the discipline in Canadian universities.

Numerous articles written from the 1920s to the '40s by Canadian historians and social scientists have a geographical bent, such as examining the geographical and cultural aspects of the major regions, geographical determinants in Canadian history, the importance of staples in the development of the country, the rôle of the Canadian Shield, and the like. This innate interest in geography from outside the discipline perhaps culminated in two textbooks by the economist A. W. Currie on the economic development of Canada using a regional approach, a synthesis of the scholarly work done on the Canadian economy, drawing heavily on Harold Innis (Currie 1942, 1945). A decade later economic historians turned from Currie's regional approach to become more purely economic and historical (Easterbrook and Aitken 1956).

European geographers on research and teaching trips to Canada contributed to the geographical investigations of Canada. Raoul Blanchard's work has already been noted. Carl Schott (1905-1990), a German geographer, wrote his Ph.D. dissertation on the evolution of settlement patterns in Ontario (Schott 1936). Thus research traditions classic in the geographical work of France and Germany were brought to the study of Canada. American geographers also contributed to the study of Canada, usually presenting their research in U. S. geographical journals, or in textbooks (e.g., Smith 1925).

4 University Departments are Finally Created

We have seen that geography courses were introduced into Canadian universities; however, it was no easy task to persuade them to establish geography departments. Trevor Lloyd (1906-1995), Professor of Geography at McGill University, told the story of how, on his first visit to Canada in 1930 as a member of a British university debating team, the chairman of the Board of Governors in a western Canadian university introduced him at a debate in these words:

> This young man tells me that he spent four years at university studying geography. I don't know what he was doing with his time. In this province we complete it in grade seven! (Lloyd 1979, 95)

Not until 1935-1936 did a Canadian university—the University of Toronto—finally establish a department of geography. Its origins and early years have been thoroughly studied (Dunbar 1985). Griffith Taylor (1880-1963), an Australian who had taught geography at the universities of Sydney and Chicago before he came to Toronto, was appointed. For one year he taught in the political economy department and then in 1936 founded and headed the new geography department.

In Québec we have seen how French Canadian and French geographers were active in the universities. They held appointments in various schools and faculties and could hold the title of Professeur de Géographie. Geography departments, however, were not created until after World War II. The department at Laval started within the Institut d'Histoire et de Géographie in 1946, became the Institut de Géographie in 1955, and then a Département in 1969. Louis-Edmond Hamelin and Henri Dorion were early leaders. At the Université de Montréal the Institut de Géographie was founded in 1947 by Pierre Dagenais, and he was the director until 1962, when it became the Département de Géographie. Dagenais, along with Benoît Brouillette, shaped the development of modern geography in Québec, and they were the first French Canadians to become professors within active departments of geography in the province. Geography courses at the Université de Sherbrooke were given in the commerce department after 1957, and the Département de Géographie was established in 1963.

In English-speaking Canada the University of Western Ontario was the second university to hire a geographer. In 1938, Edward Pleva, American-born and University of Minnesota-trained, was appointed to the geology department, and by 1948 he headed an independent department. J. Wreford Watson (1915-1990), a Scottish-born Edinburgh graduate in geography, began lecturing in geography at McMaster University in 1939, completed a Ph.D. at the University of Toronto in 1945, and began building a department after World War II. McGill University in Montréal appointed its first geographer in 1945, British-born and trained meteorologist G. H. T. Kimble, who rapidly established a department. At the University of British Columbia, in 1945, Thomas R. Weir (1912-1997), a Canadian, was the first geographer with an advanced degree (M.A.) appointed to lecture in the Department of Geology and Geography. J. Lewis Robinson, another Canadian, who had earned his Ph.D. at Clark University, went to UBC in 1946 with the expectation of establishing a separate geography department, and within seven years an independent department was functioning under his direction. Manitoba, too, started geography courses in the 1940s. Thomas Weir, who had done his doctoral work at the University of Wisconsin after teaching for two years at UBC, was appointed as a geographer in the UM Department of Economics in 1949; the next year a Department of Geography was created with Weir as head. At Queen's University, Canadian-born Donald Q. Innis was appointed lecturer in geography in the Department of Political and Economic Science in 1952. In 1960, a Department of Geography was established at Queen's, with Richard Ruggles, another Canadian, as the founding head.

The pattern is evident. The postwar period was a time of profound social change in Canada, of investment in and expansion of university education, and a realization that the social sciences included more disciplines than just political economy and psychology. A subject such as geography benefited from this cultural shift, which helped stimulate its rise in Canadian universities. At this early stage in establishing post-secondary geography in Canada, universities drew upon British-, French-, or United States-trained geographers. A few young Canadians, graduates of the new departments, soon went to the U. S., French, or British universities for advanced degrees, and then returned to Canada to found geography departments or join those already in existence. By 1950 the leaven in Canadian higher education was just beginning to work; in the following decades there were many changes.

Geography 1950-2000

1 Geographical Branch and Other Government Agencies

After World War II the federal government recognized that there was a need to look afresh at what changes were taking place in resource development, in transportation needs, and in urban centres, and in 1947 the Department of Mines and Resources created a Geographical Bureau, later called the Geographical Branch, first under the anthropologist Diamond Jenness and then under geographer Trevor Lloyd. Shortly J. Wreford Watson was recruited from McMaster University in 1949 to become director, and by 1950 a group of ten geographers had been hired. In the following decade the professional staff of the Branch included English-Canadian, French-Canadian, British-trained, and Polish-trained geographers. In summer months, geography students at Canadian universities were employed to help in geographical work in Ottawa and in the field. In 1951 Dr. Watson described the Geographical Branch's responsibilities this way:

> Its principal duties are to obtain, collate, and organize all the material relating to the geography of Canada that may be of use to the economic, commercial, and social life of this country; to prepare studies on the specific aspects of the geography of Canada for the use of those engaged in government, defense, business, scientific research, etc.; and to obtain material relating to the basic geographic facts and conditions of foreign countries, to the satisfaction of national or institutional interests. (Watson 1951, v)

It is often difficult to implement an ideal, and the Branch had a difficult time establishing a place for itself in the entrenched Ottawa bureaucracy. In a substantial reorganization of federal government services, the Department of Energy, Mines and Resources was created in 1966, and it was decided by senior bureaucrats that, rather than keeping geographers together in one agency, their geographical skills would be of greater advantage to the government if they were deployed amongst various specialized branches. As a consequence, in 1967 the Geographical Branch ceased to exist as a separate organization. Despite its somewhat indeterminate mandate, with its broad and diffuse interests, not unlike the way geography itself as a discipline is often regarded, the Branch accomplished significant work during its short existence. Quite logically, the creation of Canadian national atlases became the Branch's responsibility, and it produced one edition of the national atlas and before its demise had started another. Basic research was carried out in both northern and southern Canada. Much of this work was good descriptive and analytical geography, comprehensive in approach, and searching for relationships, similar to what was published in scholarly geographical journals of the time. Major reports were published separately as memoirs, but much of this research entered the geographical literature through the *Geographical Bulletin*, published from 1952 to 1967. These publications remain a valuable reference source on Canada. The Branch as well made a highly significant contribution to the Canadian geography profession. Of great importance was the fact that it offered a field of employment to geographers at a crucial time when honours undergraduate and advanced training in geography was getting established in Canadian universities. Thus, in the 1950s and '60s the Geographical Branch instilled confidence that there was useful professional work for geographers at a time when the profession was feeling its way in Canada. Indeed, in its early years the Branch was the largest employer of geographers in the country.

Even as the Branch was dismantled, geographical work continued under other government agencies. In the 1960s, a massive mapping of Canadian resources, named the Canada Land Inventory (CLI) project, was undertaken by the federal government. Geographers were involved in many aspects of this work. Much effort went into devising appropriate classification schemes that had to be consistent across the entire country. Under the pioneering leadership of Roger F. Tomlinson, GIS was developed as part of the CLI project to cope with mapping the enormous amount of resource information collected (McGrath and Sebert 1999, 462-490). The new innovative system of compiling and mapping spatial land information by 1966 became known as the Canada Geographic Information System (CGIS), and since then GIS has become a generic term. Provincial governments in the 1950s and '60s were newly conscious of the need to develop resources carefully, including recreational resources, and many geographers who graduated at that time found employment in various government departments. Some geographers joined consulting firms that provided resource planning services both nationally and internationally. In both the federal and provincial civil services, geographers rose to senior policy-making positions, usually in areas that demanded an integrated overview of conservation and development, for which a broad training in geography was highly appropriate.

2 Universities

By 1950, the pioneer trail had been blazed for geography in Canadian academe. That year 56 geographers met at McGill University in Montréal to form a learned society, the Canadian Association of Geographers/L'Association Canadienne des Géographes. Staff of the Geographical Branch, particularly J. Wreford Watson, provided leadership in founding the CAG/ACG, and also strong administrative support in its early years. In the first quarter century of its existence the CAG/ACG enabled practically all geographers in Canada to get acquainted with one another. The profession was small enough that no matter what the field of specialization geographers knew of one another's work, most attended all sessions at annual meetings since there was no overlap in scheduling, and went on field trips together. It was an era of friendly and rewarding camaraderie.

In the years 1950 to 1965, departments of geography were founded in universities that had missed the first wave, and, simultaneously, existing departments were expanded. This was an era when numerous new universities were founded in Canada. By now geography was well enough known and respected by academic administrators that it was commonly included in the first group of disciplines introduced in the new institutions. A regional pattern within Canada is apparent in making geography appointments. In each of the four western provinces, every institution founded as the provincial university had appointed at least one geographer by 1953, and departments developed quickly. Smaller institutions in the West lost little time in following this lead. In Ontario and Québec it was accepted without exception that all universities should have geography departments. Only Atlantic Canada stood apart. In this region with many long-established institutions, only three out of thirteen colleges and universities created geography departments. Thus it cannot be taken for granted that all Canadian colleges and universities regard geography as an essential liberal arts and sciences subject. In some institutions budgetary constraints held back geography appointments.

As new departments were established and existing ones expanded, recruitment of geographers kept department heads exceedingly busy. Many appointees were Canadians. In the 1940s, Canadian honours B.A.s had begun to do graduate work in leading American, British, and French institutions and returned to join Canadian departments. Also, beginning in the late 1940s, among the large numbers of immigrants pouring into Canada were geographers, including students from the United Kingdom and other Commonwealth countries doing graduate work and highly trained academics who were refugees from the war in Europe, especially Polish geographers. Canada benefited greatly, because some of the students stayed and joined departments, as did the refugees. In the years before 1970, faculty came from the United Kingdom, the United States, France, the Netherlands, Scandinavia, and Poland. Although some were established geographers, most were looking for their first academic opportunities in Canada. Easily the largest incoming stream was from the United Kingdom, either appointed directly to departments or via graduate programmes in Canada or the United States.

It had long been apparent that if geography were to establish strong permanent roots in Canadian universities, departments could not continue to depend on securing properly qualified instructors from other countries. Establishing high quality graduate training was vital for the long-term welfare of geography. Most departments after they had four or five members began to offer an M.A. degree, and then with greater caution, if they continued to grow, would offer the Ph.D. degree.

Geography curricula in Canadian universities and colleges changed significantly after the mid-20th century. In the 1940s and '50s, great emphasis was placed on a well-rounded liberal education in geography. In the undergraduate curriculum a balance between regional and systematic geography was sought, and students received a good grounding in both physical and human geography. Probable reasons for the regional courses were that the French school of regional geography under Vidal de la Blache had made a strong impact and was considered the pinnacle of geographical study, a model of geographical synthesis, and also because the 19th-century scientific exploration of the earth, imperialism, and a consequent curiosity about all parts of the world still had great momentum. In the years before 1960, few students in a four-year undergraduate programme graduated without having taken a course on every continent. If they were in a three-year programme the chances were that they had taken a course in world regional geography to ensure general coverage, as well as a course on either North America or Europe. An important reason for the comprehensive curriculum was to ensure that students who would become school teachers acquired a sound geographical knowledge of all parts of the world and also of the basic thematic elements of geography. Another reason was that research in the humanities and social sciences was not stressed in Canadian universities until the 1960s, so there was only limited specialization. Research funding was limited, and often dependent on U. S. foundations. It must be emphasized that many individual instructors were dedicated to research, and produced scholarly work consistently. However, Canadian universities in those years were not organized around research, but rather around teaching, as their prime responsibility.

The report of the federal Royal Commission on National Development in the Arts, Letters and Sciences, 1949-1951 (Canada 1951) recommended the creation of new funding agencies for research, including the Social Sciences and Humanities Research Council. Things began to change as research funds became available. Some professors secured large research grants for their departments from government agencies and international foundations to do research on the North. Also, students and instructors arriving from the United Kingdom and the United States, used to well equipped departments and a more dynamic research culture, had a positive impact. Instructors devoted more time to research, and soon this change was reflected in the curriculum.

The broad spectrum of departmental regional courses was quickly narrowed after the 1950s. Regional courses that remained in the curriculum were those where particular faculty members had special research interests, language skills, or field experience. Students no longer were required to take a prescribed number of regional courses. Indeed, thematic courses were triumphant, and the curriculum quickly became so flexible that there was a danger that regional courses might succumb entirely in the rush to more and more thematic specializations.

In thematic geography, courses in the 1950s covered widely accepted broad areas such as geomorphology, climatology, soils geography, economic geography, agriculture and land use, and urban geography. Those courses have held their own, introducing students to different sub-disciplines of geography, but beginning in the 1960s there was much more specialization in the undergraduate curriculum, and these broad courses were followed by advanced work on more focused problems within the sub-fields. Quantitative methods increasingly entered all geography programmes, as well as new conceptual frameworks, sometimes called theory and modeling, often adopted from other disciplines, such as economics. Just as skill in topographic map reading had been insisted upon at one time, in the 1960s all students in honours programmes were required to take courses in statistics. New areas of interest, some of which had a long history in the discipline, included historical geography, perception of landscape, environmental studies, behavioural geography, natural hazards, social geography, cultural geography, literature and geography, biogeography, and feminist geography.

3 Major Publications

Canada is a country that keeps on producing national atlases, conceived and financed by federal government agencies. Three editions of *The National Atlas of Canada/L'Atlas National du Canada* were published in this period, following a gap of many years since the preceding editions of 1906 and 1915 (Canada 1957, 1974, 1978-1995). Many geography professors feel a strong responsibility to make geographical information available to the general public, and as part of this dedication geography departments in provincial universities have produced atlases of the provinces in which they are located, with funding by interested government agencies. British Columbia led the way (British Columbia Resources Conference 1956), followed by Manitoba (Weir 1960), Alberta (Department of Geography, University of Alberta 1969), Ontario (Dean and Matthews 1969), and Saskatchewan (Richards 1969, Fung 1999). In Québec the Centre d'études en enseignement du Canada, Université Laval produced an atlas of Québec and Canada (Paradis, Raveneau, and Tessier 1986), an atlas of the Maritimes was published by a commercial house (McCalla 1988), and a Newfoundland and Labrador atlas was produced at Memorial University (McManus and Wood 1991). Almost all these atlases were comprehensive, covering physical and human geography, although the emphases varied, sometimes depending to a degree on the interest of the sponsoring agency. The atlases have become reliable mainstays for teaching the geography of Canada in university courses, and are valued as general reference works.

Much research energy has been devoted to producing historical atlases. The *Historical Atlas of Canada* was a large atlas project that engaged over a hundred and fifty scholars from across the county in many disciplines. The director of the atlas was William Dean and the cartographer/designer Geoffrey Matthews. Three volumes were published covering the time from the Ice Age to 1960 (Harris 1987, Kerr and Holdsworth 1990, Gentilcore 1993). An interpretive atlas, the maps and accompanying text depict and analyze the population, transportation, resource, industrial, rural, urban, social, and cultural development of the country. Another historical atlas of comparable significance is *Lines of Country: An Atlas of Railway and Waterway History in Canada* (Andreae 1997). Under the editorial direction of Serge Courville and Normand Séguin a multi-volume *Atlas Historique du Québec* began to be published in the 1990s, particularly noteworthy for its integration of maps, graphics, and illustrations into an extensive interpretive text (Courville, Robert, and Séguin 1995, Courville 1996, Boudreau, Courville, and Séguin 1997).

Numerous descriptions of Canada's major geological and landform divisions have been published since the last half of the 19th century, but not until the mid-20th century was a definitive classification and nomenclature of the terrain features of the entire country put together (Bostock 1964). L. J. Chapman and D. F. Putnam published a classic account of the intricate, mainly glacial, landscape of southern Ontario (Chapman and Putnam 1951).

Following the early work on the ecumene by Mark Jefferson, already mentioned, Canadian geographers expanded and refined the idea (Gajda 1960, Hamelin 1972). Canadian geographers constantly use the concept, but, strangely, it has not registered in the broader academic community. Louis-Edmond Hamelin analyzed the significance of the North in Canadian regionalization, culminating in an 'index of nordicity' (Hamelin 1978, 1980). The Canadian urban system has been extremely important in holding together a thinly populated country of immense distances, and numerous studies have examined it closely (Careless 1989, Simmons 1979, Whebell 1969). In a broadly conceived geographical interpretive work, rare for a geographer, James Lemon of the University of Toronto analyzed the conditions and development of North American society after 1600 through selected North American cities (Lemon 1996).

The Canadian Association of Geographers/L'Association Canadienne des Géographes sponsors publications that seek to provide a perspective on Canadian geography (Dean 1967, Warkentin/Beauregard 1968, 1970). In the 1990s, the CAG/ACG launched a 'Series in Canadian Geography' under the general editorship of R. Cole Harris, which presents the current state of research in major fields of Canadian geography (Bourne and Ley 1993, French and Slaymaker 1993, Britton 1996). In 1972, in connection with the International Geographical Congress in Montréal, a series of six monographs was produced on the major regions of the country (Trotier 1972).

Geographers have been giving increasing attention to the creative arts in studying Canada. J. Wreford Watson alerted Canadian geographers to the importance of literary works in understanding Canadian regions (Watson 1965). A former winner of the Governor General's award for poetry, Watson, under the name James Wreford, published a volume of poems that provide a fine insight into various regions across Canada (Wreford 1979). Ronald Rees has analyzed the landscape painting of western Canada, elucidating the inhabitants' relationships with the land, before and after agricultural settlement (Rees 1984).

4 Geography in the Schools

University-level geography changed dramatically after World War II, and so it was in Canadian schools. A narrow window of opportunity was created for geography in the schools of some provinces between 1950 and 1970, when both the geography curricula and textbooks were greatly improved. Geography was part of a wider revolution, as governments across the country made great new investments in education. Geography's stronger place at the university level, and persistent lobbying by geographers, became pivotal in advancing geography in the schools. Of greatest importance, well trained geography graduates became teachers, supported in some provinces by geography specialists working on curriculum reform, course outlines, and teaching guides. University geographers served on educational committees, initiating and implementing the changes. Canadian geographers and textbook publishers produced new geography books and reading kits. Furthermore, in 1955 the CAG/ACG established a committee—which is still in existence—to assist in improving geography teaching in schools. For many years, from the 1940s to the late '80s, a strong mutually supportive education system for geography was in place in most parts of the country. These promising developments began to weaken as more subjects were added to the school curriculum, including social studies courses. Budgets tightened, money for developing new textbooks declined, and the fights between subjects for a place in the school curriculum, and for teaching hours in the classroom, intensified.

In the early 1990s, the Canadian Association of Geographers Education Committee surveyed the status of geography and social studies in Canadian elementary and secondary schools, and found significant contrasts across the country (Baine 1996, 25-28). There was a sharp divide at the Manitoba-Saskatchewan border; geography courses figured significantly in school programmes to the east and social studies in those to the west. Of course, even if geography does not get all the designated school classroom time it requires, geographers do participate in interdisciplinary courses in social and environmental studies that sometimes take geography's traditional place in the schools. In 1993, to promote geographical education the Canadian Council for Geographic Education was launched, a joint initiative of the Royal Canadian Geographical Society and the National Geographic Society of the United States. The aim was to increase emphasis on geography in Canada's schools and colleges, improve teaching methods and classroom materials, and increase public awareness of the importance of geography.

5 Recent Trends in Universities and Colleges

The 2000 directory of the CAG/ACG lists 48 university departments and some government agencies where geographers are employed. A few colleges have only one person, but most university departments have 15 to 25 members, and the largest, in the University of Toronto, reaches 36. The total number of members in the CAG/ACG is about 750, a sufficient number to cover a wide range of research interests.

Great changes occurred in Canadian geography from the 1950s to the 1990s, parallel to international trends. Broadly speaking, there was a shift from descriptive geography, much of it regional, to quantification and modelling, and then on to interpretive and process-oriented studies. Historical analysis is unhesitatingly turned to for seeking explanations, and much openness and freedom of approach in geographical inquiry is apparent.

Models and theory from the physical and social sciences were introduced into geography in the 1950s and '60s, rather than generated within the discipline. Quantitative methods used in psychology and economics were particularly important. In the search for precision, statistics were used increasingly, and in 1963 Ian Burton christened the time of change just past the 'quantitative revolution' (Burton 1963). Positivism, a search for theory, behavioural perspectives, systems analysis, model building, more abstract generalizations, and perhaps even laws of spatial interaction seemed to be in the air in some Canadian departments. Reacting to this seemingly mechanistic turn of the discipline, a few geographers in the 1970s scrutinized and criticized the search for theory and laws in such a complex field as geography. This debate was salutary. It helped preserve geography's broad approach, serving as protection for its humanistic concerns.

Ideas continue to be brought into geography from the physical and social sciences and humanities. Discussions of new concepts and the quest for theorizing have not let up. If anything, in the 1990s they were more active than ever. This reflects lively and reflective individuals and departments. Geographers should know what goes on in neighbouring disciplines, and use it to vitalize their own. Clearly, however, in the 1990s there was great caution in adopting such ideas, and careful attempts made to evaluate their relevance and value, not the reckless and too often barren application of an earlier time. Moreover, notwithstanding the search for theory and laws extolled at various times, the Canadian research literature does not in practice reflect the exhortations. Quantitative methods, occasionally highly sophisticated ones, are frequently used but there is no concentrated positivist search for geographical laws. Partly this is because of the inherent, near-infinite complexity of the geographical worlds that are being investigated. Also, funding agencies require specific research proposals. In practice, therefore, sharply defined and limited problems are usually investigated.

At the same time, philosophical discussions concerning research frameworks and methodologies flourish, reflecting what is discussed internationally in the social and natural sciences. Spatial analysis, diffusion theory, social ecology, behavioural approaches, Marxism, post-Marxism, neoclassical economics, humanism, feminism, idealism, nationalism, modernism, post-modernism, post-colonialism, deconstruction, and structuration are or have been part of the methodological and philosophical reflections and exchanges in Canadian geography. Though it may be hard to discern how this theorizing and debate are reflected in substantive research and interpretations, some theorizing certainly has affected the kind of research published, for example work in feminist geography. Even as all this fervent discussion is going on, empirical research in geography sedately continues, often done by those also engaged in methodological controversy.

6 Research Specialization

One might ask whether any scholarly specialization has come out of special Canadian geographical conditions. This is a tricky question in the light of geography's past sensitivity to accusations of geographical determinism. One thing is clear: there is considerable interest in northern Canada. Much of the research relates to physical geography, with work carried out on northern climates, permafrost, snow and ice, sea ice conditions, vegetation, and Arctic ecosystems. With some of the world's largest areas of northern wetland, Canada is an ideal laboratory, which has spawned its own subset of research fields. Special glacial landforms and regions have been investigated, and Ross Mackay of the University of British Columbia, in his study of ice wedges and pingoes (Mackay 1963), and Roger Brown (1931-1980), of the federal civil service, in his study of permafrost (Brown 1970), achieved international reputations. Human geographers are interested in northern aboriginal peoples and the conditions in which they live including housing, and in political geography, especially with the major administrative changes in the 1990s through the creation of Nunavut. The history of Arctic exploration still holds the interest of geographers, as well as the changing perceptions of the North since the 19th century. A number of scholars, including Louis-Edmond Hamelin and William Wonders, have made the North a research specialty.

Canadians recognize an increasingly diverse population as distinctive of their country. Geographers have discussed settlement, ethnicity, and cultural landscapes since the profession reached significant numbers in the 1950s. Today, migration, refugees, and problems of integration also drive research, with huge grants supporting massive investigations of some of these themes (Hiebert 2000, 25-42). In a way, this kind of research might be considered as applied geography, with a government agenda. Researchers, however, design the studies and have full research freedom. In some ways population studies are similar to research on the North—of national interest so that funds are available.

Canada's vast territory and generally sparse settlement, with marked urban concentrations, have provided special opportunities for applied research. In the 1950s, a strong feeling prevailed that a new world was in the making, and that geographers were active participants in an exciting process. Geographers were employed in resource inventory, conservation work, planning and regional development programmes. Geographers, as civil servants or consultants, were particularly active in research that led to the establishment of provincial and national parks in many parts of Canada's highly varied natural environments. Subsequently, there has been much research into recreation and tourism, including work on ecotourism. Geographical work continues in the public service at all levels, and the idealism of involvement in public policy and service remains high.

Beyond these special circumstances it is probably fair to say that Canadian geographers' research interests arise out of the same concerns that drive their colleagues elsewhere in the world, although they will often use Canadian case studies and regions in their investigations.

7 Physical, Environmental, and Resource Geography

In physical geography the traditions of climatology and geomorphology remain strong, and hydrology and biogeography have gained strength. In the 1960s, physical geographers were warned that their fields would become second-class if descriptive work continued; they would have to analyze quantitatively in climatology, and in geomorphology use concepts from neighbouring fields such as hydrology and soil mechanics (Hare 1960, Mackay 1961). Researchers were encouraged in 'process' work, and after the 1960s this new emphasis has been evident in physical geography (Saunderson, Smith, and Woo 2000, 67-86). As funding increased from the National Research Council of Canada to improve laboratory and field instrumentation, field mapping and study of landform processes continued. In climatology there is emphasis on measurement, with particular attention given to surface micro-climates and permafrost investigations. Wetlands are being investigated, work is advancing in biogeochemical geography, and biogeography is an active research field in both temperate and tropical environments. Work in physical geography has reached the stage where general integrative works in selected areas of physical geography have been published (Bird 1980, Hare and Thomas 1979). The overarching concept of Global Change has changed the context of research and teaching in physical geography. Broad issues have come to the fore, and process-oriented specialized research addresses them. Thus two approaches, that towards process in detailed physical geographical research to gain a greater understanding of nature, and that of interpreting the natural world better to the general public and to human geographers, both go forward, with emphasis on the first.

Physical and human geography used to meet in regional geography, in attempts to understand the full complexity of a region. It still is such a meeting place, but mainly in relating broad physical patterns and human development. As sub-fields of physical and human geography dig deeper into their own specializations, drawing on applicable process and theory from other disciplines, there has been increasing separation between the two sides of geography. However, the balance is changing again. With increasing human impact on the earth a new association of physical and human geography has been thrust upon the discipline, providing opportunities for coordination in the investigation of Global Change and the sustainability of ecosystems. Now, instead of merely studying relationships through broad correlations of physical and human distribution patterns, there is a close examination of the processes of environmental change brought about by the interactions of humans and the earth. There is a long way to go yet in re-integrating physical and human geography effectively, but there is recognition of the value of doing so in the context of earthly and human needs.

8 Human Geography

Human geographical research continues in empirical ways, both qualitative and quantitative. In economic geography research proceeds across a broad front. Economic geographers continue to investigate the resource industries, manufacturing, and the service sector, and increasingly stress the adoption of advanced technologies, knowledge-intensive activities, productivity, the associated industrial restructuring from Fordist to flexible production, the segmentation of labour markets, the sustainability of enterprises, the impact of government policies, the social and cultural context in which economies exist, and regional clusters of economic growth (Barnes et al. 2000, 4-24). Canada is a consumer society, and applied work is done on the retail sector of the economy. Increasing effort is given to interviews for information gathering to supplement government statistics. Overcoming great distances is a given in Canadian life, so that organization of space, transportation, and commodity flows are continuing themes in Canadian geography. Globalization and its many related economic impacts are investigated, particularly the geographical impact of the North American Free Trade Agreement on the relationship with the United States and Mexico.

Selected aspects of social geography are not new in Canadian geography. Rural studies have long commanded loyalty, while Wreford Watson introduced ideas of urban social geography as early as the 1940s (Watson 1957). "Radical" geography appeared suddenly: in the early 1970s, William Bunge, well known for his early positivist work in theoretical geography, wrote, with Ron Bordessa, a report of a geographical exploring expedition into the heart of Toronto that investigated the human condition in a modern machine-based society and recommended possible means of survival (Bunge and Bordessa 1975). In the 1980s and '90s, social geography became one of the most exciting fields in Canadian geography, most noticeably in investigating the city. Publications on immigration, multiculturalism, and language distribution reflect the character of the country, as do studies of language, language retention, and language change in immigrant groups. The sporadic geographical research on aboriginal peoples gained strength in the 1990s (Peters 2000, 41-55). Geographers have kept abreast of contemporary social issues, including studies of geographical implications of aging at one end of the spectrum and child care at the other. Much research has been done on housing and poverty in major urban centres; in this work there is a clear consciousness of class, ethnicity, feminist issues, and the impact of public and private financing (Bourne and Ley 1993). Medical geography has become an important field of inquiry, both in historical and contemporary studies, including analyses of health care delivery. Studies in life history, in both urban and rural contexts, basically family histories taking into account ethnicity and other factors, represent a strong turn to a humanistic geography.

Consistently, attention has gone into studying the rôle of Canadian urban systems in the development of the country, on urban administration, and the efficiency of metropolitan government, and urban modeling. An important area of research continues to be the investigation of the development processes occurring at the urban-rural fringe. However, the sharpest change has been to investigate the inner city, where the research blends with that in social geography. Social structure, urban landscapes, social polarization, and the forces that drive them have been subjects of particular interest (Ley 1996).

Cultural geography is very closely associated with historical geography; much of the research is linked to Canadian topics. Using early explorers' accounts and Hudson's Bay Company archives, important work has been done on aboriginal peoples in Canada, including their relationships with Europeans, particularly in the fur trade, the diffusion of epidemics, and their experiences in resource development in the 20th century. James Gibson has built an impressive body of historical geographical work on the Pacific Northwest, from the time of the first international interactions on that coast to the early settlement period (Gibson 1991). The history of Canadian mapping has long been an active field of research for historical geographers. The rural landscape produced by different ethnic groups has also been a vital theme, and cultural symbols in landscapes are studied. In many cultural fields work has gone beyond detailed local studies to examining the broader picture. In the 1990s, books were published on the heritage landscape of Ontario, and on building styles in Canada (McIlwraith 1997, Ennals and Holdsworth 1998). Increasing attention is being given to the use of literature and painting in interpreting the character of Canadian regions. Historical geography in Canada is part of a larger academic panorama because many disciplines strive to understand Canadian development, and historical geography is an integral contributor to this endeavour. To this end the Historical Atlas of Canada project was a great bringing together of scholars from many disciplines. Furthermore, geographers from many sub-fields plunge happily into historical geography in the common search for explanation and historical perspective.

Theory in human geography, much of it brought in from other social sciences, is used to define and clarify issues, but most research is not so driven, although there is some highly technical model testing. Critical social and economic theory and critical inquiry have certainly affected geographical thinking in Canada. Interest in the interrelations of capitalism, social structure and the state continues, in some ways reminiscent of investigations by Canadian political economists in the first half of the 20th century before their broad area of study broke into technical sub-fields.

9 Regional Geography and International Studies

Canadian geographers rarely mention regional geography in published lists of their research interests. Despite lacking broad appreciation amongst Canadian geographers in the 1980s and '90s, the fact remains that regional geography, as well as being an academic and intellectual challenge to achieve high quality work, is very much a practical subject, essential in geographical undergraduate instruction, on which so much of the well-being of regional geography teaching in the schools ultimately depends. Regional geographies of Canada, produced for university and college, range in approach from historical interpretations, to thematic analyses, to regional synthesis. In the 1990s, books of geographical essays appeared on different parts of the country. Some were essentially contemporary regional texts; others had a strong historical bent (Harris 1997).

It is worth noting that since the 1960s the Canadian media and general public have become much more conscious of the importance of regions in Canadian life and politics. Arising partly out of the high media profile of the secessionist movement in Québec, called the nationalist movement by many Québécois, it also reflects the increasing sophistication of television news, especially through satellite transmission, that has made the economic and social problems of different parts of the country much more apparent to Canadians. Generally, geographers have not addressed these broad Canadian regional issues, except as a minor component of some thematic discussions or in textbooks, while regionalism has become an important subject for discussion in Canadian political science.

Relative to those scholars working on Canadian topics, foreign area geography specialists are few. Serious scholarly work by at least a few geographers is being done in many parts of the world--the former Soviet Union, Latin America, the Caribbean, Africa, Asia, Oceania, and parts of Europe. Some of this results from the support of the Canadian International Development Agency, the World Bank, and the United Nations, with geographers often attached to large research teams. Individual geographers or departments in Canadian universities have research links with colleagues and departments in many other countries. Going in the other direction, geographers from other countries do significant work in Canada, with German geographers in particular producing an important range of research monographs on Canada since the 1960s, following the example of Carl Schott.

Perspective

It is exasperating that some subjects always seem to have to prove themselves, but this does force them to bring out their best qualities. Certainly, Canadian geography has never been able to rest on its laurels. In universities the Canadian geography community has been large enough in the last quarter of the 20th century to keep abreast of the main areas of international geographical scholarship, both in contributing research and in the kinds of courses offered. Despite reversals in a few departments for external budgetary reasons, the subject in most institutions has held its own or advanced. However, schools are particularly susceptible to social pressures on what is deemed educationally important, especially in an era when international comparisons are made on performance in the sciences, thus demanding more time for certain subjects. As well, educational theorists have power to influence what is stressed in school curricula. Consequently, eternal vigilance and effort are required in schools to ensure that both the content of geography and its identity as a subject are maintained. It is harder to do this today than ever before, and the supporting rôle of the university departments is vital.

Geography reflects on very broad human and natural problems. Although they may be of direct importance in many geographical inquiries, geographers do not usually study power, individuals, or human motives for their own sake. Likely this is why geography has taken a subordinate rôle, at its expense, in some of the great issues involving policy and geographical change in the world. This is true of geography in Canada, although in the period from the late 1940s into the '60s, at a time when Canada was changing rapidly through renewed immigration and industrial growth, it appeared that Canadian geographers were heading toward a stronger rôle in public affairs, whether in local or national issues. It was a period in which geographers were trained in both physical and human geographical approaches to the subject, and a spirit of comprehensive geographical thinking still prevailed, even as research specialization increasingly characterized geography and marked its advance in universities.

Renewed world environmental problems at the beginning of the new millennium wake us to a realization of the continued relevance of geography in investigating the highly complex intersections of humans and nature at both local and global scales. Nor are environmental problems alone of increasing concern; multiculturalism, the fruits of imperialism and its effect on aboriginal peoples, and sharp regional disparities, all these are geographical issues in which Canadians have long experience, and are problems in many other parts of the world as well. The increasing specialization in the sub-fields of geography—and the resulting expertise—provide the needed methodology to study these areas and achieve a better understanding of them, but the broad problems—the context—must be borne in mind. It is likely that in the future, an implied ethical component will be introduced into geographical studies to provide a perspective on issues confronting geographers. Geographical ways of thinking have practical applications when they affect citizens and those who have power in a society. In some ways this is a call for the return of the spirit of Canadian geography of the mid-20th century, when it was a fresh, relatively new discipline aware of big problems to be investigated and with which to become engaged. There is a long way to go in understanding geographical reality and conveying this understanding to students and the general public, but today an arsenal of new analytical research approaches is available to help in this great endeavour.

Professional geographers in Canada are few in number relative to other disciplines. Thus it is common for Canadian geographers to work in collaboration with specialists in other subjects. There is a critical need for cross-disciplinary work; the traditional synthesizing rôle of geography is essential, and calls for stronger implementation in research and in the university geography curriculum.

Acknowledgments

We particularly wish to thank Pierre Deslauriers for assiduously and cheerfully providing us with information on the development of geography in Québec, and Ian Brookes for a close critical reading and editing of the manuscript, and for his suggestions on physical geography. We express appreciation as well to Alan Baker, the late William Baker, J. Keith Fraser, Brian Osborne, Edward Pleva, J. Lewis Robinson, Peter Speak, and Suzanne Zeller for information, cogent criticisms, and ideas. We, of course, remain responsible for all content and interpretations.

References

Ami, H. M. (1915) *North America*, Vol. I, *Canada and Newfoundland*, London.
Andreae, Christopher (1997) *Lines of Country: An Atlas of Railway and Waterway History in Canada*, Erin.
Anonymous (1922) *Ontario Public School Geography*, Toronto.
Baine, Richard P. (1996) A Survey of the Status of Geography and Social Studies in Canadian Elementary and Secondary Schools, *The Monograph* 47, no. 2, 25-28.
Barnes, Trevor J., John Britton, William Coffey, David Edgington, Meric Gertler, and Glen Norcliffe (2000), Canadian Economic Geography at the Millennium, *The Canadian Geographer* 44, no. 1, 4-24.
Bird, J. Brian (1980) *The Natural Landscapes of Canada*, 2nd ed., Toronto.
Blanchard, Raoul (1930) *La Presqu'île de Gaspé*, Grenoble.
Blanchard, Raoul (1935) *L'est du Canada Français: Province de Québec*, Paris.
Blanchard, Raoul (1938) *Les Laurentides*, Montréal.
Blanchard, Raoul (1947) *Le Centre du Canada Français, Province de Québec*, Montréal.
Blanchard, Raoul (1949) *Le Québec par l'Image*, Montréal.
Blanchard, Raoul (1954) *L'Ouest du Canada Français, les Pays de l'Ottawa, l'Abitibi, Témiscamingue*, Montréal.
Blanchard, Raoul (1964) *Le Canada Français*, Paris.
Bostock, H. S. (1964) *A Provisional Physiographic Map of Canada*, Geological Survey of Canada.

Boudreau, Claude, Serge Courville, and Normand Séguin (1997), *Le Territoire*, Saint-Foye.
Bourne, Larry S. and David F. Ley, eds. (1993) *The Changing Social Geography of Canadian Cities*, Montréal.
Britton, John H., ed. (1996) *Canada and the Global Economy: The Geography of Structural and Technological Change*, Montréal.
British Columbia Natural Resources Conference (1956) *British Columbia Atlas of Resources.*
Brooke, Rupert (1919) *Letters from America*, London.
Brown, Roger J. E. (1970) *Permafrost in Canada: Its Influence on Northern Development*, Toronto.
Brouillette, Benoît (1950) Un Pionnier de la Géographie au Canada Français, Émile Miller, *Revue Canadiene de Géographie* 4, 94-97.
Buies, Arthur (1880) *Le Saguenay et la Vallée du Lac St. Jean*, Québec.
Buies, Arthur (1889) *L'Outaouais Supérieur*, Québec.
Buies, Arthur (1890) *La Région du Lac Saint-Jean, Grenier de la Province de Québec, Guide des Colons*, Québec.
Buies, Arthur (1895) *La Vallée de la Matapédia, Ouvrage Historique et Descriptive*, Québec.
Bunge, W. W. and R. Bordessa (1975) *The Canadian Alternative: Survival, Expeditions and Urban Change*, Toronto.
Burton, Ian (1963) The Quantitative Revolution and Theoretical Geography, *The Canadian Geographer* 7, 151-162.
Canada (1951) *Report of the Royal Commission on National Development in the Arts, Letters and Sciences, 1949-1951*, Ottawa.
Canada, Department of the Interior (1906) *National Atlas of Canada*, Ottawa.
Canada, Department of the Interior (1915) *National Atlas of Canada*, Ottawa.
Canada, Department of Mines and Technical Surveys, Geographical Branch (1957) *The National Atlas of Canada/L'Atlas National du Canada*, Third edition, ed. by Norman Nicholson, Ottawa.
Canada, Department of Energy, Mines and Resources and Information Canada (1974) *The National Atlas of Canada/L'Atlas National du Canada*, Fourth edition, ed. by Gerald Fremlin, Ottawa.
Careless, J. M. S. (1989) *Frontier and Metropolis: Regions, Cities, and Identities in Canada before 1914*, Toronto.
Chapman, L. J. and Donald Putnam (1951) *The Physiography of Southern Ontario*, Toronto.
Courville, Serge, Jean-Claude Robert, and Normand Séguin (1995) *Atlas Historique du Québec: Le Pays Laurentian au XIXe Siècle: Les Morphologies de Base*, Sainte-Foye.
Courville, Serge, ed. (1996) *Population et Territoire*, Sainte-Foye.
Creighton, Donald (1937) *The Commercial Empire of the St. Lawrence, 1760-1850*, Toronto.
Currie, A. W. (1942) *Canadian Economic Development*, Toronto.
Currie, A. W. (1945) *Economic Geography of Canada*, Toronto.
Dawson, G. M. and A. Sutherland (1892) *Elementary Geography of the British Colonies*, London.
Dawson, S. E. (1897) *North America*, Vol. I, *Canada and Newfoundland*, London.
Dean, William and Geoffrey Matthews (1969) *The Economic Atlas of Ontario*, Toronto.
Dean, William G., ed. (1967) *The Canadian Geographer* ('Centennial Publication') 11, no. 4.
Department of Geography, University of Alberta (1969) *Atlas of Alberta*, Edmonton.

Dunbar, Gary S. (1985) Harold Innis and Canadian Geography, *The Canadian Geographer* 29, 159-163.
Duncan, Sara Jeannette (1904) *The Imperialist*, Toronto.
Dunham, Mabel (1924) *The Trail of the Conestoga*, Toronto.
Easterbrook, W. T. and H. J. Aitken (1956) *Economic History of Canada*, Toronto.
Ennals, Peter and Deryck Holdsworth (1998) *Homeplace: The Making of the Canadian Dwelling over Three Centuries*, Toronto.
French, Hugh M. and Olav Slaymaker, eds. (1993) *Canada's Cold Environments*, Montréal.
Fung, Ka-iu (1999) *Atlas of Saskatchewan*, Second edition, Saskatoon.
Gajda, Roman T. (1960) The Canadian Ecumene - Inhabited and Uninhabited Areas, *Geographical Bulletin* 15, 5-18.
Gentilcore, R. Louis, ed. (1993) *Historical Atlas of Canada*, Vol. II, *The Land Transformed, 1800-1891*, Toronto.
Gibson, James R. (1991) *Otter Skins, Boston Ships, and China Goods: The Maritime Fur Trade and the Northwest Coast, 1785-1841*, Montréal.
Hamelin, Louis-Edmond (1972) Types of Canadian Ecumene, *Readings in Canadian Geography*, ed. by Robert Irving, pp. 20-29, Toronto.
Hamelin, Louis-Edmond (1978) *Nordicité Canadienne*, La Salle.
Hamelin, Louis-Edmond (1980) *Canadian Nordicity: It's Your North Too*, Montréal.
Hare, F. Kenneth (1960) Review Article - Climatology and the Geography Student, *The Canadian Geographer*, 4, no. 16, 36-38.
Hare, F. Kenneth and Morley K. Thomas (1979) *Climate Canada*, Second edition, Toronto.
Harris, R. Cole (1997) *The Resettlement of British Columbia: Essays on Colonialism and Geographical Change*, Vancouver.
Harris, R. Cole, ed. (1987) *Historical Atlas of Canada*, Vol. I, *From the Beginning to 1800*, Toronto.
Hayden, F. V. and A. R. C. Selwyn (1883) *North America*, London.
Hémon, Louis (1916) *Maria Chapdelaine*, Montréal.
Herbertson, A. J. and O. J. R. Howarth, eds. (1914) *The Oxford Survey of the British Empire*, Vol. V, *America*, Oxford.
Hiebert, Daniel (2000) Immigration and the Changing Canadian City, *The Canadian Geographer* 44, 25-43.
Innis, H. A. (1930) *The Fur Trade in Canada*, New Haven, Connecticut.
Jefferson, Mark (1934) The Problem of the Ecumene: The Case of Canada, *Geografiska Annaler* 16, 146-159.
Kerr, Donald and Deryck Holdsworth, eds. (1990) *Historical Atlas of Canada*, Vol. III, *Addressing the Twentieth Century, 1891-1961*, Toronto.
Laureys, Henry (1913) La Géographie et les Hautes Études Commerciales, *Revue Économique Canadienne* 3, 68-79.
Laureys, Henry (1914) *Essai de Géographie Économique du Canada*, Bruxelles.
Laureys, Henry (1926) *La Conquête des Marchés Extérieurs*, Montréal.
Laureys, Henry (1929) *The Foreign Trade of Canada*, Toronto.
Lemon, James T. (1996) *Liberal Dreams and Nature's Limits: Great Cities of North America since 1600*, Toronto.
Ley, David (1996) *The New Middle Class and the Remaking of the Central City*, Oxford.
Lloyd, Trevor (1979) Comments on the Occasion of the 25th Anniversary of the CAG, *Retrospective 1951-1976*, ed. by L.-E. Hamelin and L. Beauregard, Montréal, 95-100.
Mackay, J. Ross (1961) Review Article - Geomorphology and the Geography Student, *The Canadian Geographer* 5, no. 3, 30-33.

Mackay, J. Ross (1963) The Mackenzie Delta, North West Territories, *Geographical Branch Memoir* 8, Ottawa.
Mackintosh, W. A. and W. L. G. Joerg, eds. (1934-1940) *Canadian Frontiers of Settlement*, 9 vols., Toronto.
McCalla, Robert J. (1988) *The Maritime Provinces Atlas*, Halifax.
McGrath, Gerald and Louis M. Sebert (1999) *Mapping a Northern Land: The Survey of Canada, 1947-1994*, Montréal.
McIlwraith, Thomas F. (1997) *Looking for Old Ontario: Two Centuries of Landscape Change*, Toronto.
McIntyre, Alexander (1904) *The Canadian West: A Geography of Manitoba and the North-West Territories*, Toronto.
McManus, Gary E. and Clifford H. Wood (1991) *Atlas of Newfoundland and Labrador*, St. John's.
Mellen, Peter (1970) *The Group of Seven*, Toronto.
Miller, Émile (1912) *Terres et Peuples du Canada*, Montréal.
Miller, Émile (1921) *Pour qu'On Aime la Géographie*, Montréal.
Newbigin,Marion (1926) *Canada: The Great River, the Lands and the Men*, New York.
Nicholson, Norman L. and L. M. Sebert (1981) *The Maps of Canada*, Folkestone.
Paradis, Pierre, Jean Raveneau, and Yves Tessier (1986) *L'interAtlas Les Ressources du Québec et du Canada*, Montréal.
Parkin, G. R. (1895) *The Great Dominion: Studies of Canada*, London.
Peters, Evelyn J. (2000) Aboriginal People and Canadian Geography: A Review of the Recent Literature, *The Canadian Geographer* 44, 44-55.
Rees, Ronald (1984) *Land of Earth and Sky: Landscape Painting of Western Canada*, Saskatoon.
Reid, Dennis (1990), *Lucius R. O'Brien: Visions of Victorian Canada.*
Richards, J. Howard and K. I. Fung (1969) *Atlas of Saskatchewan*, Saskatoon.
Rogers, J. D. (1911) *A Historical Geography of the British Colonies*, Vol. V, *Canada*, Part III, *Geographical*, and Part IV, *Newfoundland*, Oxford.
Saunderson, Houston C., Dan Smith, and Ming-Ko Woo (2000) Progress in Canadian Geomorphology and Hydrology, *The Canadian Geographer* 44, 67-86.
Seton, Ernest Thompson (1911) *The Arctic Prairies* , New York.
Schott, Carl (1936) *Landaufnahme und Kolonisation in Canada am Beispiel Südontarios*, Kiel, Germany.
Service, Robert (1940) *The Complete Poems of Robert Service*, New York.
Shadbolt, Doris (1979) *The Art of Emily Carr*, Toronto.
Siegfried, André (1906) *La Canada, les deux races: problèmes politiques contemporains*, Paris.
Simmons, J. S. (1979) The Canadian Urban System: An Overview, *Research Paper* 104, University of Toronto, Centre for Urban and Community Studies.
Simpson-Housley, Paul and Glen Norcliffe, eds. (1992) *A Few Acres of Snow: Literary and Artistic Images of Canada*, Toronto.
Smith, Goldwin (1891) *Canada and the Canadian Question*, Toronto.
Smith, J. Russell (1925) *North America*, Boston.
Stefansson, Vilhjalmur (1921) *The Friendly Arctic, Five Years of Polar Exploration*, New York.
Sverdrup, Otto (1904) *New Land: Four Years in the Arctic Regions*, 2 vols., London.
Tackabury's Atlas of the Dominion of Canada (1875), Montréal, Toronto and London.

Thompson, David (1962) *David Thompson's Narrative 1784-1812*, ed. by Richard Glover, Toronto.
Trotier, Louis, gen. ed. (1972) *Studies in Canadian Geography*, 6 vols., Toronto.
Walker and Miles New Standard Atlas of the Dominion of Canada (1875), Montréal and Toronto.
Warkentin, John, ed. (1968) *Canada: A Geographical Interpretation*, Toronto.
Watson, J. W. (1951) Preface, *Geographical Bulletin*, no. 1, v.
Watson, J. W. (1965) Canadian Regionalism in Life and Letters, *Geographical Journal* 131, 21-33.
Watson, J. W. (1957) The Sociological Aspects of Geography, *Geography in the Twentieth Century*, ed. by Griffith Taylor, Third edition, 463-499.
Weir, Thomas R. (1960) *Economic Atlas of Manitoba*, Winnipeg.
Whebell, C. F. J. (1969) Corridors: A Theory of Urban Systems, *Annals of the Association of American Geographers* 59, 1-26.
Wreford, James (1979) *Countryside Canada*, Fredericton.
Zaslow, Morris (1975) *Reading the Rocks: The Story of the Geological Survey of Canada, 1842-1971*, Toronto.

John Warkentin and Paul Simpson-Housley
Department of Geography
York University
North York, Ontario
Canada M3J 1P3

Index of Personal Names

The GeoJournal Library

1. B. Currey and G. Hugo (eds.): *Famine as Geographical Phenomenon.* 1984 ISBN 90-277-1762-1
2. S.H.U. Bowie, F.R.S. and I. Thornton (eds.): *Environmental Geochemistry and Health.* Report of the Royal Society's British National Committee for Problems of the Environment. 1985 ISBN 90-277-1879-2
3. L.A. Kosiński and K.M. Elahi (eds.): *Population Redistribution and Development in South Asia.* 1985 ISBN 90-277-1938-1
4. Y. Gradus (ed.): *Desert Development.* Man and Technology in Sparselands. 1985 ISBN 90-277-2043-6
5. F.J. Calzonetti and B.D. Solomon (eds.): *Geographical Dimensions of Energy.* 1985 ISBN 90-277-2061-4
6. J. Lundqvist, U. Lohm and M. Falkenmark (eds.): *Strategies for River Basin Management.* Environmental Integration of Land and Water in River Basin. 1985 ISBN 90-277-2111-4
7. A. Rogers and F.J. Willekens (eds.): *Migration and Settlement.* A Multiregional Comparative Study. 1986 ISBN 90-277-2119-X
8. R. Laulajainen: *Spatial Strategies in Retailing.* 1987 ISBN 90-277-2595-0
9. T.H. Lee, H.R. Linden, D.A. Dreyfus and T. Vasko (eds.): *The Methane Age.* 1988 ISBN 90-277-2745-7
10. H.J. Walker (ed.): *Artificial Structures and Shorelines.* 1988 ISBN 90-277-2746-5
11. A. Kellerman: *Time, Space, and Society.* Geographical Societal Perspectives. 1989 ISBN 0-7923-0123-4
12. P. Fabbri (ed.): *Recreational Uses of Coastal Areas.* A Research Project of the Commission on the Coastal Environment, International Geographical Union. 1990 ISBN 0-7923-0279-6
13. L.M. Brush, M.G. Wolman and Huang Bing-Wei (eds.): *Taming the Yellow River: Silt and Floods.* Proceedings of a Bilateral Seminar on Problems in the Lower Reaches of the Yellow River, China. 1989 ISBN 0-7923-0416-0
14. J. Stillwell and H.J. Scholten (eds.): *Contemporary Research in Population Geography.* A Comparison of the United Kingdom and the Netherlands. 1990 ISBN 0-7923-0431-4
15. M.S. Kenzer (ed.): *Applied Geography.* Issues, Questions, and Concerns. 1989 ISBN 0-7923-0438-1
16. D. Nir: *Region as a Socio-environmental System.* An Introduction to a Systemic Regional Geography. 1990 ISBN 0-7923-0516-7
17. H.J. Scholten and J.C.H. Stillwell (eds.): *Geographical Information Systems for Urban and Regional Planning.* 1990 ISBN 0-7923-0793-3
18. F.M. Brouwer, A.J. Thomas and M.J. Chadwick (eds.): *Land Use Changes in Europe.* Processes of Change, Environmental Transformations and Future Patterns. 1991 ISBN 0-7923-1099-3

The GeoJournal Library

19. C.J. Campbell: *The Golden Century of Oil 1950–2050.* The Depletion of a Resource. 1991 ISBN 0-7923-1442-5
20. F.M. Dieleman and S. Musterd (eds.): *The Randstad: A Research and Policy Laboratory.* 1992 ISBN 0-7923-1649-5
21. V.I. Ilyichev and V.V. Anikiev (eds.): *Oceanic and Anthropogenic Controls of Life in the Pacific Ocean.* 1992 ISBN 0-7923-1854-4
22. A.K. Dutt and F.J. Costa (eds.): *Perspectives on Planning and Urban Development in Belgium.* 1992 ISBN 0-7923-1885-4
23. J. Portugali: *Implicate Relations.* Society and Space in the Israeli-Palestinian Conflict. 1993 ISBN 0-7923-1886-2
24. M.J.C. de Lepper, H.J. Scholten and R.M. Stern (eds.): *The Added Value of Geographical Information Systems in Public and Environmental Health.* 1995 ISBN 0-7923-1887-0
25. J.P. Dorian, P.A. Minakir and V.T. Borisovich (eds.): *CIS Energy and Minerals Development.* Prospects, Problems and Opportunities for International Cooperation. 1993 ISBN 0-7923-2323-8
26. P.P. Wong (ed.): *Tourism vs Environment: The Case for Coastal Areas.* 1993 ISBN 0-7923-2404-8
27. G.B. Benko and U. Strohmayer (eds.): *Geography, History and Social Sciences.* 1995 ISBN 0-7923-2543-5
28. A. Faludi and A. der Valk: *Rule and Order. Dutch Planning Doctrine in the Twentieth Century.* 1994 ISBN 0-7923-2619-9
29. B.C. Hewitson and R.G. Crane (eds.): *Neural Nets: Applications in Geography.* 1994 ISBN 0-7923-2746-2
30. A.K. Dutt, F.J. Costa, S. Aggarwal and A.G. Noble (eds.): *The Asian City: Processes of Development, Characteristics and Planning.* 1994 ISBN 0-7923-3135-4
31. R. Laulajainen and H.A. Stafford: *Corporate Geography.* Business Location Principles and Cases. 1995 ISBN 0-7923-3326-8
32. J. Portugali (ed.): *The Construction of Cognitive Maps.* 1996 ISBN 0-7923-3949-5
33. E. Biagini: *Northern Ireland and Beyond.* Social and Geographical Issues. 1996 ISBN 0-7923-4046-9
34. A.K. Dutt (ed.): *Southeast Asia: A Ten Nation Region.* 1996 ISBN 0-7923-4171-6
35. J. Settele, C. Margules, P. Poschlod and K. Henle (eds.): *Species Survival in Fragmented Landscapes.* 1996 ISBN 0-7923-4239-9
36. M. Yoshino, M. Domrös, A. Douguédroit, J. Paszynski and L.D. Nkemdirim (eds.): *Climates and Societies – A Climatological Perspective.* A Contribution on Global Change and Related Problems Prepared by the Commission on Climatology of the International Geographical Union. 1997 ISBN 0-7923-4324-7
37. D. Borri, A. Khakee and C. Lacirignola (eds.): *Evaluating Theory-Practice and Urban-Rural Interplay in Planning.* 1997 ISBN 0-7923-4326-3

The GeoJournal Library

38. J.A.A. Jones, C. Liu, M-K. Woo and H-T. Kung (eds.): *Regional Hydrological Response to Climate Change.* 1996 ISBN 0-7923-4329-8
39. R. Lloyd: *Spatial Cognition.* Geographic Environments. 1997 ISBN 0-7923-4375-1
40. I. Lyons Murphy: *The Danube: A River Basin in Transition.* 1997 ISBN 0-7923-4558-4
41. H.J. Bruins and H. Lithwick (eds.): *The Arid Frontier.* Interactive Management of Environment and Development. 1998 ISBN 0-7923-4227-5
42. G. Lipshitz: *Country on the Move: Migration to and within Israel, 1948–1995.* 1998 ISBN 0-7923-4850-8
43. S. Musterd, W. Ostendorf and M. Breebaart: *Multi-Ethnic Metropolis: Patterns and Policies.* 1998 ISBN 0-7923-4854-0
44. B.K. Maloney (ed.): *Human Activities and the Tropical Rainforest.* Past, Present and Possible Future. 1998 ISBN 0-7923-4858-3
45. H. van der Wusten (ed.): *The Urban University and its Identity.* Roots, Location, Roles. 1998 ISBN 0-7923-4870-2
46. J. Kalvoda and C.L. Rosenfeld (eds.): *Geomorphological Hazards in High Mountain Areas.* 1998 ISBN 0-7923-4961-X
47. N. Lichfield, A. Barbanente, D. Borri, A. Khakee and A. Prat (eds.): *Evaluation in Planning.* Facing the Challenge of Complexity. 1998 ISBN 0-7923-4870-2
48. A. Buttimer and L. Wallin (eds.): *Nature and Identity in Cross-Cultural Perspective.* 1999 ISBN 0-7923-5651-9
49. A. Vallega: *Fundamentals of Integrated Coastal Management.* 1999 ISBN 0-7923-5875-9
50. D. Rumley: *The Geopolitics of Australia's Regional Relations.* 1999 ISBN 0-7923-5916-X
51. H. Stevens: *The Institutional Position of Seaports.* An International Comparison. 1999 ISBN 0-7923-5979-8
52. H. Lithwick and Y. Gradus (eds.): *Developing Frontier Cities.* Global Perspectives – Regional Contexts. 2000 ISBN 0-7923-6061-3
53. H. Knippenberg and J. Markusse (eds.): *Nationalising and Denationalising European Border Regions, 1800–2000.* Views from Geography and History. 2000 ISBN 0-7923-6066-4
54. R. Gerber and G.K. Chuan (eds.): *Fieldwork in Geography: Reflections, Perspectives and Actions.* 2000 ISBN 0-7923-6329-9
55. M. Dobry (ed.): *Democratic and Capitalist Transitions in Eastern Europe.* Lessons for the Social Sciences. 2000 ISBN 0-7923-6331-0
56. Y. Murayama: *Japanese Urban System.* 2000 ISBN 0-7923-6600-X
57. D. Zheng, Q. Zhang and S. Wu (eds.): *Mountain Geoecology and Sustainable Development of the Tibetan Plateau.* 2000 ISBN 0-7923-6688-3

The GeoJournal Library

58. A.J. Conacher (ed.): *Land Degradation.* Papers selected from Contributions to the Sixth Meeting of the International Geographical Union's Commission on Land Degradation and Desertification, Perth, Western Australia, 20–28 September 1999. 2001 ISBN 0-7923-6770-7
59. S. Conti and P. Giaccaria: *Local Development and Competitiveness.* 2001 ISBN 0-7923-6829-0
60. P. Miao (ed.): *Public Places in Asia Pacific Cities.* Current Issues and Strategies. 2001 ISBN 0-7923-7083-X
61. N. Maiellaro (ed.): *Towards Sustainable Buiding.* 2001 ISBN 1-4020-0012-X
62. G.S. Dunbar (ed.): *Geography: Discipline, Profession and Subject since 1870.* An International Survey. 2001 ISBN 1-4020-0019-7

KLUWER ACADEMIC PUBLISHERS – DORDRECHT / BOSTON / LONDON

GPSR Compliance

The European Union's (EU) General Product Safety Regulation (GPSR) is a set of rules that requires consumer products to be safe and our obligations to ensure this.

If you have any concerns about our products, you can contact us on ProductSafety@springernature.com

In case Publisher is established outside the EU, the EU authorized representative is:

Springer Nature Customer Service Center GmbH
Europaplatz 3
69115 Heidelberg, Germany

Batch number: 10755284

Printed by Printforce, the Netherlands